Advances in Asian Human-Environmental Research

The *Advances in Asian Human-Environmental Research* series aims at fostering the discussion on the complex relationships between physical landscapes, natural resources, and their modification by human land use in various environments of Asia. It is widely acknowledged that human-environmental interactions become increasingly important in Area Studies and development research, taking into account regional differences as well as bio-physical, socio-economic and cultural particularities. The book series seeks to explore theoretic and conceptual reflection on dynamic human-environment systems applying advanced methodology and innovative research perspectives. The main themes of the series cover urban and rural landscapes in Asia. Examples include topics such as land and forest degradation, vulnerability and mitigation strategies, natural hazards and risk management concepts, environmental change, impact studies and consequences for local communities. The relevant themes of the series are mainly focused on geographical research perspectives of Area Studies, however there is scope for interdisciplinary contributions from various spheres within the natural sciences, social sciences and humanities. Key themes: Human-Environment Interaction - Asian regional studies - Asian geography - Impact studies - Landscape - Society - Land use and land cover change - Natural resources *Submit a proposal* : Proposals for the series will be considered by the Series Editor and International Editorial Board. An initial author/ editor questionnaire and instructions for authors can be obtained from the Publisher, Dr. Robert K. Doe (robert.doe@springer.com).

Anwesha Borthakur • Pardeep Singh
Editors

The Resilience of Traditional Knowledge Systems for a Sustainable Future

A Focus on Agriculture and Food Practices in the Himalayas

Springer

Editors
Anwesha Borthakur
KU Leuven
Leuven, Belgium

Pardeep Singh
Department of Environmental Studies
PGDAV College, University of Delhi
New Delhi, Delhi, India

ISSN 1879-7180 ISSN 1879-7199 (electronic)
Advances in Asian Human-Environmental Research
ISBN 978-3-031-56857-2 ISBN 978-3-031-56858-9 (eBook)
https://doi.org/10.1007/978-3-031-56858-9

This Springer imprint is published by the registered company Springer Nature Switzerland AG
The registered company address is: Gewerbestrasse 11, 6330 Cham, Switzerland

Acknowledgments

We, the editors, are grateful to the University of Leuven (KU Leuven), Belgium and Indian Council of Social Science Research (ICSSR), India (ICSSR/RPD/MJ/2023-24/ST/98) for the financial support provided to conduct our research in the Indian Himalaya. This book is a result of our experiences working in this majestic mountain range. We would also like to thank our respective current institutions—Rachel Carson Center at Ludwig Maximilian University of Munich (LMU), Germany and PGDAV College, University of Delhi, India—for their support. Further, we acknowledge with thanks all the authors who contributed to this book along with the peer reviewers who were kind enough to comment on the chapters.

Introduction

With debates on climate change and sustainable development holding a central place in the current global environmental narratives, the focus on traditional knowledge systems emerges as a major outcome. For centuries, these knowledge systems have been offering insights and instigating practices that have the potential to significantly contribute towards environmental sustainability. Today, there are ever-increasing evidence that local communities and indigenous people observe, experience and report numerous indicators of climate change and existing environmental impacts (Reyes-García et al. 2024). All across the globe, local and indigenous communities own systematic knowledge and in-depth understandings of their surroundings including the local ecosystems. They are not only well-linked with their local ecosystems but also shoulder responsibilities, as custodians, for their maintenance and care (Sangha et al. 2019). Aspects associated with sustainable management of natural resources, the interconnectedness among diverse species, climate/weather phenomenon and so on play a major role in their culture and cultural interpretations. For instance, in the Indian Trans-Himalaya of Ladakh, social norms supervised by local institutions guide a considerate use of resources, underlining the significance of community-led management practices in nourishing the spirit of the traditional agropastoral structure (Ladon et al. 2023).

The traditional and indigenous know-how and understandings are developed over several centuries through the lived-experiences of the indigenous and local communities and their active experiments with nature and natural processes. Overall, this knowledge is immensely diverse, grounded in their territories and is constantly evolving through interaction of various types of knowledge such as oral, written, visual, gendered, tacit, scientific, practical, etc., experiences and innovations (Hill et al. 2020). Consequently, many of these local communities are often equipped with knowledge about how to respond to, adapt and alleviate natural disaster risks and climate change challenges, how to fight local diseases of humans, plants and animals and how to conserve and promote biodiversity (including agricultural diversity) in their respective regions. Unfortunately, with larger implications for protection of biocultural heritage, factors such as urbanization could result in the loss of knowledge of both the management and use of local wild species, for instance (Arjona-García et al. 2021). Thus, traditional knowledge must be safeguarded towards providing valuable inputs in efforts to protect the environment in today's world.

In this book, we are focusing on the resilience of traditional knowledge systems for a sustainable future with a particular focus on agriculture and food practices in the Himalaya. The Himalayan region is characterized by its overall diversity in terms of linguistic, cultural, socio-ecological among others. Agricultural diversity is an integral part of it. This agricultural diversity results in a wide range of food processes and practices, providing the local populations with the essential nutrients and vital energy to endure and survive in challenging weather and climatic conditions. Further, it provides a unique gastronomy culture across the Himalayan mountain range, many of which are recognized for both their diversity and quality. Today, some of the traditional Himalayan gastronomy are able to travel outside their places of origins— to countries all over the globe and find a place in the popular cuisine and culture. Together, both agriculture and food systems in the Himalayas contribute to a resilient structure where not only the health of the humans is ensured but also wellbeing of the overall environment is ascertained. These systems provide some essential suggestions towards a sustainable future in the current distressed world. Accordingly, in the present-day global environmental circumstances, such ideas are instrumental for tacking the ecological and climate change challenges. Further in the Himalayan region, local communities, beyond their cultural and ethnic specificities, aid in safeguarding their livelihoods by adopting measures such as 'altitudinal diversification across agroecological zones and by decreasing the threat that natural hazards pose to any of these environmental resources' (Nüsser and Baghel 2016). Unfortunately, a major challenge confronted by the international community today is how to reverse or slow down the current trend of agrobiodiversity loss towards safeguarding the future food provisions (Rasul et al. 2022).

The book starts with chapters on the traditional agricultural systems in the Himalayan mountain region in Part I. Traditional farming practices are often unique, tribe/community-specific and thus, uphold cultural and agrobiodiversity in a region (Giri et al. 2020). Accordingly, considering instances from the central Himalayan region (primarily Nepal), North-Western and North-Eastern India, Chapter "Traditional/Indigenous Knowledge and Technologies for Sustainable Agriculture in the Himalayas" provides an overview of traditional/indigenous knowledge and technologies for sustainable agriculture in the Himalayas. Traditional agricultural practices have a significant role not only in offering financial resilience to the farmers or pastoralist but also in ensuring agrobiodiversity, food security and nutrition quality (Patel et al. 2020). The contributions of these systems in achieving sustainable agricultural practices are addressed in detail in the book. Moving a step further, Chapter "Traditional Farming in the Indian Himalayan Region: Preserving Agro-Biodiversity and Future Food Security" focuses on traditional farming practices in the Indian Himalayan Region and their role in preserving agrobiodiversity and ensuring future food security. In a similar effort, Chapter "Traditional Agri/Horticultural Crops, Farming Practices and Conservation in the Himalayan Region" emphasizes on traditional agricultural/horticultural crops, farming practices and conservation in the Himalayan region, taking examples from different parts of the mountain range. In the North-Eastern part of India in the Eastern Himalaya, shifting cultivation or jhum cultivation is a major system of farming which involves a cyclical progression of 'slash and burn' of community-owned natural forest areas with the farming of food crops thereafter (Giri et al. 2020). Chapter "Shifting Cultivation: An Analysis of Jhum Cultivation in Karbi Society of Assam" provides an analysis of shifting cultivation in indigenous Karbi society in the North-East Indian state of Assam. Part I ends with a chapter on the relevance of traditional knowledge for sustainable agriculture in the Indian Himalayas Region (Chapter "Relevance of Traditional Agriculture Knowledge for Sustainable Agriculture in the Indian Himalayas Region"). An attention to the importance of traditional agricultural knowledge is crucial considering the fact that such practices are often threatened with extinction owing to inadequate consideration for protection.

Section I is followed by a focus on the traditional food systems and the delights of local cuisines in Section II. Culturally accepted ingredients from the native environment composed traditional foods of the local indigenous population and the Indian Himalayan region, with its rich tribal populace, is no exception as each of this tribe displays a distinct food culture (Bhatt et al. 2023). Unfortunately, climate change-induced precipitation and temperature diversion in the Himalayan region with all the associated changes in cropping patterns and agrobiodiversity are currently responsible for putting a lot of pressure on traditional eating or food practices (Das & Mishra 2022). Accordingly, it becomes essential to document the existing mountain cuisines before they disappear due to external environmental circumstances. Chapter "Mountain Cuisine in a Changing Climate: The Resilience of Traditional Himalayan Food Systems" of our book provides an account of mountain cuisine in a changing climate, particularly focusing on the resilience of traditional Himalayan food systems. With an emphasis on the Eastern Himalaya (on the state of Nagaland in India), Chapter "Bearing the Brunt of Climate Change in Eastern Himalayas: Conserving the Traditional Food Systems of Nagaland, India" gives an overview of how traditional food systems are forced to bear the brunt of climate change and argues for the urgent need to conserving these systems. Traditional and indigenous food systems, with their incredible cultural and biological diversity, have tremendous potential towards ensuring food and nutritional security to the local indigenous communities (Pandey et al. 2022). Therefore, in a world characterized by a growing population and environmental challenges, these food systems could provide important clues for a sustainable future. Many of the traditional food practices throughout the Himalayan region are at the brink of extinction. Accordingly, the purpose of this book is to document those practices and list them down so that it could act as a reference to both the current and future academicians and decision-makers.

The book further endeavours to report the challenges encountered by the Himalayan region which affect all the traditional practices and ecological balance of the fragile mountain ecosystems. In reality, no book on the Himalayas is probably complete without addressing the anthropogenic challenges constantly faced by this majestic mountain range. These challenges are observed to be amplified manifolds in the recent years. More often than not, the man-made challenges both intensify and complicate natural environmental challenges. Accordingly, we decided to dedicate a few chapters of the book to the current trials and tribulations the Himalayas are forced to pass through. Accordingly, Chapter "Traditional Knowledge Systems in the Himalayas: Why Moving Away from Biomass Cooking Fuel Is So Difficult?", Chapter "Unorganized Tourism Practices and the Allied Challenges in the Himalayas" and Chapter "Unveiling Anthropogenic Environmental Burdens Impacting Ecosystem Services in the Himalayas" of the book in Part III examine, respectively, (i) the socio-cultural context involving the interaction between culture, livelihood and social practices about clean cooking fuel preferences in the Nepal Himalaya, (ii) challenges of unorganized tourist activities and related repercussions on the Himalayan ecosystems and (iii) anthropogenic environmental burdens impacting ecosystem services in the Himalayas. The objective is to provide the readers a glimpse of the current challenges in one of the most expansive, delicate and remarkably diverse mountain ecosystems in the world. This will help them comprehend the relevance of the 'traditional knowledge systems' in the region better, which is the central theme of majority of the chapters in the book. The final two chapters of the book in Part IV deliver food for thought on possible way forwards. That is, Chapter "The Himalayan Ecosystem and Development Paradigm: A Sustainability Perspective" addresses the conservations of Himalayan ecosystem and development paradigm from a sustainability perspective. For instance, anthropogenic actions such as deforestation, unrestrained forest fires along with developmental activities are responsible for putting enormous pressure on spring-fed groundwater sources in the Central Himalaya. In such contexts, it becomes imperative to uninterruptedly engage in environment and/vs. development-related debates in the contemporary global environmental conditions. The book ends with a chapter on the significance of harnessing the traditional socio-cultural institutions for sustainable development in the Himalayas (Chapter "Revitalizing Roots: Harnessing Traditional Socio-Cultural Institutions for Sustainable Development of the Himalayas"). Overall, along with acknowledging the significance and contributions of traditional knowledge systems in the present-day world, the book aims to further the debates on environment vs. development and the associations among traditional knowledge, climate science and environmental management, especially in the Global South.

The book is initiated by us, the editors, with our own experiences of carrying out research projects in and on the Indian Himalayas. The editor, Dr. Anwesha Borthakur, has been involved in her individual research project on public engagement through knowledge co-production for environmental sustainability in the Indian Himalayan Region. Funded by the University of Leuven (KU Leuven) in Belgium, the insights attained during the conduct of the project have been instrumental in shaping the current book. Earlier this year, the editor Dr. Pardeep Singh also received a major grant from the Indian Council of Social Science Research (ICSSR) to carry out a project in the Indian Himalayas for a period of two years. The experiences and understandings during the development of his project ideas and subsequent implementation strategies help streamlining the book further. Moreover, we, both the editors are natives of the Indian Himalayas (of Assam and Jammu and Kashmir, respectively). This, in turn, has been a supporting factor because of our own lived-experiences and socio-political-cultural-economic-environmental understandings of the Himalayan region.

The book primarily aims to document the existing traditional agricultural and food practices in the Himalayas. Further, it aspires to chart directions towards integrating such practices in the modern scientific knowledge systems for a sustainable future. Today, it is well accepted that both the significance and efficacy of environmental sustainability initiatives could be enhanced by efforts to incorporate traditional knowledge into modern environmental policies and practices. For example, any real-world resolution for environmental resource management calls for respect of the sanctity and integrity of traditional ecological knowledge and provides appropriate value of this know-how by any agroecological systems (Saylor et al. 2017). In general, decision-making processes that actively involve local and indigenous communities tend to be more effective than top-down policy approaches. Moreover, such processes are often considerate of the long-term impacts on the surrounding environment and aid in an inclusive community-based governance system. It is imperative, for instance, to ensure participatory land use planning in the Indian Trans-Himalaya considering the locale-specific characteristics of the mountain communities and local sub-regions (Schmidt and Nüsser 2023). A creative collaboration and a proactive partnership (respecting the rights of the indigenous/local communities) between indigenous/local communities and scientific organizations are essential to harness the wisdom of both traditional and modern knowledge for greater environmental sustainability.

References

Arjona-García C, Blancas J, Beltrán-Rodríguez L, López Binnqüist C, Colín Bahena H, Moreno-Calles AI, Sierra-Huelsz JA, López-Medellín X (2021) How does urbanization affect perceptions and traditional knowledge of medicinal plants? J Ethnobiol Ethnomed, 17(1), 48. https://doi.org/10.1186/s13002-021-00473-w

Bhatt S, Purohit MV, Balodi KN, Kumar U, Sharma A, Shridhar V, Arunachalam A, Arunachalam K (2023) The ethnic food system of Jaad Bhotiya tribal community in the Gangotri landscape of western Himalaya: a transition in traditional practices. AlterNative: Int J Indigenous People, 19(2), 250-260. https://doi.org/10.1177/11771801231173396

Das S, Mishra AJ (2022) Dynamics of indigenous community's food and culture in the time of climate change in the Himalayan region. J Ethnic Food, 9(1), 1. https://doi.org/10.1186/s42779-022-00118-7

Giri K, Mishra G, Rawat M, Pandey S, Bhattacharyya R, Bora N, Rai JPN (2020) Traditional Farming Systems and Agro-biodiversity in Eastern Himalayan Region of India. In R. Goel, R. Soni, & D. C. Suyal (Eds.), Microbiological Advancements for Higher Altitude Agro-Ecosystems & Sustainability (pp. 71-89). Springer Singapore. https://doi.org/10.1007/978-981-15-1902-4_5

Hill R, Adem Ç, Alangui WV, Molnár Z, Aumeeruddy-Thomas,Y, Bridgewater P, Tengö M, Thaman R, Adou Yao CY, Berkes F, Carino J, Carneiro da Cunha M, Diaw MC, Díaz S, Figueroa VE, Fisher J, Hardison P, Ichikawa K, Kariuki P, . . . Xue D (2020) Working with Indigenous, local and scientific knowledge in assessments of nature and nature's linkages with people. Curr Opin Environ Sustain, 43, 8-20. https://doi.org/https://doi.org/10.1016/j.cosust.2019.12.006

Ladon P, Nüsser M, Garkoti SC (2023) Mountain agropastoralism: traditional practices, institutions and pressures in the Indian Trans-Himalaya of Ladakh. Pastoralism, 13(1), 30. https://doi.org/10.1186/s13570-023-00289-1

Nüsser M, Baghel R (2016) Local Knowledge and Global Concerns: Artificial Glaciers as a Focus of Environmental Knowledge and Development Interventions. In P. Meusburger, T. Freytag, & L. Suarsana (Eds.), Ethnic and Cultural Dimensions of Knowledge (pp. 191-209). Springer International Publishing. https://doi.org/10.1007/978-3-319-21900-4_9

Pandey DK, Momin KC, Dubey SK, Adhiguru P (2022) Biodiversity in agricultural and food systems of jhum landscape in the West Garo Hills, North-eastern India. Food Security, 14(3), 791-804. https://doi.org/10.1007/s12571-021-01251-y

Patel SK, Sharma A, Singh GS (2020) Traditional agricultural practices in India: an approach for environmental sustainability and food security. Energy, Ecol Environ, 5(4), 253-271. https://doi.org/10.1007/s40974-020-00158-2

Rasul G, Hussain A, Adhikari L, Molden DJ (2022) Conserving agrobiodiversity for sustainable food systems in the Hindu Kush Himalaya. Int J Agric Sustain, 20(6), 1117-1135. https://doi.org/10.1080/14735903.2022.2057642

Reyes-García V, García-del-Amo D, Álvarez-Fernández S, Benyei P, Calvet-Mir L, Junqueira AB, Labeyrie V, Li X, Miñarro S, Porcher V, Porcuna-Ferrer A, Schlingmann A, Schunko C, Soleymani R, Tofighi-Niaki A, Abazeri M, Attoh EMNAN, Ayanlade A. Ávila JVDC, . . . Zakari IS (2024) Indigenous Peoples and local communities report ongoing and widespread climate change impacts on local social-ecological systems. Comm Earth Environ, 5(1), 29. https://doi.org/10.1038/s43247-023-01164-y

Sangha KK, Russell-Smith J, Costanza R (2019) Mainstreaming indigenous and local communities' connections with nature for policy decision-making. Global Ecol Conserv, 19, e00668. https://doi.org/https://doi.org/10.1016/j.gecco.2019.e00668

Saylor CR, Alsharif KA, Torres H (2017) The importance of traditional ecological knowledge in agroecological systems in Peru. Int J Biodiver Sci, Ecosys Ser Manage, 13(1), 150-161. https://doi.org/10.1080/21513732.2017.1285814

Schmidt S, Nüsser M (2023) Glaciers of Central Ladakh: Distribution, Changes and Relevance in the Indian Trans-Himalaya. In B. Humbert-Droz, J. Dame, & T. Morup (Eds.), Environmental Change and Development in Ladakh, Indian Trans-Himalaya (pp. 11-30). Springer Nature Switzerland. https://doi.org/10.1007/978-3-031-42494-6_2

Contents

Part I

Traditional Knowledge Systems: Agriculture and Livelihood

Traditional/Indigenous Knowledge and Technologies for Sustainable Agriculture in the Himalayas

Roshan M. Bajracharya

1 Introduction

As humanity steps forward into the twenty-first Century, it has become evident that the socio-economic and development activities of modern societies over the past two centuries are creating a wide range of crises having global ramifications. The ever-growing world population, along with advancements in technologies that touch upon all facets of our lives, has placed tremendous pressures upon the environment, natural ecosystems and the resource base. This is due to the insatiable demand of societies for food, water, energy and a whole host of raw materials required for manufacturing and industries. In fact, most of the approaches, processes and technologies that modern societies have come to depend upon for everything from food production, energy generation, transportation and communication to manufacturing and construction, are clearly unsustainable (Bajracharya 2021; Harwood 1990). The realization has already dawned upon the scientific community that unless we adopt a paradigm shift in the way we conduct many of the day-to-day activities needed for virtually every aspect of our lives, the very survival of the human race may be jeopardized.

For millennia, in the past, the human race had existed in relative balance with the natural environment. Prior to the Industrial Revolution, which began in the latter part of the eighteenth Century, people lived in fairly small communities without sophisticated technologies and relied mainly on small-scale manual or semi-mechanized techniques for farming, transportation and goods manufacturing. However, with the invention of the steam engine and the use of coal and subsequently petroleum-based fuels, large-scale production processes became possible.

R. M. Bajracharya (✉)
Department of Environmental Science and Engineering, Kathmandu University, Dhulikhel, Kavrepalanchok, Nepal

A. Borthakur, P. Singh (eds.), *The Resilience of Traditional Knowledge Systems for a Sustainable Future*, Advances in Asian Human-Environmental Research,
https://doi.org/10.1007/978-3-031-56858-9_1

This further escalated with the development of modern machinery used for land and sea transportation, agriculture, aviation, and so forth. Moreover, with the advancement of modern medicine, human longevity and life expectancy increased significantly and the global population began to grow exponentially, thereby, exerting evermore demands upon the natural resource base. During the mid-twentieth Century, after the Second World War, the Green Revolution introduced chemical fertilizers along with plant breeding and genetic manipulation, which promised to substantially enhance food production and alleviate hunger worldwide. However, it soon became evident, by the late 1900s, that modern agriculture with its large-scale monoculture cropping, excessive reliance on chemical fertilizers and pesticides, and over-grazing of rangelands, could not be sustained indefinitely (Harwood 1990; Firebaugh 1990).

With technological advancement, modern societies came to rely more and more upon the exploitation of natural resources, with evermore extractive processes for everything from fossil fuels to water, minerals, ores, forests, plants and other biological resources. Furthermore, the modern techniques used in agriculture, heavy industries, transportation, and energy generation, all produced a variety of wastes and pollutants that began to contaminate our land, water bodies and air. The consequences of land, water and air pollution with toxic substances, such as, heavy metals, inorganic compounds and persistent organic chemicals, entail initial uptake and accumulation in plants and organisms at the lower end of the food chain, and subsequent bio-magnification at the higher trophic levels. This ultimately adversely affects human health as people are at the top of the food web, thus leading to life-threatening diseases such as cancer, endocrine disruption, and birth defects. In addition pollution from industries and the transport sector, the large-scale deforestation of tropical and temperate forests, drainage of wetlands, conversion of naturally vegetated land to monoculture farming, and the ever-growing livestock populations are all contributing to global warming and climate change, which unless drastic measures are taken, will spiral in a positive feedback loop to accelerate in the not too distant future (Bajracharya 2021; Bajracharya et al. 2014).

2 Traditional and Indigenous Knowledge in the Modern Context

With the growing recognition of the detrimental impacts caused by the development and economic activities that humans were engaged in, the environmental movement began in the early 1970s. The United Nations Organization also put forth proposals for moving towards a sustainable and more balanced development approach. As the agriculture sector is a major contributor to global climate change, sustainable and climate friendly approaches to food production have become an urgent necessity. To this end, incorporating traditional and indigenous farming practices and techniques into modern agricultural systems may offer environmentally sound, sustainable and climate neutral means of producing a significant proportion of the food and fiber that human societies demand (Bajracharya et al. 2014; Firebaugh 1990).

From the term 'traditional', it is generally understood that such a practice or knowledge is typically "old" or passed down through the generations. On the other hand, the term 'indigenous' refers to something that is of local origin and developed locally, although the inspiration or motivation for doing so may be from elsewhere. Indigenous practices may, in fact, be relatively new and still undergoing evolution (Gill 1993). A practice may be traditional, indigenous, or even both, and because they are often of local origin, practices tend to be well adapted to the local conditions and more likely to be accepted by the community than technology transferred from elsewhere (Gill 1993). Indeed traditional or indigenous knowledge often builds upon the vast experience and wisdom of past generations that have developed the techniques or practices over centuries (Chaudhary et al. 2017). These knowledge or technical systems tend to be in tune with the natural environment and circumstances of the location and, hence, may be more sustainable, in most cases, than modern, imported technologies (Tamang 1993).

3 Selected Traditional Practices of the Central Himalaya

The ensuing sections provide brief descriptions of the salient aspects of some of the common indigenous and/or traditional agricultural practices observed in the hill areas of the central Himalayan region, mainly in Nepal, northwestern and northeastern India. The practices are sub-divided into six key categories, namely, traditional land preparation, soil management, livestock rearing, crop production, water management, and pest control practices. The list is not exhaustive, rather, only those practices that are deemed to be sustainable and likely to contribute to adaptation to or mitigation of the impacts of climate change are included.

3.1 Land Preparation Practices

3.1.1 In most of the lower and middle hill areas of the central Himalayas, farmers have traditionally used the age-old practice of tilling the land with an ox-drawn wooden plough that loosens the soil to a depth of about 15 cm. The wooden plough is locally fabricated from hardwoods, such as, Sal (*Shorea robusta*), and having the tip fitted with a flat iron strip of 20–30 cm in length. The plough is drawn by two oxen and steered by the farmer with a long wooden handle connected to the back part of the plough (Fig. 1). With this handle, the farmer is able to control the direction and depth of tillage. This tillage technique is the most commonly used method of land preparation in both lowland, paddy cultivation areas (valleys and near stream banks), and upland farms where maize and millet are usually grown. It is used on sloping lands of up to 50 or 60% slopes, as well as, on hill terraces as narrow as 2 m in width, which is the minimum width required to turn the oxen around. This has been the most affordable, practical and efficient method for tilling their fields for most farmers in the mostly fragile, inaccessible and marginal rural hill farms (Bajracharya 2001; Singh 1998).

Fig. 1 Wooden plough drawn by oxen. (Source: Author)

Fig. 2 Different kinds of farm hand-tools. (Source: Author)

3.1.2 Manual cultivation using a hand hoe is the next most common method of land preparation used in the hills and valleys of the central Himalaya. This is especially so for farmers who own very small parcels (typically less than 0.1 ha) of land and for those who do not have oxen, as well as, for tilling narrow terraces of less than 2 m width on steep slopes or for use in small kitchen gardens. Hand-held hoes come in variety of blade widths, ranging from about 5 to 15 cm, and varying handle lengths from 30 cm to 1 m long. They are used by both men and women for primary, as well as, secondary tillage operations (Figs. 2 and 3). Hoes are also employed for clod-breaking after oxen plowing and even for weeding purposes. This type of tillage loosens the soil down to

Fig. 3 Hand hoes and wooden clod-breaker. (Source: Author)

a depth of 10 cm or less and does not involve overturning of the soil (Bajracharya 2001).

3.1.3 Secondary tillage operations are done by farmers for the purposes of clod-breaking, leveling or smoothing the soil surface, or for post-emergence weeding. The timing of such operations, number of tillage operations, and types of tools used depend mainly on the soil texture, land type, slope steepness, terrace width, climate, irrigation availability, cropping patter, and religious or socio-cultural factors (Amatya et al. 1997; Turton et al. 1995). On broad fields and gently sloping areas, post-tillage clod-breaking, leveling and surface smoothing operations are usually done by means of a wooden, locally-made, ox-drawn harrow with wooden tines, or simply, a wooden plank (Fig. 3). On steep slopes, narrow terraces or other small parcels of land, such as kitchen gardens, these operations, along with weed removal, are usually done with a hand-held wooden clod-breaker or hand-hoe, frequently by women (Adhikari and Karmacharya 1988; Bajracharya 2001).

A different kind of secondary tillage that is done primarily on upland fields cropped to maize and millet, is the "earthing-up" or ridging of soil in-between crop rows or for individual plants. The practice is locally known as '*dohry-aune*' and for crops planted in rows on broad terraces or gently sloping areas, it is done by means of an ox-drawn wooden plough or tine harrow. However, for narrow terraces and steep slopes, it is done by hand hoe for individual plants. This practice serves the dual purpose of providing support to easily lodged plants, such as maize, and for the removal of unwanted weeds (Bajracharya 2001).

3.1.4 For millennia, farming in hill regions have involved the preparation and meticulous maintenance of terraces. Different kinds of terraces are made by

Fig. 4 Level bench terraces (left) and outward sloping terraces on steep hills (right). (Source: Author)

Fig. 5 The "Bukma" system of preparing seedbed for growing potatoes and millet on marginal lands at high elevations. (Source: Author)

farmers, with the most common in the mid-hills being level bench terraces and outward-sloping terraces (Fig. 4). Level bench terraces are most often made for the cultivation of paddy with the soil under submerged conditions. Where the land or soil type is unstable and easily eroded, the terrace risers may be supported with stone walls. These are found in the lowlands of valleys and along stream banks where adequate water can be diverted for flood irrigation of the fields. On the other hand, outward-sloping terraces are generally prepared on sloping land with often shallow soils (50–60 cm or less soil depth). The main reason for maintaining a slope across the terrace is due to the shallow depth of soil which would lead to exposure of rocky and low fertility subsoil or parent material at the back part if the terraces were made level.

3.1.5 The "*Bukma*" system of cultivating finger millet, barley and potatoes in some high (1800–3500 m) hill areas of the central Himalaya can be considered both a land preparation and soil management practice that has the added advantage of incorporating a type of locally produced biochar into the soil. This cultivation technique involves scraping or slicing thin layers of sod (soil with a mat of grassy vegetation) on sloping lands and piling them into elongated row configurations to dry for several days (Fig. 5). Once the overturned grassy vegetation and roots are dry, the piles are burned and left to smolder for a few days. The crop is then sown into the soil row piles containing ash and biochar without adding any manure of fertilizers. Occasionally, finger millet or barley is planted in the soil following harvest of the potatoes and subsequent tillage. After cultivating one or two crops in this manner, the land is left fallow for 4–6 years to allow for the re-growth of local grassy vegetation (Bajracharya et al. 2013; Dhakal 1993).

3.2 Soil Management Practices

3.2.1 Application of locally prepared farm yard manure (FYM) on fields is a ubiquitous traditional method of fertilizing the soil to replenish essential crop nutrients and soil organic matter (SOM) in the Himalaya. This local compost or FYM is generally made from any available forest leaf litter, crop residues (grain husks and straw), weeds, animal manure and spent animal bedding material from livestock stalls. The traditional method for composting the waste organic matter is to simply mix of the materials together and pile them up in a heap (Fig. 6) to decompose over a period of a few to several months, depending on the climate and season (Bajracharya 2001; Sharma et al. 2009). The ripening period is shorter at low elevations and during the summer months, whereas, it is longer at higher elevations and in the winter due to cool conditions. During the ripening period, heap of decomposing organic material may or may not be mixed occasionally. Often times, the FYM compost is applied to the crop fields in a partially decomposed state 2–4 weeks prior to sowing. The FYM is spread out and incorporated into the soil during the tillage operations.

3.2.2 On hill and valley farms where bench terraces can be established for crop cultivation, a practice that has been passed down through the generations is terrace-riser slicing. This practice involves the removal of a thin (2–5 cm) layer of soil with weedy vegetation from the face of the terrace riser and

Fig. 6 Farmyard manure compost pile near livestock shed (left); application of FYM on farm fields prior to tillage (right). (Source: Author)

Fig. 7 Terrace riser slicing. (Source: Author)

incorporating the sliced or scraped soil along with vegetation into the topsoil of the adjacent terrace below (Fig. 7). The main reasons claimed for doing this are to eliminate the growth of weeds and to control pests and diseases that may occur if a weedy vegetative growth is kept on the terrace risers. Another minor benefit of this technique is the addition of small amounts of organic matter to the adjacent terrace from the incorporation of the weeks and grass into the topsoil as a kind of green manure (Bajracharya 2001).

3.2.3 A practice that has not been traditionally used widely by farmer in the central Himalayas is mulching. It is, however, used on a small-scale mainly for vegetable nursery beds or fore specialty high-value crops (Fig. 8). Mulching has gained importance in recent decades due to changes in cropping patterns to include commercial cash crops like vegetables and horticultural plants. For instance, the use of straw mulching is now commonly used in the cultivation of crops like ginger, strawberries, as well as, around fruit trees and to cover nursery seedbeds prepared for vegetable production.

3.2.4 Another traditional soil fertilization technique that is both a soil, as well as, livestock management practice is in-situ manuring. This involves the grazing of livestock, such as, goats, sheep, cows and oxen, on fallow farm fields after the crop has been harvest. As the animals roam about the fields grazing on the crop stubble, they disperse their droppings across the surface of the soil (Fig. 9). It is usually done between successive crops in the spring and autumn or during the fallow winter months when cropping is not economical (Buddhathoki et al. 1992). If in-situ manuring is to be carried out for extended periods (2–3 weeks), farmers may construct temporary shelters on the fields for the animals and tether them at different locations across the field (Subedi et al. 1995). The practice allows for the dung deposited by the animals to be

Fig. 8 Straw mulching on vegetable seed bed. (Source: Author)

Fig. 9 In-situ manuring with livestock on fallow fields. (Source: Author)

spread across the farm field with a minimum of labor and it is subsequently incorporated into the topsoil during tillage. It, thereby, enhances the fertility of the soil by replacing or supplementing the use of FYM.

3.3 Livestock Rearing (Animal Husbandry)

3.3.1 The use of forest leaf-litter and other waste organic materials, such as, grasses and weeds as bedding materials for animals in the livestock stalls is an age-old and common traditional practice (Fig. 10). For protection from predation by wild animals (like leopards and wolves), the livestock stalls are constructed

Fig. 10 Leaf-litter and crop residue pile for animal bedding (left); Livestock shed with animal bedding material (right). (Source: Author)

Fig. 11 Stall feeding of livestock due to limited areas for free grazing. (Source: Author)

adjacent to the farm house or, in many instances, kept on the ground from of the main house with the farmers and their families living above. Once the animal bedding is soiled, it is removed and used in the preparation of local compost FYM. Such bedding material is replaced typically every week or two as required.

3.3.2 As suitable areas for free-grazing of livestock has become limited and with increased livestock numbers in most areas of the lower and middle hills, most farmers are obliged to stall-feed their animals. For this purpose, fodder is gathered from nearby forests and shrub lands. However, in recent decades, fodder collection has become restricted or limited to certain times of the year, and hence, is often inadequate for year-round feeding of the livestock. Therefore, many farmers have planted local fodder species of trees and shrubs on their farm lands to supplement forest fodder collection for stall-feeding (Fig. 11). This type of traditional agroforestry practice can, however, be traced back for generations in the hill regions, as trees on farms have been a mainstay with multiple uses and benefits, such as, fire wood, timber, poles and other construction materials apart from merely fodder for livestock (Sharma et al. 2022; Sherpa 2005).

3.3.3 Transhumance pastoral system of livestock rearing has been a long-standing traditional practice in the high hills of the Himalayan region. It is a free-grazing method that involves leaving the animals to roam free across a large expanse of common mountain highlands over a period of 3–5 months during the summer, and then herding them back to the farmsteads at lower elevations during the winter months (Tiwari et al. 2020). Farmers using transhumance pastoral systems in the high hill regions generally keep sheep, mountain goats and yak. The animals graze on rangelands across altitudes ranging from 1500 to about 5000 m elevations (Fig. 12). This practice has, however, come under pressure from climate change impacts on rangelands, as well as, government policies that allocate more mountain areas to protected areas and restrict free grazing to limited areas (Chaudhary et al. 2017; Tiwari et al. 2020).

3.4 Cropping Systems

3.4.1 Mixed cropping patterns have been a prominent feature of hill farming for millennia, due in part to the characteristics of the land (topography and climate), as well as, the fact that much of the farming is subsistence in nature. Few farmers in the central Himalaya (northern India and Nepal) have the capacity or land holding size to allow for commercial-scale cropping with yield surpluses for sale. Thus, the main strategy has been to meet household dietary requirements through diversified copping by growing both grains and

Fig. 12 The practice of transhumance pastoralism in the Himalayan region. (Source: K.R. Tiwari)

some vegetables and pulses (Aich et al. 2022; Bajracharya et al. 2015a). Nonetheless, this practice also has the advantage of being climate adaptive, and even climate change mitigation potential, if perennial tree crops are included. Various combinations of mixed annual crops are planted by farmers to suit the local topographic and climatic conditions. These include grains like maize, millet and wheat (at lower elevations), and barley, oats and buckwheat (at higher elevations) along with vegetable, tuber and leguminous crops. Many farmers also incorporate fodder and fruit trees along the field borders and between terraces (Fig. 13).

3.4.2 Planting of legumes or pulses like rice beans or mung beans on terrace riser bunds is a traditional practice seen mainly in valleys or gently sloping hill areas with broad, level terraces (Fig. 14). These crops help to stabilize the terrace risers and also contribute some nitrogen and organic matter back to the soil as they are leguminous plants and symbiotically fix nitrogen via Rhizobium bacteria. However, a more tangible benefit for farming households is the dietary supplemental role they play as a protein-rich food source, while also enhancing the crop biodiversity of the hill farms.

3.4.3 Planting of beans in fields with standing maize stalks after harvest of the maize is a traditional practice found mainly in the upland farm areas where maize-millet or maize-mustard systems of cropping dominate (Fig. 15). With this practice, the maize stalks are not cut down after the harvest, rather they serve as support to the climbing bean stalks. This practice also takes advantage of the benefits of growing a leguminous crop between the main growing season (summer) and the dry season (winter) when, at high elevations, the fields are left fallow. The cropping system is not only environmentally friendly, but also helps to provide both cropping diversity, as well as, dietary diversity to the local farmers' meals with the nutritious, protein-rich beans.

3.4.4 Kitchen gardens with a mixed crop of vegetables and greens are a common local practice maintained by most farming and even non-farm households. These gardens are usually established on small (20–50 m^2) plots of land adjacent to the residential homes (Fig. 16). This facilitates the ready utilization of

Fig. 13 Mixed cropping in hill regions (left); mixed farming with fodder trees on farm border (right). (Source: Author)

Fig. 14 Leguminous crop (beans) planted on the terrace riser. (Source: Author)

Fig. 15 Climbing crop (beans or squash) planted into field with standing maize stalks. (Source: Author)

Fig. 16 Kitchen gardens with mixed vegetable crops. (Source: Author)

Fig. 17 Traditional agroforestry systems with locally available trees, perennial and annual crops (left); coffee agroforestry (right). (Source: Author)

waste water from the homes and livestock waste materials for production of the crops. The kitchen gardens are a source of fresh seasonal vegetables and greens that help to augment the diets of the local communities. The most common crops grown on these small plots include squash, pumpkin, gourds of various types, spinach, bok choy, carrots, radish, onions, garlic, chilies, broad beans among others (Aryal et al. 2017).

3.4.5 Traditional hill agroforestry systems with mixed cropping are found throughout the mountainous areas of the Himalaya and have been practiced for centuries. This type of local agroforestry practice consists of a random mix of horticultural and fodder trees with or without annual crops interspersed between the tree crops in open spaces (Aich et al. 2022; Bajracharya et al. 2015b). The practice is typically used on marginal and steep lands where conventional annual cropping is not economical (Fig. 17). The choice of crops and trees depends largely upon the land characteristics like soil depth and rockiness, climate and season length (governed by elevation). The western Himalayas are somewhat drier with less rainfall than the eastern mountains, which generally receive higher precipitation, and this influences the types of perennial crops that perform well. In the central Himalayan region, common fruit trees grown are Asian pear, apple, orange, peach, plum, etc. At lower elevations (1000–1500 m), banana, papaya and coffee (Arabica) have been planted in recent decades (Dahal et al. 2018; Gautam et al. 2017). This type of agroforestry practice is also done by landless tenants through a leasehold forestry scheme promoted by the Nepal government.

3.5 Agricultural Water Management

3.5.1 A stone dam or diversion, locally known as '*bhakari*', on streams is an age old method used by farmers along with canals for the irrigation of crops in agricultural fields (Fig. 18). This practice is the predominant technique for irrigation of crops across the Himalayan region. It is a simple and efficient

Fig. 18 Stone dam to divert water for irrigation (left); canal to transfer water to the agricultural field (right). (Source: Author)

Fig. 19 Wooden aqueduct used for irrigation (left); split-bamboo aqueduct (right). (Source: Author)

method where feasible as it relies only upon gravity to conduct the water to the farm fields (Kandel and Wagley 1999; Shrestha 1998; Shukla and Khanal 1997). It does require, however, that the irrigation canals and water ways be maintained regularly to ensure availability of water when needed and to avoid channel failure which could result in landslides and erosion of the hilly terrain. It should be noted that only a limited proportion of agricultural land in the region has access to such irrigation as it is suitable only for areas downstream of the water abstraction points.

3.5.2 Bamboo or carved wooden aqueducts are indigenous techniques for the delivery of irrigation water across gullies and ravines to the farm fields where it is required (Fig. 19). The choice and type of aqueduct is determined by the availability of resources or materials. In the case of bamboo aqueducts, a length of bamboo of 10–15 cm in diameter is split longitudinally and notches removed to create a dual conduit for the water to flow across the length of the bamboo. Where bamboo is not readily available, a tree trunk can be hollowed out along its length to create a channel for the water flow from one end to the other (Sharma et al. 2009; Shukla and Khanal 1997).

3.5.3 Farm ponds and water harvesting for irrigation and livestock watering (Fig. 20) are indigenous water collection and distribution used in many places throughout the Himalayan region (Shrestha 1998; Sharma et al. 2022). Traditionally, such ponds have been made simply by digging out an area of soil, usually close to the farm house, and the sides and bottom lined with fine clay soil to minimize seepage of the water. In recent decades however, these farm ponds are lined with plastic sheeting, where available, to prevent water seepage. Farm ponds vary in size and may collect water from the roofs of houses and livestock sheds or from the surrounding hill slopes in which case the water is channeled or piped to the pond. In some places, water is transferred from a dug-well or a spring (Fig. 21) to the pond by means of pipes or channels.

3.5.4 The construction of brushwood or stone check-dams for stabilizing gullies to prevent head-ward erosion is a long practices indigenous technique (Fig. 22). Hill areas are particularly susceptible to gully formation and erosion during the rainy season when intense rainfall events or large amounts rain (1000–2000 mm) is deposited over a 4 to 5-month period. The materials used to construct the check-dams depend mostly on local availability. Traditionally, either living or dead branches or bamboo may be used for brushwood check-dams. Where stones are more readily available, stone check-dams may be constructed to trap eroding soil and reducing the velocity of flowing water (Kandel and Wagley 1999; Sharma et al. 2009). In recent decades, with the intervention of government watershed conservation initiatives, such check-dams are often constructed using wire gabion boxes to keep the stones in place or with cement and concrete masonry structures.

3.5.5 Prior to gaining access to electricity, local farming communities have relied on traditional water mills along mountain streams for grinding grains (Fig. 23). Remote areas still use this indigenous technique for making flour from wheat, maize, millet, buckwheat, oats, or other grains (Kandel and Wagley 1999). The water mills have two large stone wheels held within a wooden frame over or adjacent to a small stream. Water from the stream is

Fig. 20 Farm ponds used for irrigating crops, watering animals, or aquaculture. (Source: Author)

Fig. 21 A mountain spring enclosed and covered for protection (left); a dug well (right). (Source: Author)

Fig. 22 Bamboo check dam used for preventing gully erosion (left); brush-wood check dam (right). (Source: Author)

directed through a channel to a wooden paddle-wheel that is connected to the upper stone. As the water turns the paddle-wheel, it rotates the stone grinding wheel to crush the grain. The structure is typically housed inside a shed for convenience and protection from the elements.

3.6 Pest Control

3.6.1 Use of biological measures for grain and seed preservation during storage is an ancient indigenous practice going back many generations. A variety of materials are used, such as, ash, camphor, neem (*Azadiracta indica*) leaves and oil, castor (*Ricinus communis*) oil, chili (*Capsicum sp.*) powder, mint (*Mentha sp.*) leaves, etc. (Sharma et al. 2022; Aich et al. 2022). These materials are usually mixed in with the stored grains, which are kept in a cool dark place in the farm house. Another common practice that is used to repel insect pests from the storage room is to paint the mud floor with a slurry of cow

Fig. 23 Traditional water-mill along mountain stream for grinding grains. (Source: Author)

dung and red clay soil (Sherpa 2005; Sharma et al. 2009). This is believed to keep many types of insects (like termites, ants, cockroaches, etc.) away. Salt mixed in with paddy seeds for long-term storage and the use of coriander seeds with salt for the preservation of oil (for instance groundnut or soybean oil) have also been noted. Additionally, coating seeds with a mix of ghee, honey and cow dung, and mixing chili and neem or mint leaf powder with seeds for long-term storage are other traditional methods used in the region (Sharma et al. 2022).

3.6.2 The use of ash, crushed leaves or plant extracts as an insect repellant spread over the ground or sprayed over crops is a long-practiced traditional method in northwest and northeast India, as well as, in the Nepal Himalaya (Sharma et al. 2009, 2022; Sherpa 2005). Ash spread around plants and maize or fruit trees is often used to control grub worms and ants. Depending on local availability, many different plant extracts or leaves are used, such as, neem, titepati (*Artemisia vulgaris*), tobacco, *Citronella*, khira (*Cucumis sativus*), lemon grass (*Cymbopogon citratus*), etc. The specific combinations of plant extracts or crushed leaves varies locally, however, neem leaf extract or tobacco powder mixed with cattle urine has been reported to be used in India (Sharma et al. 2022), while khira leaf pulp spread around wheat crops is known to be practiced for the control of insect pests in Nepal (Sharma et al. 2009). In addition, lime juice mixed with onion or garlic juice is thought to be used for the control of grub worms in Nepal and northeastern India (Sherpa 2005). Other

Fig. 24 Sun-drying of grains prior to storage (left); hanging of corn cobs on veranda below the farm house roof (right). (Source: Author)

Fig. 25 Different types of bio-fences used to protect crop fields and homesteads. (Source: Author)

combinations of plant extracts include titepati, banmara (*Eupatorium sp.*), and *Lantana camara*; *Citronella* leaves, *Vitex negundo* leaves and lemon grass for the control of diseases.

3.6.3 Another common traditional method of curing or preserving grains like maize, wheat and rice for storage involves sun-drying (Sharma et al. 2009). The loose grains are often spread out over local straw mats for sun-drying before storage in clay containers in a cool dark part of the farm house. For harvested maize crop, the cobs are frequently hung out on the upper balcony of the house just below the roof for drying and storage until they can be processed for grinding into corn meal (Fig. 24).

3.6.4 An environmentally friendly indigenous practice for protecting crops, livestock or the homestead in general from wild animals involves bio-fencing (Fig. 25). Such bio-fences may be constructed from live vegetation, like thorny shrubs and locally found cactus plants, or from brushwood and bamboo (Sharma et al. 2009). Such measures help to protect crops and livestock in farmsteads from wild animal depredation like monkeys, wolves or leopards.

4 Discussion

Indigenous and most traditional approaches for agriculture, land and water management all have in common a local focus and reliance on locally available materials and resources. Thus, they tend to be more in harmony with the natural environment of the place and usually blend in rather than contrast with the ecosystem in which they are practiced. Many of the practices are typically done on a small-scale and with manual or semi-mechanized implements, and hence are less destructive than the large-scale modern agricultural practices. Moreover, as they do not rely on the import or transfer of materials or technology, they may be better suited to the circumstances and existing local conditions (Sharma et al. 2009; Tamang 1993). While some of the practices, such as the choice of crops or agroforestry species, may be site specific and depend on factors like climate, land characteristics and soil types, others may be more suited to scaling-out or replication in different locations.

In view of the adverse impacts of rapidly accelerating global climate change and the anticipated future consequences, there is an urgent need to take drastic action and make a paradigm shift in the way human societies carry out a variety of activities including agriculture, which is a major contributor to atmospheric greenhouse gases. By adopting practices that minimize the use of heavy farm equipment like tractors and mechanized harvesters that use fossil fuels and emit carbon, a significant portion of the GHG emissions to the atmosphere could be avoided. This can be done by using manual or animal draught power and reducing the number of operations. Furthermore, incorporation of perennial with annual crops in mixed farming and agroforestry systems would not only reduce GHG emissions, but also help in carbon capture and sequestration (Bajracharya 2021; Bajracharya et al. 2015a, b; Gautam et al. 2017). Planting of fodder trees and horticultural species like fruit trees have to the advantage of reducing the pressure on nearby forests as much of the needed materials, such as, fodder, leaf-litter and timber can be obtained on-farm. It also avoids the added labor and time costs spent, usually by women, on collecting these materials from the forest. The application of local farmyard manure/compost along with locally prepared biochar offers benefits of enhancing crop production while simultaneously contributing to increase SOC contents and subsequent carbon accumulation (Gautam et al. 2017; Pandit et al. 2018).

Sound and efficient water management is also a crucial component of farm productivity and is affected by the changing and unpredictable climatic patterns in recent decades. Therefore, the development and use of low-cost, local methods for harvesting water and using efficient techniques for irrigating farm land are needed. Indigenous approaches to water management, including harvesting, transferring, storage and irrigation, developed to suit the local situation and requirements offer the best possibilities to be sustainable (Shukla and Khanal 1997; Sharma and Gurung 2005). Adopting traditional pest control measures using plant extracts and other natural substances to repel insects or weeds instead of using chemical pesticides would reduce air, water and land pollution, thereby, avoiding the detrimental environmental and health impacts of persistent toxic chemicals. Bio-fencing and the use of vegetative measures like bamboo and brushwood check-dams to stabilize hill slopes and gullies minimize dependence on external inputs while being low-cost and effective. The adoption of environmentally friendly and sustainable traditional and indigenous practices that are economically viable and locally adapted offers the benefits of being climate resilient (Bajracharya et al. 2015a; LWF Nepal 2017) while also contributing to reducing the carbon footprint and other harmful or polluting consequences of agriculture.

5 Conclusions

Farming communities in the Himalayan region have used traditional and indigenous practices passed down through the generations for millennia. These practices related to tillage, soil management, crop cultivation, animal husbandry, water management, and pest control, have evolved through both local innovations and external initiatives as per the needs of the times. As such indigenously adapted agricultural practices are generally well suited to the local situation and often based on low-cost, efficient and locally available materials. Hence, from a sustainability standpoint they are preferable to imported technologies. Furthermore, these types of practices and technologies are likely to have lower environmental impacts, low carbon footprints, as well as being beneficial for climate change adaptation and mitigation. There is undoubtedly an urgent need for societies to rapidly adopt ways and means to minimize the adverse effects caused by production and development activities, including agriculture which is a major contributor to greenhouse gas emissions and global climate change.

References

Adhikari LK, Karmacharya BB (1988) Trainer's manual: farm implements and machinery. Department of Agriculture, Manpower Development Agricultural Project, Kathmandu, 185pp

Aich A, Dey D, Roy A (2022) Climate change resilient agricultural practices: a learning experience from indigenous communities over India. PLOS Sustain Transform 1(7):e0000022. https://doi.org/10.1371/journal.pstr.0000022, 14 p

Amatya LK, Thapa DR, Sapkota IB (1997) Study of agricultural systems in mid-hills of Nepal through RRA/PRA approach. LARC Occasional Paper No. 97/2. Lumle Agricultural Research Centre, Kaski, Nepal, 29pp

Aryal K, Poudel S, Chaudhary RP, Chhetri N, Ning W, Shaoliang Y, Kotru R (2017) Conservation and management practices of traditional crop genetic diversity by the farmers: a case from Kailash Sacred landscape, Nepal. J Agric Environ 18:15–28

Bajracharya RM (2001) Land preparation: an integral part of farming systems in the mid-hills of Nepal. Nepal J Sci Technol 3:15–24

Bajracharya RM (2021) Regenerative approach for sustainability and climate resilience of mountain agroecosystems. In: Bajracharya RM, Sitaula BK, Raut N, Gurung S (eds) Sustainable natural resource management in the Himalayan region: livelihood and climate change. Nova Science Publ. Inc., New York, pp 42–56

Bajracharya RM, Dahal NR, Bajracharya SB, Gautam N (2013) Biochar as a soil amendment and its indigenous use for crop production on marginal lands in the mountains of Nepal. Paper presented at the Int'l. Conf. on Forest, People & Climate: Changing Paradigm, 28–30 August 2013, Pokhara

Bajracharya RM, Sherchan DP, Dahal BM, Raut N (2014) Soil management for sustainable agricultural intensification in the Himalayan region. In: Lal R, Stewart BA (eds) Soil management of smallholder agriculture. Taylor & Francis Publ, Boca Raton, pp 143–164

Bajracharya RM, Atreya K, Raut N, Shrestha HL, Gautam DK, Dahal NR (2015a) Sustainable diversified agriculture and land management in the Himalaya: implications for climate change adaptation and mitigation. J Mountain Area Res 1:20–52. Publ. Online

Bajracharya RM, Shrestha HL, Shakya R, Sitaula BK (2015b) Agro-forestry systems as a means to achieve carbon co-benefits in Nepal. J Forest Livelihood 13(1):59–68

Buddhathoki K, Gurung GB, Lokar DP (1992) Vegetable crop indigenous knowledge and technology in the western hills of Nepal. LARC Seminar Paper No. 92/13. Lumle Agricultural Research Centre, Kaski, Nepal, 25pp

Chaudhary RP, Bhattarai SH, Basnet G, Bhatta KP, Uprety Y, Bhatta LD, Kotru R, Oli BN, Sharma LN, Khanal S, Sharma UR (2017) Traditional practice and knowledge of indigenous and local communities in Kailash Sacred Landscape, Nepal. Int'l. Center for Integrated Mountain Development, ICIMOD working paper 2017/1, Kathmandu, 30 p

Dahal N, Bajracharya RM, Wagle LM (2018) Biochar effects on carbon stocks in the coffee agroforestry systems of the Himalayas. Sustain Agric Res 7(4):103–114

Dhakal B (1993) The *Buk* method of potato farming in the high hills of Nepal. In: Tamang D, Gill GJ, Thapa GB (eds) Indegenous management of natural resources in Nepal. HMG/Ministry of Agriculture and Winrock International, Kathmandu, Nepal, pp 120–125

Firebaugh FM (1990) Sustainable agricultural systems: a concluding view. In: Edwards CA, Lal R, Madden P, Miller RH, House G (eds) Sustainable agriculture systems. Soil and Water Conservation Society, Ankeny, pp 674–676

Gautam DK, Bajracharya RM, Sitaula BK (2017) Effects of biochar and farm yard manure on soil properties and crop growth in an agroforestry system in the Himalaya. Sustain Agric Res 6(4):74–82

Gill GJ (1993) Indigenous systems in agriculture and natural resource management – an overview. In: Tamang D, Gill GJ, Thapa GB (eds) Indigenous management of natural resources in Nepal. HMG/Ministry of Agriculture and Winrock International, Kathmandu, pp 24–51

Harwood RR (1990) A history of sustainable agriculture. In: Edwards CA, Lal R, Madden P, Miller RH, House G (eds) Sustainable agriculture systems. Soil and Water Conservation Society, Ankeny, pp 3–19

Kandel DD, Wagley MP (1999) Some salient indigenous technology practices for watershed management in Nepal. Participatory Watershed Management Training in Asia, Field Document No. 19

LWF Nepal (2017) Identification of indigenous, eco-friendly and climate-resilient agricultural practices in Nepal, A Field Study Report, The Lutheran World Federation Nepal, Kathmandu, 30 p

Pandit NR, Mulder J, Hale SE, Martinsen V, Schmidt HP, Cornelissen G (2018) Biochar improves maize growth by alleviation of nutrient stress in a moderately acidic low-input Nepalese soil. Sci Total Environ 625:1380–1389

Sharma S, Gurung B (2005) Indigenous water quality management practices in Dhading district, Nepal. Environ J Environ 9(10):61–67. Min Envir Sci & Tech HMG/Nepal

Sharma S, Bajracharya RM, Sitaula B (2009) Indigenous technology knowledge (ITK) in watershed management – a review of studies from Nepal. Indian J Tradit Knowl 8(4):569–576

Sharma A, Manpoong C, Pandey H, Gupta CK, Baja Y, Singh MS, Mounglang CC (2022) Practices of indigenous agricultural knowledge of farmers in India. www.preprints.org. https://doi.org/10.20944/preprints202206.0071.v1, 83 p

Sherpa NS (2005) Indigenous peoples of Nepal and traditional knowledge. UNDESA Int'l Workshop on Traditional Knowledge, Panama City, 21–23 September 2005

Shrestha S (1998) A compilation of people's indigenous knowledge (ITK) for watershed management in Nepal. B.Sc. Project Report, Kathmandu University

Shukla A, Khanal PR (1997) Indigenous knowledge in irrigation management: Some lessons from farmer managed irrigation systems in Nepal. Paper presented at the Nat'l. Workshop on Indigenous Technology Knowledge for Watershed Management. Dep. Soil Conser. & Watershed Mgmt., HMG/Nepal, 25–27 December, 1997, Institute of Forestry, Tribhuvan Univ., Pokhara

Singh SP (1998) Chronic disturbance, a principal cause of environmental degradation in developing countries. Environ Conserv 25:1–2

Subedi KD, Sthapit BR, Suwal MRS, Joshi KD (1995) Search and utilization of indigenous green manure and bio-fertilizers in the western hills of Nepal. In: Joshi KD, Vaidya AK, Tripathi BP, Pound B (eds) Formulating a strategy for soil fertility research in the hills of Nepal. Worshop proceedings, 17-18 August 1995. Lumle Agricutural Research Centre, Kaski, Nepal, pp 55–62

Tamang D (1993) Challenges and opportunities in farm and community resource management in Nepal. In: Tamang D, Gill G, Thapa GB (eds) Indigenous management of natural resources in Nepal. HMG/Ministry of Agriculture and Winrock International, Kathmandu, pp 13–23

Tiwari KR, Sitaula BK, Bajracharya RM, Raut N, Bhusal P, Sengel M (2020) Vulnerability of pastorialism: a case study from the high mountains of Nepal. Sustainability 2020(12):2737. https://doi.org/10.3390/su12072737

Turton CNR, Gurung GB, Sherchan DP (1995) Traditional farming systems of red soil areas: the eastern hills of Nepal. PAC Technical Paper No. 164, Pakhribas Agricultural Research Centre, Dhankuta, Nepal, 19pp

Traditional Farming in the Indian Himalayan Region: Preserving Agro-Biodiversity and Future Food Security

Sanoj Kumar Patel, Anil Sharma, Anil Barla, and Gopal Shankar Singh

1 Introduction

The agricultural system provides a vast proportion of the global food supply and potentially ensures the majority of ecosystem services. Agriculture is imperative for worldwide food security and an initiative to fulfil the United Nations Sustainable Development Goals (UNSDGs) (Viana et al. 2022). The expansion of agriculture is important because approximately 820 million people worldwide suffer from chronic starvation, more than 149 million children are undernourished, and approximately 2 billion people lack sufficient micronutrients in their daily diet (Altieri and Nicholls 2020). Thus, agricultural expansion serves as the most potent tool for eradicating poverty and growing the global economy by feeding the expected 9.7 billion people by 2050 (Duro et al. 2020). According to projections, approximately 670 million people will still be hungry in 2030, accounting for 8% of the global population, which is the same as when the 2030 Agenda was introduced in 2015 (FAO 2022). At the global scale, starvation affected more than 425 million people in Asia, 278 million in Africa and 56.5 million in Latin America in 2021 (FAO 2022). This indicated that the world's people with the most hunger were in Asia, followed by Africa. Nearly 40% of people worldwide are experiencing moderate or severe food hardship and food insecurity. The prevalence of severe food insecurity and hunger

S. K. Patel · A. Sharma · A. Barla · G. S. Singh (✉)
Institute of Environment and Sustainable Development, Banaras Hindu University, Varanasi, India

A. Borthakur, P. Singh (eds.), *The Resilience of Traditional Knowledge Systems for a Sustainable Future*, Advances in Asian Human-Environmental Research,
https://doi.org/10.1007/978-3-031-56858-9_2

increased from 9.3% in 2012 to 11.7% in 2019, which is equivalent to 207 million people in both years. Therefore, agriculture is important for human life and survival.

Agriculture is the ultimate source of food, producing a range of food products that contribute to the dietary needs of the global population. It including food, fruit, vegetables, pharmaceuticals, livestock and fish for human feed and survival. This diversity in food production also diversifies people's dietary systems and provides balanced nutritive feed that contains various essential nutrients. A diverse farming system strengthens the supply of grains, vegetables and fruits, which are vital for sustaining good health and wealth (Patel et al. 2022). Agriculture production directly or indirectly contributes nearly 90% of food calories and 80% of protein and fat requirements (Cassidy et al. 2013; Steinfeld et al. 2006). According to the current evaluation, 7039 plant species are edible, and most of these species have multiple uses (Ulian et al. 2020). The most common use is medicinal (70%), followed by material (59%) and environmental use (40%). The agriculture industry is intricately linked to socioeconomic factors and food security (Patel et al. 2023a). Smallholders play an important role in global food production by producing 50–70% of the total global food production. However, small farms (less than 2 ha) are essential and have local relevance in South Asia, Sub-Saharan Africa and Southeast Asia, where they produce approximately 30% of most food commodities (Herrero et al. 2017).

The expanding world population and climate change have created a shortage of agricultural production, resulting in hunger and nutrient imbalances (Patel et al. 2019a, b; Ulian et al. 2020). The increase in agricultural production on currently available cultivated land is a critical task (Godfray et al. 2010). These challenges become more complicated by the addition of various other factors, such as resilience to few crops, decreased cultivated land and global climate change (Sharma et al. 2023). To address these issues, we need to improve food production worldwide. This enhancement cannot be accomplished by the simple expansion of modern cultivation, which occurs at the expense of the surrounding ecosystem and biodiversity decrease (Ulian et al. 2020). Therefore, it is necessary to rehabilitate degraded modern cultivation systems by diversifying the farming of highly nutritive and climate-adaptive traditional crops in a way that increases crop production through conventional cultivation practices. Biodiversity is critical for tackling this dual dilemma, which requires improved knowledge of the global state of food resources (Ulian et al. 2020).

Traditional agriculture is based on the principle of sustainable utilization of natural resources management and depends on full organic input. Indigenous farming creates a new pathway for sustainable production as well as new local business opportunities for local farmers to produce and sell traditional crops for social and economic well-being. These advantages, when combined, support livelihoods and social justice, climate change mitigation and adaptation in local farmers. Traditional agricultural practices provide a highly flexible strategy for enhancing previous Green Revolution investments in ways that meet both environmental and social restrictions, limiting both mainstream agriculture and sustainable agriculture.

Indigenous agricultural knowledge is very broad and diverse in nature, which has been acquired, developed, and polished over millennia. The indigenous knowledge of agriculture is very effective in meeting food production requirements, especially for land preparation, seed sowing, weeding and crop harvesting under severe cultivation conditions. This knowledge is widely available to smallholder farmers, who often cultivate less than 2 ha of cultivated land. These small-holder farmers are important for global food production because they produce 70–80% of food worldwide (FAO 2014). Indigenous crop varieties from a certain region play an important role in food security, where mixed cropping and crop rotation produce a tremendous harvest (Pandey et al. 2017). The Himalayan region of India is renowned for its unique and varied agroecosystems, which have been formed by centuries's traditional knowledge and methods (Nüsser and Baghel 2016). These diverse and tribe-specific traditional farming methods can help to preserve rich cultural and agro-biodiversity in the region (Patel et al. 2020b). Thus, present chapter reviews and describes the unique traditional agricultural practices of the Himalayan region of India. This section further describes the traditional techniques and sustainable production potential of these farming systems.

2 Traditional Knowledge: Principles and Objectives

Traditional knowledge may be crucial in the development of sustainable agricultural systems, increasing the likelihood that rural people will embrace, develop and maintain through innovations and interventions. It may be defined as the collection of experiences and knowledge of an indigenous community that serves as the foundation for decision-making in the face of both familiar and unknown difficulties and obstacles (Warren and Cashman 1988). This information has accumulated over many years and is considered a key and significant element of society's culture and technology. Indigenous knowledge operates under the economic and physical constraints of society and actively contributes to the cultural issues of society. This knowledge is conveyed and conserved by community members and used to accomplish certain goals in conjunction with productive activity within the community (Bell 1979). The transfer of traditional knowledge from one generation to another occurs through the use of communicated stories, legends, folklore, rituals, music, art, and even legislation. Contrary to popular belief, indigenous knowledge is not a static phenomenon; rather, it continually develops in response to changes in the internal and external environments of the community. Knowledge of resource and ecosystem dynamics, as well as associated management practices, exists among community members. These interactions with ecosystems for benefits and acquired livelihoods are developed on a daily basis as well as over long periods of time (Fabricius and Koch 2004; Patel et al. 2018).

Numerous scientific publications have documented, how indigenous knowledge systems may help management of environment, conserve biodiversity, and enhance the sustainable use of natural resources (Sharma et al. 2021; Patel et al. 2020a). Indigenous knowledge also influences wildlife conservation, maintenance of rural setups, promotion of traditional ethno-medicinal systems and preparedness for natural disasters and responses (IPBES 2013; Patel et al. 2020b). India has a long and rich history of sustainable cultivation knowledge and culture, which are abundant in every part of the country, especially in the Himalayan region (Ulian et al. 2020).

3 Traditional Agricultural Knowledge: A Sustainable Approach for Food Security

Indigenous knowledge is holistic in nature because of its numerous uses in different disciplines such as diverse agriculture, soil health, water systems, climate, animals, forestry, and human health (Howes and Chambers 1980; Jungerius 1985; Wilken 1987; Agrawal 1995; Pulido and Bocco 2003). Traditional farming plays an important role in the development of scientific agricultural knowledge (Sandor and Furbee 1996; Singh et al. 1997). Traditional farming comprises biodiverse, resilient, energetically efficient, socially justifiable forms of agriculture. This practice serves as an energy-efficient production, climate-smart agricultural strategy that has fed much of the world's population from centuries. Alley cropping is an example of successful cultivation of traditional agriculture, where crops are cultivated between the alleys formed by hedgerows of leguminous trees. This cultivation approach may be a more lucrative and environmentally stable alternative to the destructive effect of shifting cultivation. Traditional agriculture has the ability to significantly increase food production by moderating the environmental stress in Africa (Kang et al. 1984). Most of the small farmers in Asian countries, usually cultivate local cultivars of rice in highland environments under rain-fed circumstances for efficient production (Koohafkan and Altieri 2010) (Fig. 1).

4 Traditional Knowledge of Sustainable Agriculture in India

The diverse agroclimatic zone of India comprises a mosaic of traditional agricultural techniques that created a heterogeneous farming structure with varying soil and vegetation cover (Sharma et al. 2018). This diverse farming provides ecological adaptability, economic stability, and survival capability of farm with supporting different cultural and religious knowledge systems of society (Patel et al. 2020a, b). The eight states of the northeastern region of India comprise a mountainous landscape interspersed with plain and basin topography. Farmers execute various types of rainfed cultivation on these landscapes that conserve soil erosion and biodiversity together with preservation of various native crop varieties. The most ancient

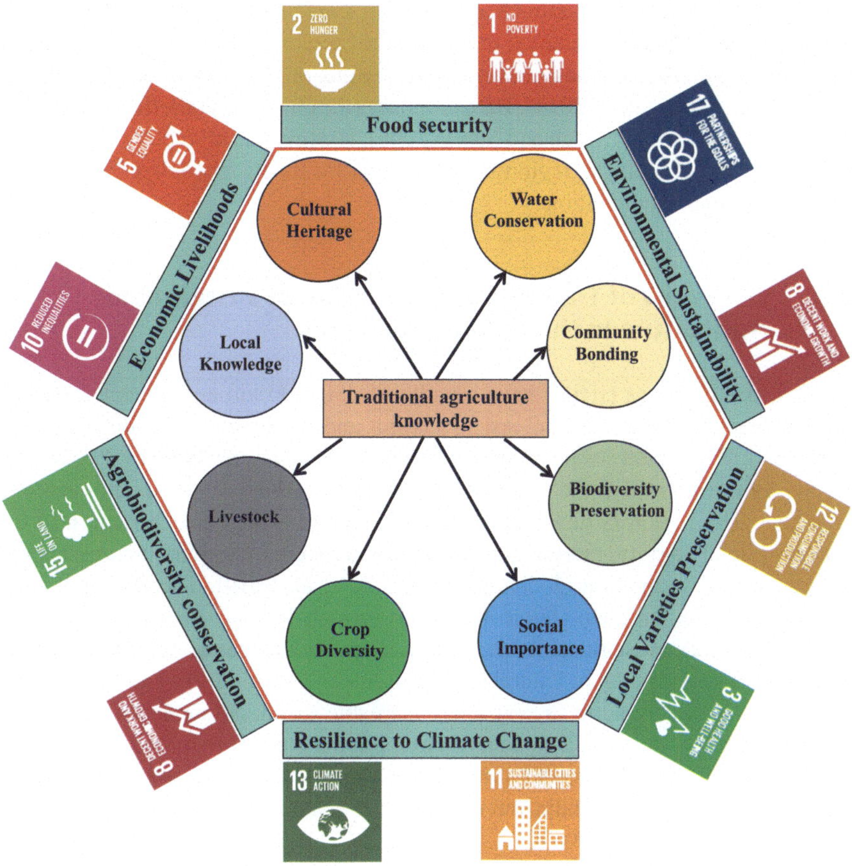

Fig. 1 Representing the sustainability approach of traditional agricultural knowledge and its contribution to SDGs. Source: The authors

cultivation practices, including shifting cultivation (jhum farming), Alder (*Alnus nepalensis*)-based farming (Ramakrishnan 1992; Kehie et al. 2017), Zabo cultivation (Singh and Munda 2008), the Bun shifting method (Upadhaya et al. 2020), drip irrigation of bamboo (Van der Kooij et al. 2013), cattle shed rotation in upland and pond-based farming systems (Singh et al. 2016), have been recorded in the northeast region of India. The western Himalayan region of the country is dominated by mountain farming (Singh et al. 1997), traditional agroforestry systems (Bijalwan 2012; Phondani et al. 2020; Vishvakarma et al. 1998), homesteads (Kyaroo) (Kashyap et al. 2014), cow dung composting (Singh et al. 1997, 2016) and diversification of cultivation by traditional crops (Semwal and Maikhuri 2015). On the other hand, the coastal and Deccan regions of southern India include the Aquaforestry (Dagar et al. 1991), Kaipad (rice–fish farming) (Vanaja 2013) and Pannendu Pantalu (Bhushan et al. 2001) agricultural practices. However, Walora or Jhumming (Patel et al. 2020a), dry land farming (Venkateswarlu and Shanker 2012), Kanabandi farming (Mathur 1995), Khadin (Kolarkar et al. 1983) and Raab (Choudhary et al. 2015) practices were documented in the dry-arid region of Gujarat and Rajasthan. The central Indian and Vindhyan highlands regions of the country are famous for the Taungya system (Basu 2014), Badi cropping (Singh et al. 2006), livestock tracking and Utera cropping (Singh and Sureja 2008). However, in all reported regions of the country, the Himalayan region is well known for its diverse indigenous knowledge of sustainable cultivation.

5 Overview of Traditional Knowledge of Sustainable Agriculture in the Indian Himalayas

The Indian Himalayas (IHRs) have a width of 250–300 km and stretch 2500 km from Arunachal Pradesh in the east to Jammu and Kashmir in the south (Nüsser et al. 2019). This region covers an area of 533,604 km^2 and is inhabited by 39,628,311 people. Physiologically, the IHR is divided into eastern, central and western Himalaya. The region is distinguished by mountains, which are characterized by inaccessibility, fragility, marginality, variety (heterogeneity), niche (natural appropriateness) and adaptation (Jodha 1992; Nüsser et al. 2019). The IHR is spread across the 13 Indian state/Union Territories, which include Arunachal Pradesh, Sikkim, Manipur, West Bengal, Meghalaya, Tripura, Nagaland, Assam, Mizoram, Himachal Pradesh, Ladakh, Uttarakhand, Jammu and Kashmir. In addition to having a diverse population, the area also has adaptable social, political, economic, and environmental institutions. It sustains an exceptional ethnic group that has a varied society and culture and a wealth of indigenous knowledge. There are a total of 573 designated tribes living in India, out of which 173 inhabited in the IHR (Samal et al. 2000; Bergmann et al. 2008).

Physiographically, the mountain range of the IHR starts from the foothills of the south, called Siwalik, and extends to the Tibetan Plateau in the north, known as the Trans-Himalaya. Geographically, it divides into Himadri (greater Himalayas), Himachal (lesser Himalayas) and Siwalik (outer Himalayas), which are uninterrupted throughout its entire length. The climate of the IHR greatly depends on the altitudinal gradient, which ranges from a height of <800 m (subtropical) to >3600 m (arctic zone) (Nüsser and Gerwin 2008). According to the Champion and Seth (1968) classification, the subtropical semidesert type of forest is situated in the foothills of Jammu & Kashmir as well as western Himachal Pradesh, whereas the Montane subtropical vegetation recorded at an altitude of 500–1500 m is characterized as broad-leaved vegetation of the eastern Himalayas. The pine forest is ruled in the lower part of Siwalik together with the chir pine forest, whereas dry evergreen vegetation is ruled in the Terai-Bhabar tract. Characteristic montane wet temperate forests dominated at altitudes of 1800–3000 m in Darjeeling, Arunachal Pradesh and some parts of West Bengal. The hill parts of the western and eastern IHR have more biodiversity because of high rainfall and good soil conditions. However, the temperate climate of the cold desert Ladakh in the western IHR and eastern IHR of Sikkim, Darjeeling, is limited by insufficient moisture and climatic conditions.

The Himalayan Mountains feature a dense forest with more than a thousand different types of tree, shrub and herb vegetation (Singh 2004). It supported approximately 50% of the flowering flora of India with 8000 angiosperm species and 44 species of gymnosperms. Compared with commercial agriculture, mountain terrain and inaccessibility have contributed to the preservation of traditional agri-biodiversity. For a very long time, the Himalayan people practiced integrated agriculture, which involves balance farming, agroforestry, animal husbandry, and forestry. This diverse farming occurs due to the presence of varied land use patterns at different altitudes in landscapes with various climatic conditions.

5.1 Traditional Knowledge of Sustainable Cultivation in IHR and Its Environmental Sustainability

A significant proportion of the mountainous people of the Himalayan region are entirely dependent upon agrarian as well as forest-based natural resources for their survival and livelihood (Ramakrishnan 1997). However, these resources are managed sustainably through traditional ecological knowledge systems (Farooquee et al. 2007). The traditional Himalayan cultivation system is a complex network of agriculture, livestock rearing, forestry and horticulture (Ramakrishnan 1993; Dollo et al. 2006). The agriculture of mountain regions is diverse in nature, ranging from slash and burn cultivation (Ramakrishnan and Patnaik 1992), field crop production (De 2021), terrace or bun cultivation (De 2021), and mountain farming (Singh et al. 1997) to sedentary practices such as Apatani (Chaudhary et al. 1993), Bari (Saravanan 2010) and pakho khet (Mishra et al. 2020) (Table 1). Multiple cropping (De 2021), Barahnaja cultivation (Gururani et al. 2021), and Tar Bari (Sharma and Dhakal 2011) are renowned for multiple crop cultivation, where various crop species are cultivated in a single patch of land. However, Dhankheti (Subba 2009), Alainchi Bari (Negi et al. 2018) and Alder-based agriculture (Rathore et al. 2010) are low-diversity cultivation systems that depend on few crops. Jhum/shifting/slash and burn, field crop production and large cardamom plantations are forest-based cultivation, where agricultural land is shifted from one place to another place. These phenomena maintain soil fertility with high crop production, while the rotation of cultivated land helps forest conservation. This type of cultivation was mostly performed by the Nishis, Karbis, and Kacharis tribes. However, alder-based agriculture, the Zabo system, large cardamom plantations and diversified hill horticulture are forestry-based cultivation methods (Table 1).

5.2 Traditional Knowledge of Soil Management

The loss of fertile soil from agricultural land is a common phenomenon in which the upper layer of soil erodes due to water flow. More than 80% of global agriculture suffers from soil erosion, which causes land degradation and moisture loss (Pimentel

Table 1 List of traditional knowledge-based sustainable agriculture practices in the Indian-Himalayan region

S.N.	Traditional agriculture practices	Feature	Tribal	State	References
1	Jhum/ Shifting/Slash and burn	This type of cultivation is mostly performed by hill farmers that involves cleaning and burning of vegetation in order to increase yield and minimize the negative impact of agriculture on the environment	Most of the tribal communities	Entire Himalayan region	Singh (2007)
2	Field crop production	It is an alternative of shifting cultivation where pulses and rice grow on upland, while lowland utilized for rice cultivation	Most of the tribal communities	Entire Himalayan region	De (2021)
3	Double cropping of rice	In the Northeastern region, prekharif rice cultivated by adjusting the planting date and selecting early varieties such as Vivek Dhan-82, Megha Rice1, and VL Dhan 61 before the main rice crop cultivation	Most of the tribal communities	Entire Himalayan region	De (2021)
4	Multiple cropping-based production system	It is prevalent in shifting farming, where farmers planted 8–35 crop species on a small plot of 2–2.5 hectares with simultaneous sowing and sequential harvesting	Most of the tribal communities	Entire Himalayan region	De (2021)
5	Terrace or Bun cultivation	This kind of farming is particularly common in Meghalaya, where crops are grown on a series of raised beds called "*Bun*" in the local language which are made down the hillside's slope	Khasis, Jaintias and Garos	Meghalaya	Jeeva et al. (2006)

(continued)

Table 1 (continued)

S.N.	Traditional agriculture practices	Feature	Tribal	State	References
6	Rice-fish system of Apatani plateau	It is a multipurpose water management technique that integrates land, water, and farming by preventing soil erosion, saving water for irrigation, and cultivating paddy-cum-fish	Apatani	Arunachal Pradesh, Manipur and Nagaland	Roy (1986), Chaudhary et al. (1993)
7	Alder based agriculture	Alder (*Alnus nepalansis*) is a legume tree that fixes atmospheric nitrogen for nutrition and its leaf-litter helps to conserve soil moisture in jhum production	Angami, Chakhesang, Chang, Yimchunger and Konyak	Nagaland	Rathore et al. (2010)
8	Zabo system	The term "Zabo" implies for water impoundment that combines forest, agricultural, and animal husbandry which create a solid foundation in soil and water conservation	Chakhesang tribe	Nagaland	Singh et al. (2014)
9	Large Cardamom Plantation	The farmers of Lepcha and Limbu community collect large cardamom from wild. This collected cardamom domesticated in the shade of alder tree. Where alder plant fixes atmospheric nitrogen for soil fertility in sloppy land and cardamom plant produces large production under the shade of this tree	Lepcha and Limbu	Sikkim	Singh et al. (1989)

(continued)

Table 1 (continued)

S.N.	Traditional agriculture practices	Feature	Tribal	State	References
10	Diversified Hill Horticulture	The terrain of Eastern Himalaya provides an ideal setting to diversify agriculture which dominated by horticulture. This system generated an excellent production options for sustaining livelihoods of hill farmers	Most of the tribes	Arunachal Pradesh	De et al. (2007)
11	Bari	The Adi tribal of Siang valley in the North–East India has practiced a traditional homestead agroforestry system that demonstrating sustainability, productivity, food security, environmental protection, biodiversity conservation, and natural resource management	Adi tribe	Entire North east region	Saravanan (2010)
12	Barahnaja cultivation	It is a traditional orphan crop-based mixed cropping system that guaranteed a sustainable crop yield assurance against variable weather circumstances	Most of tribes	Uttarakhand	Gururania et al. (2021)
13	Rachyan ghuryan (Home Garden)	A plot of land that generally situated around a home and mostly used for producing seasonal vegetables, fruits and certain traditional crops	Most of tribes	Sikkim	Sharma et al. (2009)

(continued)

Table 1 (continued)

S.N.	Traditional agriculture practices	Feature	Tribal	State	References
14	Pakho Khet (Terrace paddy field in a sloppy land)	It is terraced rice fields with a slope of >30 in sloppy region, where ground fodder is produced along the high terrace raisers	Most of tribes	Sikkim	Mishra et al. (2020)
15	Byansi (Flat paddy field)	The flat plains along with riverbeds also known as Thang, are typically watered by neighboring streams that accumulated a nutrient-rich soil collects from upstream, making an extremely fruitful and productive system	Most of tribes	Sikkim	Sharma et al. (2009)
16	Shim/ Kholyang (Paddy field with a natural spring)	Shim is a term for a tiny type of dug pond used for water storage. It is used to prevent seepages by planting appropriate tree species in their catchments, which is important for seasonal drinking water supply to residence, grazing cattle and wild animals in dry season	Most of tribes	Sikkim	Sharma et al. (2009)
17	Bagarey (Paddy field at the bank of a river)	Bagarey is a high productivity system formed at river bank, where rice, wheat, lentils, buckwheat and soybeans are grown	Most of tribes	Sikkim	Upadhyay and Rai (2023)
18	Birauto (Newly terraced paddy field)	The Birauto is a freshly terraced rice field, in which farmers continue to preserve farms, repairing and managing slopes through terracing	Most of tribes	Sikkim	Sharma et al. (2009)

(continued)

Table 1 (continued)

S.N.	Traditional agriculture practices	Feature	Tribal	State	References
19	Tar bari (Flat dry field)	Tar bari is a non-irrigated flat dry field used for producing maize, vegetables, ginger, orange, wheat, barley, beans, pulses, oilseeds, millets, buckwheat and other crops	Most of tribes	Sikkim	Sharma and Dhakal (2011)
20	Kothe bari (Homestead Garden)	It is a terraced arid field with slope of more than 30 degrees, where farmers cultivate maize, millets, pulses, and grass fodder. Land left over from the dry season are managed by slash and burn techniques throughout the winter	Most of tribes	Sikkim	Kumar et al. (2019)
21	Bhasmey/ Khoria (Slopy land, slashed and burn for cultivation)	These pieces of land are generally unsuitable for terracing or other agricultural practices. Thus, cattle's are grazed in winter on those field followed by cultivation of mixed landraces of pulses, maize and horse gram	Most of tribes	Sikkim	Kumar et al. (2019)
22	Siru bari (Dry field for growing thatch grass)	Khar is used by small farmers to thatching the roofs of livestock shelters, where farmers plant lumber, fodder and fuelwood tree species in a dry infertile land	Most of tribes	Sikkim	Sharma et al. (2009)
23	Alainchi bari (Large cardamom agroforestry)	In this process, large cardamom has been cultivated under *Alnus nepalensis* in traditional agroforestry system	Most of tribes	Sikkim	Negi et al. (2018)

(continued)

Table 1 (continued)

S.N.	Traditional agriculture practices	Feature	Tribal	State	References
24	Bansghari + Jungle	The Bans-ghari is a collection of trees near the edge of unfertile sloppy agricultural land which are unsuitable for cultivation. This practice avoids landslide and soil erosion while also meeting household needs	Most of tribes	Sikkim	Mishra et al. (2019)
25	Panikheti system	The farmers of Nagaland and Sikkim have performed cultivation on bench terraces since millennia, that contains >100% slope with stony fields with just 10–15 cm of soil depth	Abor, Ahom, Aka, Bhutia, Lepcha, Limboo	Nagaland, Sikkim, Manipur	Das et al. (2012)
26	Cattle shed rotation in upland	A temporary cattle house is constructed in the field to collect in situ urine, cow manure and organic litters	Bhutia, Lepcha, Limboo	Sikkim	Das et al. (2012)
27	Pond based farming systems	It is a man engineered pond near to the houses for water harvesting, where fish + cattle + Areca nut + vegetable gardens + rice are grown. The irrigation performed using harvest pond water and farmyard manure	Kukis, Nagas	Tripura, Assam and Manipur	Das et al. (2012)
28	Mountain farming	Natural resources-based crop was cultivated on high slope mountain	Most of the farmers	Himachal Pradesh	Singh et al. (1997)
29	Kyaroo (homesteads)	A number of tree species is growing for fuel, fodder and timber along with livestock, poultry and fish to satisfying the basic need of the farmer	Most of the farmers	Himachal Pradesh	Kashyap et al. (2014)

(continued)

Table 1 (continued)

S.N.	Traditional agriculture practices	Feature	Tribal	State	References
30	Dhankheti	It is a direct sowing of paddy without irrigation in upland of sloppy region	Bhutia, Lepcha, Limboo	Sikkim	Subba (2009)
31	Terrace farming	It is the narrow strips of land, carved out across the hill slopes for cultivation, where cereals crop mainly cultivated with soil conservation	Most of the farmers	Sikkim	Mishra and Rai (2013)
32	Alley cropping	In the watershed landscape of the hilly region, tree and shrubs are planted in the row to create alley along with agriculture crops	Most of the farmers	Sikkim	Mishra and Rai (2013)
33	Bench terrace farming	In this process, hilly terrain region is changed into graduated step across the slope	Most of the farmers	Sikkim	Chapagain and Raizada (2017)
34	Traditional tree-based farming	Trees are integrated in the crop cultivation according to the climatic condition	Entire Himalayan region	Most of the farmers	Sharma et al. (2007)
35	Pit farming for seedling	Seedling is sowing in a 1 m depth pit and top is covered with colored polythene for moisture	Most of the tribal community	Entire Himalayan region	Ansari et al. (2021)
36	Intercropping	Cultivation of two or more crop on the similar field where each crop is separately seeded or planted in separated row	Most of the tribal community	Entire Himalayan region	Singh and Singh (2017)
37	Multiple farming	Farmers grow 10–30 species on a same plot with simultaneous sowing and sequential harvesting	Most of the tribal community	Entire Himalayan region	De (2021), Maikhuri et al. (2019)

Source: The authors

and Kounang 1998). The degradation of fertile soil through soil erosion is a major problem in the Himalayan region. Soil erosion has a negative impact not only on agricultural output but also on the environmental and ecological functions of soil. The local population of the Himalayan region has developed a series of soil and water conservation techniques based on indigenous knowledge. Terracing is one of the oldest endogenous practices for soil conservation in Sikkim (Mishra and Rai 2013). This included agroforestry and cereal cultivation on a narrow strip of land carved across hill slopes. The Kebang tribes of Arunachal Pradesh implemented land-following periods in a specific time period of each year cycle to maintain soil health and fertility (Table 2). Farmers commonly plant tree species on the border of fields to create vegetation barriers, while some common practices, such as mixed cropping, crop rotation, reduced tillage, green manure and agroforestry, are used to conserve soil health in hilly regions (Patel et al. 2020a, b). Mulching is one of the most effective practices for sustainable moisture conservation and protection of early growing buds of crops (Negi and Kandpal 2003). However, a stone barrier is used to prevent the eroded soil from leaving the field and entering the river. However, a series of local techniques, including bamboo culms, mind/bidd cultivation, bunds and water, have been applied to reduce the flow of running water as well as decrease the strength of slopes to prevent soil erosion.

5.3 Traditional Knowledge of Irrigation and Water Management

Water is one of the most valuable and common natural resources. The conservation and management of water resources in the Himalayan region are typically unique. They applied various indigenous knowledge in the irrigation of the field as well as the conservation of water resources (Nüsser 2017). Through experience with local conditions, the local community of the Himalayan region developed a network of Khals along the face of hill slides, which work to harvest rainwater (Lal and Verma 2008). This system is similar to the traditional pond system and is built equidistant in a symmetrical line along the land slope. This practice generally occurs in low-rainfall areas, where rainwater harvesting is performed to store rainwater for further irrigation (Table 3). In regions with high hills and steep slopes, bamboo drip irrigation, bench terrace irrigation and Cheo-ozihi practices have been used to channelized water to accommodate a series of terraces (Jeeva et al. 2006). However, kuhl/gul practices are used to channel water from the hilltop to the foot plain of the land along the slope gradient of gravity (Negi and Kandpal 2003; Nüsser and Schmidt 2018).

In the case of water conservation, Dongs is used to collect rainwater, which is further redistributed to the field by an instrument called *lahoni* to prevent the loss of water (Dhiman and Gupta 2011). However, Zing is an indigenous practice for collecting water from melting glaciers in the cold desert region of Ladakh (Namtak and Sharma 2018). Gools harvest rainwater for storage, which is redistributed to the field through the use of canal systems (Table 2). In the Choyas and Khatris techniques, seepage water from agricultural fields and hard rock that is insufficient for

Table 2 Collection of important traditional knowledge-based soil conservation, water management and irrigation practices in the Indian Himalayan region

S.N.	Practices	Feature	Tribal	State	References
1	Fallow period	Kebang tribes preformed directives fallowing of land with offered grazing in each year to preserve soil health and nutrient status	Kebang	Arunachal Pradesh	Saravanan (2010)
2	Bamboo culms	Bamboo culms are planted at the bottom of the slope to slow down the flow of running water in order to reduce erosion caused by water flow	Most of the farmers	Sikkim	Jeeva et al. (2006)
3	Mind/Bidd cultivation	Farmers used a variety of methods to flatten the slopy terrain in this practice, such as lowering the slope of hills, creating a tiered agricultural landscape	Most of the farmers	Himachal Pradesh	Sarkar et al. (2015)
4	Mulching	Crop residue and dry fodder are used for mulch in open land, which works as to retain the soil moisture	Most of the farmers	Uttarakhand	Negi and Kandpal (2003)
5	Bunds	It is the small embankments created at outer edge of the terraces to control the water flow and prevents soil erosion	Most of the farmers	Sikkim	Mishra and Rai (2013)
6	Water ways	It is a crucial component of steep terraced farming, which helps control run-off to minimize soil erosion and aids in land management	Most of the farmers	Sikkim	Mishra and Rai (2013)
7	Stone barriers	Some farmers constructed a stone barrier with the help of pebbles and stone particles at the field boundary to check the soil erosion	Most of the farmers	Sikkim	Mishra and Rai (2013)
8	Farm yard mannure	Decomposed mixture of crop litter, animal excreta and fodder are applied in the agricultural land to enhance the organic matter and soil fertility	Most of the farmers	All the Himalayan state	Maikhuri et al. (2019)

(continued)

Table 2 (continued)

S.N.	Practices	Feature	Tribal	State	References
9	In situ manure	Cattles are allowed to graze directly into agricultural field in winter season, where animal excreta is straight incorporated with soil	Most of the farmers	All the Himalayan state	Maikhuri et al. (2001)
10	Crop residue manure	Burnt residues of crops are directly ploughed into the farms	Most of the farmers	All the Himalayan state	Upadhyay and Rai (2022)
11	Green manure	Non cereal and short-term crops such as *alfa–alfa* and grass clover are grown and integrated with soil for improving soil fertility	Most of the farmers	All the Himalayan state	Rodríguez et al. (2022)
12	Use of stones, gunny bags filled with soil	Stones and gunny bags filled with soils are generally used for prevention of erosion in terrace farm	Most of the farmers	All the Himalayan state	Jeeva et al. (2006)
13	Gools	It is also called Gaad or rivulets named *Gadheram*, where water from a rainfed small rivers to channeled in agricultural fields using a network of canals based on gravity flow	Most of the farmers	Central Himalayan	Rao et al. (2003)
14	Choyas	Choyas are small seasonal seepages near to the farm that discharge insufficient water for canalization. The store water is redirected to a terrace which has higher slope compared to the main farm	Most of the farmers	Central Himalayan	Rao et al. (2003)
15	Dhara	Dharas are thought to be underground water conduits that revealed on lower or mid-slopes in wet season. They provide essential water needed in dry season	Most of the farmers	Central Himalayan	Rao et al. (2003)
16	Dongs	It is a constructed pond used for rainwater harvesting and irrigation. The harvested water redistributed into field through the instrument called *lahoni*	Bodo tribes	Assam	Dhiman and Gupta (2011)

(continued)

Table 2 (continued)

S.N.	Practices	Feature	Tribal	State	References
17	Khatris	It is a rectangular deep pit on the hard rocks of hill slope, where water is collected through seepage from the rock	Most of the farmers	Himachal Pradesh	Sharma and Kanwar (2009)
18	Zings	Zings are water collection devices frequently used in Ladakh. These are little tanks that catch water from melting glacier	Most of the farmers	Ladakh	Namtak and Sharma (2018)
19	Kools	These are ancient irrigation channels system, locally known as k*ools*, which help to water and soil conservation through indigenously developed management practices	Jammu and Kashmir	Most of the farmers	Slathia et al. (2016)
21	Bamboo drip irrigation	The bamboo drip irrigation system is performed using locally available bamboo species for irrigation, where hill slopes are quite steep with low soil depth	Meghalaya	Khasis, Jaintias and Garos	Jeeva et al. (2006)
22	Bench terrace irrigation practice	The water of hill streams are channeled to sequence of terraces to support the retention of water	Meghalaya	Khasis, Jaintias and Garos	Jeeva et al. (2006)
23	Kuhl/gul	It is the mud line canal carved, that out flow from the hill slope along the gradient of gravity	Most of the farmers	Uttarakhand	Negi and Kandpal (2003)
24	Khals	It is similar to traditional pond system, which is generally found in low rainfall area that is used to collect rain water for further irrigation	Most of the farmers	Uttarakhand	Negi and Kandpal (2003)
25	Cheo-ozihi	It is similar to bamboo drip irrigation system, where main channel of Mezii river brought down terraces to 8–10 km by the bamboo pipe channel	Most of the farmers	Nagaland	Agarwal and Narain (1997)

Source: The authors

Table 3 Traditional knowledge of sustainable pest crop management in Indian Himalayan region

S.N.	Indigenous knowledge	Description	References
1.	Covered head of maize cob with paper or cloths	Head of maize cobs are covered with paper or cloths to keep away birds	Tyagi et al. (2018)
2.	Use of *Artemisia parviflora*	Fresh leaves/branches of *Artemisia parviflora* used into soil as well as around the paddy field to avoid pest infestation in crops	Ansari et al. (2021)
3.	Use of Chinese mustard	Chinese mustard plants were planted on the border of cabbage field to draw pests and lessen that would harm the main crops	Ansari et al. (2021)
4.	Bamboo tree planting near parika (tree pea) plant	Bamboo shoot have antibacterial and antifungal properties. If they grow in the field of main crops it will be lessen fungal assault to the crops	Chongtham et al. (2011)
5.	Use of *clerodendrum* species	The plant of *clerodendrum* species is grown around the paddy field. Which provide space for birds to perch and hunt insects in paddy field	Chandra et al. (2017)
6.	Duck rearing in paddy field	Duck rearing is used in paddy field for pest control at seedling stage to pre flowering stage	Men et al. (1999)
7.	Use of *Vitex negundo* L.	The twig of *Vitex negundo* is used in paddy field that produced foul smells, which act as repellent for pest	Bhattacharjee and Ray (2010)
8.	Use of rotten crab and fermented fish	Rotten crab and fermented fish is used to lure the gundhi insects into the trap	Chandra et al. (2017)
9.	Use of *Datura stramonium*	*D. stramonium* plants are grown on the bunds of paddy fields. Because mature seeds have bitter taste which acts as a poison for rates	Tyagi et al. (2018)
10.	Use of ash	Rice straw's ash is used in the agricultural fields to maintain moisture level in the field, which also acts as a pest repellent fertilizers source in field	Bidyalakshmi et al. (2023)

Source: The authors

canalization are retained and rechannelled to agricultural fields for irrigation (Rao et al. 2003).

5.4 Traditional Knowledge for the Conservation of Local Crop Varieties

The distinctive mountainous terrain of the Himalayas consists of different physiographic and topographic elements that allow locality-specific microclimatic conditions, giving rise to a variety of ecosystems that are rich in agrobiodiversity (Sharma and Dhakal 2011; Singh et al. 2018). The region is an agro-biodiversity hotspot dominated by traditional agriculture that provides more than 75% of the population employment (Gururani et al. 2021). Uttarakhand's Barahnaja practice is characterized as a 12-grain system that includes not only cereal but also oilseeds, spices, fibre, and vegetable crops (Gururani et al. 2021). The most commonly cultivated crops in this system are finger millet (*Eleusine coracana),* Buckwheat *(Fagopyrum esculentum),* Maize *(Zea mays),* Horsegram *(Macrotyloma uniflorum),* Amaranth *(Amaranthus* spp.) and Perilla *(Perilla frutescens).* Multiple cropping systems of Arunachal Pradesh include rice + maize, maize + French bean, rice + banana, rice + cucurbits, and rice + citrus systems to prevent nutrient loss from soil and improve soil conditions and nutrient cycling (De 2021). In the northeastern Himalayan region, jhum cultivation is the dominant practice, comprises 5–45 traditional crop species. On the other hand, the terraced cultivation of Sikkim consists of agricultural land in the foothill region, horticulture in the middle portion and a silvipastoral system at the top portion of the hill slope. High-value cash crop-based agroforestry, including cardamom (*Amomum subulatum*), mandarin orange (*Citrus reticulata*), and bamboo cultivation dominate the slope-based cultivation of Sikkim.

The Himalayan region is characterized by valley rice cultivation practices along the riverbank, as well as on terraced slopes and lower hills. These rice systems contributed a rich gene pool of native rice varieties that were adapted to local climate conditions, such as Ghyya-Dhan, Marshi, Takmari and Bhuindhan, which grow on dry land. Attey, Nuniya, Kataka Timmurey, Krishnabhog, Mansaro, Baghey-Tulashi, Champasari, Bachhi, Sikrey and Taprey are the irrigated rice varieties. Among these indigenous paddy varieties, Krishnabhog, Kataka and Nuniya are well known for their aroma, fine grain quality and medicinal significance, which considered equivalent to Indian basmati rice (Sharma and Liang 2006). The Lepcha community of Sikkim State is also famous for the cultivation of large indigenous varieties of rice, including Tukmorzho, Kalo nunia (small Basmati), Birinful and Attey (Chettri and Sathyanarayana 2021). Apart from rice cultivation in valleys, farmers cultivate a number of pulses, beans and vegetables crops along elevated bunds and terrace edges (Sharma and Dhakal 2011). In the winter season, maize, vegetables, oil seeds, wheat, buckwheat and many other crops are cultivated together with rice in these systems. Cardamom-based agroforestry practices are performed between the terraced opening of paddy fields along slopes (Sharma and Dhakal 2011). Most intercropping systems, viz., ginger-maize, pulse-turmeric, maize-potato, paddy-soybean, cardamom, and mandarin-based intercropping, are cultivated as understorey crops.

The Apatani Plateau is known as the bowl of rice due to the cultivation of a number of rice landraces via fish culture performed by Apatani tribes (Das et al. 2012). They also engaged in dry hill farming, cultivated millet, maize and vegetables and, to some extent jhuming (Maikhuri and Ramakrishanan 1990). However, the local people of the central Himalaya also perform dry hill farming and cultivate traditional crops, including *Glycine max, Hordeum himalyens, Echinochloa frumentacea, Fagopyrum botrydis, Setaria italic, Fagopyrum tataricum* and *Panicum miliaceum,* at all altitudes (Saxena et al. 2005). Home gardens are complex and diverse systems that represent an intriguing agroecosystem for the sustainable management of crop and wild edible plant resources (Majumder et al. 2011). A total of 59 wild edible plants were recorded from various cultivation systems in Chakmas village of Arunachal Pradesh. Nearly 51 species grow in home gardens, while 33 plant species grow in various jhum fields (Majumder et al. 2011).

5.5 Traditional Knowledge of Sustainable Pest Management in Crop Systems

The farmers of the IHR follow a number of traditional practices for pest management and disease control for various crops. Diverse knowledge was recorded for different crops, which also varied from one tribe to another and from one region to another region (Table 3). The plant residue of Titeypati (*Artemisia vulgaris*), which has good larvicidal, insecticidal, antibacterial and antifungal properties, is used as mulch for the nursery preparation of Ragi (*Eleusine coracana*) and Ginger (*Zingiber officinale*) crops in Sikkim (Ghosh et al. 2013).

However, plants of *Datura stramonium* is planted on the bunds of paddies to prevent attack by mice. The *Datura* plant is bitter according to tests, and seeds act as poisonants (Tyagi et al. 2018). The heads of maize cobs are covered with paper or cloth to protect birds, while fresh leaves/branches of *Artemisia parviflora* spread in the field to avoid pest infestation (Ansari et al. 2021). Wood ash is generally used for the storage of cereals and pulses, and rice straw ash is applied in agricultural fields to maintain moisture and pest repellency (Bidyalakshmi et al. 2023). In which ash kills insects by desiccating them or by obstructing their movement and appears by filling the intergranular gaps (Chandola et al. 2011). Bamboo shoots have antibacterial and antifungal properties that protect rice and tea plants from bacterial and fungal assault (Ansari et al. 2021; Chongtham et al. 2011). On the other hand, Chinese mustard is planted on the border of cabbage fields to attract pests, while the twigs of *Vitex negundo* plants are used to repel pests (Chongtham et al. 2011; Bhattacharjee and Ray 2010). In a few cases, farmers have applied duck rearing to reduce the impact of pests during flowering in paddy fields, whereas rotten crab and fermented fish are used to lure gundhi insects into traps (Men et al. 1999; Chandra et al. 2017).

5.6 Traditional Farmer Groups Supporting Sustainable Farming

The Indian Himalayas are globally recognized for their rich eco-cultural legacy and richness of traditional ecological expertise among farmers. Traditional farmer organizations play a crucial role in establishing and maintaining sustainable outputs in certain agroecosystems. The Dokpas and Dzumsa nomadic tribal communities are agro-pastoralists in nature and survive in drought and cold-stricken climates through mobile livestock production. They rear dzos, sheep, yards, and goats in plateaus and meadows of the Greater and Trans-Himalayan regions (Sharma and Dhakal 2011).

Agroecosystems below the Trans-Himalayan range are subsistence in nature practiced by the Lepchas, Bhutias, and Sherpas communities, which grow a range of crops and fruits, including barley, maize, wheat, beans potato, cabbage, apple, and peas. The occurrence of high agrobiodiversity is attributable to differences in cultural and traditional eating preferences among ethnic populations. The cultivation of crop types and the ethnic dietary habits of mountain societies are ingrained in their culture, customs, and customary rites. Indigenous groups protect domesticated agricultural landraces, wild cuisine, and underutilized crops for traditional and cultural rites and festivals.

Apatani are well known for their rice and fish agriculture in the valley, which produces sufficient rice to export after fulfilling local requirements. The Apatani tribes of Arunachal Pradesh are recognized for modified wet rice cultivation and are locally referred to as *Aji*. This system involves rice and fish cultivation with millet on bunds that separate plants and is recognized as one of the most productive and energy-efficient agricultural systems in the region (Ramakrishnan 1994). Monpas and Sherdukpens of the state of Arunachal Pradesh and various other tribes of the plain region of Assam, Tripura, also performed this modified form of Apatani (Majumder et al. 2011). A small group of the Chakma population of Arunachal Pradesh, famous for rice cultivation. They cultivated 41 local varieties of rice in diverse agriculture, 22 of which were dry land varieties and 18 of which were wetland varieties (Majumder 2007).

Lepcha and Limbu tribes of Sikkim collected large cardamoms from forest regions that were domesticated under the shade of forestry plants such as alder (Sharma and Dhakal 2011). The Galo tribes are acknowledged for cultivating major crops such as maize, mandarin orange, tuber crops and turmeric with rice (Bhagawati et al. 2017). The understanding of crop-soil interactions, ethno-pedology and nutrient management in traditional agricultural practices cannot be managed by a single person in the tribe community. With the implementation of an indigenous cultivation system, the tribal community constructs different groups or subtribes that take care of the functioning and execution of the practice. The Adi tribes organized a village council called Kebang, which takes care operation and function of the practice (Raj 2010). For example, different groups and working categories of Apatani tribes that operate rice-fish cultivation are described below (Table 4).

5.7 Policy Framework and Recommendations for the Future

Indigenous knowledge is a crucial component in the creation of sustainable agricultural systems that are organic in nature, increasing the possibility of tribal community development and creating and sustaining innovations and interventions. Most of the developmental activities fail in these remote areas do not considering the community's preference for skills, culture and knowledge. Thus, there is a need to implement plans and techniques that are familiar to indigenous communities and involve participation by local populations. The indigenous knowledge contains a collection of techniques and principles for sustainable cultivation. During the 2002

Table 4 Types of tribal subgroup organization and their work in the Apatani tribal for rice-fish cultivation

S. N.	Local name	Description	Group manager	Task
1	Bogo	An elected farmer group, who take-care sharing of common water resource in community	Bogo Ahtoh (male)	Construction and maintenance of water supply
2	Aji Lenda	A group of 50–350 household farmers, whose land within same area	Lenda Kagenee (male/ female)	Contraction and maintenance of foot-path that reached to the field
3	Sulu-sikhii	A group of 50–350 farmers, whose land fall within the same area	Sulu Kagenee (male)	Construction and maintenance of fencing to protect the field from wild animal
4	Tanser Patang	It is a group of 5–15 family, who help in crop sowing	Patang Ahtoh (female)	They help in field preparation, nursery farming and weeding
5	Konchi Patang	It is a group of 5–10 farmers' family, who help in field preparation but works from 5–8 am	Patang Ahtoh (female)	Field preparation, transplantation and weeding activities
6	Halying Patang	A group of 5–10 household who help in field seedling transplantation	Patang Ahtoh (female)	Transplanting seedlings of rice and millets
7	Enthee Patang	It is 8–10 members group, help during the crop harvesting	Patang Ahtoh (male/ female)	They carried out agricultural harvesting and field transportation
8	Bijee Lenda	A group of 70–300 household, having bamboo garden	Lenda Kagenee (male)	Construction and maintenance of foot-path, which is used for carrying bamboo, fuelwood and timber

Source: Dollo (2007), Dollo et al. (2009)

World Summit on Sustainable Development, the United Nations Food and Agriculture Organization (FAO) launched an international partnership program for the conservation and management of traditional agricultural practices called Globally Important Agricultural Heritage Systems (GIAHS). The main objective of the GIAHS Program is to identify and protect globally significant agricultural heritage systems, as well as the accompanying landscapes, agricultural biodiversity, knowledge systems, and cultures. The three GIAHS sites are located in India, where the Pampore Saffron Heritage of Jammu & Kashmir falls within the HRI. Smallholder Innovation for Resilience (SIFOR), a plan incited to promote the indigenous knowledge-based revolution system for the generation of sustainable food security in the changing climate era, with the participation of India, Kenya, China and Peru supported by International Institute for Environment and Development (European Union and UK). The International Centre for Integrated Mountain Development

(ICIMOD) is a UNFCCC-sponsored intergovernmental learning and knowledge sharing programme that includes eight countries of the Hindu Kush Himalayas, namely, India, Pakistan, China, Nepal, Bangladesh, Afghanistan, Myanmar, and Bhutan. The preservation of this indigenous knowledge is also encouraged by the Convention on Biological Diversity (CBD), FAO, World Health Organization (WHO), UNESCO, United Nations Commissions on Human Rights (UNCHR) and other organizations.

In the Indian context, the government of India has impelled a number of judicial and administrative initiatives to conserve and document indigenous knowledge for sustainable production. The Traditional Knowledge Digital Library (TKDL), which is a collaborative plan of the Council of Scientific and Industrial Research (CSIR), Department of Ayurveda, Ministry of Science and Technology and Ministry of Health and Family Welfare, aims to develop a database of traditional knowledge. This database helps preserve and prevent the misuse of indigenous knowledge systems. The National Action Plan on Climate Change (NAPCC) is India's national strategy for implementing environmentally sustainable development using its own resources. The C.P. Ramaswami Environmental Education Centre (CPREEC) and Environmental Information System (ENVIS) work to conserve indigenous ecological knowledge and the heritage of sacred natural sites in the Indian Himalayas. The cooperative management of forest resources in Van Panchayat, which is a village community in Uttarakhand, is the oldest practice (Negi et al. 2023). Several local farmer movements, such as 'Save the Seed from Tehri Garhwal district' (Juyal and Sati 2010), also work to conserve indigenous knowledge systems for sustainable production.

6 Conclusion and Recommendations

The traditional Himalayan knowledge system is highly effective and viable for sustainable cultivation, where abundant organic inputs are added via nature-based cultivation techniques. This indigenous knowledge is limited not only to cultivation but also to crop variety selection, water harvesting and irrigation, soil management and crop protection. The knowledge of sustainable agriculture also varies among the different tribal communities as well as the different altitudes of hilly landscapes. The diversification of hill agriculture conserves a range of traditional cultures, social associations, habitats and native crop gene pools that are capable of increasing the productivity and profitability of farmers in a sustainable way. Through their diverse agroecological features, traditional agricultural techniques have the capacity to adapt to and reduce climate change. Additionally, these methods are based on low-cost and energy-efficient techniques that use locally available resources. Traditional agriculture not only contributes to sustainable agriculture but also helps to protect human health, manage natural resources, save energy, and maintain socio-ecological integrity. Agroforestry, rice-fish systems, alder-based agriculture, pakho khet, bench terrace farming and livestock-based cultivation can be used as models for climate-smart and sustainable production systems.

Therefore, expanding the present production system through the conservation of indigenous community customs, traditions, beliefs and heritage can be achieved through traditional knowledge-based cultivation. A noble option for mitigating and adapting to climate change would be to combine traditional practices with contemporary sustainable agricultural operations. The government must fund more research in these fields for documentation and conservation of indigenous knowledge of the marginal community. The digital documentation of this knowledge can be improved through app-based tactics for rapid popularization and promotion in society. A regional database of SDGs fulfilled by these indigenous knowledge systems was constructed to determine the full potential of these practices for sustainable socio-ecological profit. The Indian scientific community should generate geographical indicators (GI) and intellectual property rights (IPRs) for best indigenous practices on the basis of their application. Superior knowledge management and capacity building through exchanges between scientific communities, work-based learning, and developing the capabilities of educational institutions, government and nongovernment organizations, and local groups.

References

Agarwal A, Narain S (1997) Dying wisdom: Rise, fall and potential of India's traditional water harvesting systems. (State of India's environment – a citizens' report, No. 4). Centre for Science & Environment (CSE), New Delhi, pp 11–12

Agrawal A (1995) Indigenous and scientific knowledge: some critical comments. Indigenous Knowl Dev Monitor 3:3–5

Altieri MA, Nicholls CI (2020) Agroecology: challenges and opportunities for farming in the Anthropocene. Cienciae investigaciónagraria: revista latinoamericana deciencias de la agricultura 47(3):204–215

Ansari MA, Sharmaa SK, Roy SS, Ramakrishna Y, Datt S, Ningombama A, Prakash N (2021) Documenting the agriculture based indigenous traditional knowledge in Manipur State of North Eastern India. Indian J Tradit Knowl (IJTK) 20(4):1065–1074

Basu JP (2014) Agroforestry, climate change mitigation and livelihood security in India. N Z J For Sci 44(1):1–10

Bell M (1979) The exploitation of indigenous knowledge or the indigenous exploitation of knowledge: whose use of what for what? IDS Bull 10:44–50

Bergmann C, Gerwin M, Nüsser M, Sax WS (2008) Living in a high mountain border region: the case of the 'Bhotiyas' of the Indo-Chinese border region. J Mt Sci 5:209–217

Bhagawati K, Sen A, Shukla KK (2017) Seasonal calendar and gender disaggregated daily activities of indigenous Galo farmers of eastern Himalayan region of India. Curr Agric Res J 5(3):325

Bhattacharjee PP, Ray DC (2010) Pest management beliefs and practices of Manipuri rice farmers in Barak Valley, Assam. Indian J Tradit Knowl 9(4):673–676

Bhushan C, Srinidhi A, Kumar V, Singh G (2001) Rising to the Call: good practices of climate change adaptation in India. Centre for Science and Environment, New Delhi

Bidyalakshmi T, Bembem K, Narsaiah K, Singh HJ, Singh HD (2023) Traditional knowledge system on paddy straw management in North-East India. Indian J Tradit Knowl 22(2):255–263. https://doi.org/10.56042/ijtk.v22i2.49062

Bijalwan A (2012) Structure, composition and diversity of horticulture trees and agricultural crops productivity under traditional agri-horticulture system in mid hill situation of Garhwal Himalaya, India. Am J Plant Sci 3(4):18702,9

Cassidy ES, West PC, Gerber JS, Foley JA (2013) Redefining agricultural yields: from tonnes to people nourished per hectare. Environ Res Lett 8:034015. https://doi.org/10.1088/1748-9326/8/3/034015

Champion HG, Seth SK (1968) A revised survey of the forests types in India. Government of India Publ, New Delhi

Chandola M, Rathore S, Kumar B (2011) Indigenous pest management practices prevalent among hill farmers of Uttarakhand. Indian J Tradit Knowl 10(2):311–315

Chandra A, Das R, Singh B, Pait B, Singh R, Bhagawati R, Gogoi J (2017) Puntius spp. as a potential lure against Gundhi bug in North East region of India. Environ Ecol 35:863–866

Chapagain T, Raizada MN (2017) Agronomic challenges and opportunities for smallholder terrace agriculture in developing countries. Front Plant Sci 8:331. https://doi.org/10.3389/fpls.2017.00331

Chaudhary RG, Dwivedi RN, Dutta KK, Sarma BK, Patel CS, Prasad RN (1993) Rice based farming of Apatani-an efficient indigenous system of hill farming. Indian Journal of Hill Farming 6(1):93–102

Chettri D, Sathyanarayana N (2021) Traditional rice landraces and Indigenous knowledge associated with the rice cultivation in Sikkim Himalaya: an analysis based on local farming practices. Indian J Hill Farming 34(1):1–15

Chongtham N, Bisht MS, Haorongbam S (2011) Nutritional properties of bamboo shoots: potential and prospects for utilization as a health food. Compr Rev Food Sci Food Saf 10(3):153–168. https://doi.org/10.1111/j.1541-4337.2011.00147

Choudhary R, Viradia HM, Desai CS (2015) Raab-The traditional agricultural practice in Dang district of Gujarat. Indian Council Agric Res:19–4786

Dagar JC, Mongia AD, Bandyopadhyay AK (1991) Mangroves of Andaman and Nicobar Islands. Oxford & IBH Publication Co., Oxford, pp 157–166

Das A, Ramkrushna GI, Choudhury BU, Munda GC, Patel DP, Ngachan SV, Kumar M (2012) Natural resource conservation through indigenous farming systems: wisdom alive in North East India. Indian J Tradit Knowl 11(3):505–513

De LC (2021) Traditional knowledge practices of North East India for sustainable agriculture. J Pharmacognosy Phytochem 10(1):549–556

De LC, Sanwal SK, Singh, Akath BKD, Kumar R, Yadav RK (2007) Organic input management in important horticultural crops of NEH region. In: Munda GC, Ghosh PK, Anup Das, Ngachan SV, Bujarbarua KM (eds) Advances in organic farming technology in India, published by the Director, ICAR Research Complex for NEH Region, Meghalaya, pp 247–258

Dhiman SC, Gupta S (2011) Rainwater harvesting and artificial recharge. Central Ground Water Board, Ministry of Water Resources, New Delhi

Dollo M (2007) Traditional farmers' groups supporting sustainable farming. Leisa-Leusden 23(1):22

Dollo M, Chaudhury S, Sundriyal R (2006) Traditional farming and land tenure systems in West Kameng district, Arunachal Pradesh. In: Ramakrishnan PS, Sexena KG, Rao KS (eds) Shifting agriculture and sustainable development of North-Eastern India, UNESCO-MAB series. Oxford & IBH, New Delhi, pp 293–315

Dollo M, Samal PK, Sundriyal RC, Kumar K (2009) Environmentally sustainable traditional natural resource management and conservation in Ziro Valley, Arunachal Himalaya, India. J Am Sci 5(5):41–52

Duro JA, Lauk C, Kastner T, Erb KH, Haberl H (2020) Global inequalities in food consumption, cropland demand and land-use efficiency: a decomposition analysis. Glob Environ Change 64:102124

Fabricius C, Koch E (2004) Rights, resources and rural development: community-based natural resource management in Southern Africa. Earthscan, London

FAO (2014) The state of food and agriculture innovation in family farming. Food and Agriculture Organization of The United Nations Rome

FAO (2022) The State of Food Security and Nutrition in the World. Repurposing food and agricultural policies to make healthy diets more affordable. FAO, Rome. https://doi.org/10.4060/cc0639en

Farooquee NA, Dollo M, Kala CP (2007) Traditional wisdom of Apatani Community in the management and sharing of natural resources in North Eastern India. In: Misra KK (ed) Traditional knowledge in contemporary societies: challenges and opportunities. Pratibha Prakashan Shakti Nagar, Delhi, pp 110–126

Ghosh T, Mitra P, Mitra PK (2013) Effect of leaves of *Artemisia vulgaris* L. on growth of rats. Int J Herbal Med 1(2):30–34

Godfray HCJ, Beddington JR, Crute IR, Haddad L, Lawrence D, Muir JF, Pretty J, Robinson S, Thomas SM, Toulmin C (2010) Food security: the challenge of feeding 9 billion people. Science 327:812–818

Gururani K, Sood S, Kumar A, Joshi DC, Pandey D, Sharma AR (2021) Mainstreaming Barahnaja cultivation for food and nutritional security in the Himalayan region. Biodivers Conserv 30:551–574

Herrero M, Thornton PK, Power B, Bogard JR, Remans R, Fritz S, Gerber JS, Nelson G, See L, Waha K, Watson RA, West PC, Samberg LH, van de Steeg J, Stephenson E, van Wijk M, Havlík P (2017) Farming and the geography of nutrient production for human use: a transdisciplinary analysis. Lancet Planetary Health *1*(1):e33–e42. https://doi.org/10.1016/S2542-5196(17)30007-4

Howes M, Chambers R (1980) Indigenous knowledge: analysis, implications and issues. In: Brokensha D (ed) Indigenous knowledge and development. University Press of America, Lanham, pp 323–334

IPBES (Intergovernmental Platform on Biodiversity and Ecosystem Services) (2013) Initial elements of an IPBES approach: towards principles and procedures for working with Indigenous and Local Knowledge (ILK) systems. Available: http://www.missouribotanicalgarden.org/Portals/0/wlbc/PDF's/IPBES%20ILK%20Report%20Tokyo%202013.pdf

Jeeva SRDN, Laloo RC, Mishra BP (2006) Traditional agricultural practices in Meghalaya, North East India. Indian J Tradit Knowl 5(1):7–18

Jodha NS (1992) Mountain perspective and its utility: a framework for development strategies (unpublished). Paper presented in the workshop at Manali on 'Approaches to sustainable development of Indian Himalayas', August 1–4, Manali

Jungerius PD (1985) Perception and use of the physical environment in peasant societies. Geographical Papers, vol 93. University of Reading, England, Department of Geography, p 19

Juyal RP, Sati MC (2010) Natural-cultural practices in conservation of traditional crop diversity in mountain: a study of Uttarakhand state, Indian Himalayas. J Biodivers 1(2):103–110

Kang BT, Wilson GF, Lawson TL (1984) Alley cropping: a stable alternative to shifting cultivation. International Institute of Tropical Agriculture, Ibadan

Kashyap SD, Dagar JC, Pant KS, Yewale AG (2014) Soil conservation and ecosystem stability: natural resource management through agroforestry in Northwestern Himalayan region. In: Dagar JC, Singh AK, Arunachalam A (eds) Agroforestry systems in India: livelihood security & ecosystem services. Springer, New Delhi, pp 21–55

Kehie M, Khamu S, Kehie P (2017) Indigenous alder-based farming practices in Nagaland, India: a sustainable agricultural model. J Tradit Folk Pract 5(2):82–152

Kolarkar AS, Murthy KNK, Singh N (1983) 'Khadin' A method of harvesting water for agriculture in the Thar Desert. J Arid Environ 6(1):59–66

Koohafkan P, Altieri MA (2010) Globally important agricultural heritage systems: a legacy for the future. UNFAO, Rome

Kumar D, Chavan S, Ram A, Dev I, Newaj R (2019) Management of hill and mountain soil of North Eastern India through agroforestry interventions. Indian Farming 69(10):47–48

Lal C, Verma LR (2008) Indigenous technological knowledge on soil and water management from Himachal Himalaya. Indian J Tradit Knowl 7(3):485–493

Maikhuri RK, Ramakrishanan PS (1990) Ecological analysis of cluster of villages emphasizing land use of different tribes in Meghalaya in north-east India. Agric Ecosyst Environ 31:17–37

Maikhuri RK, Semwal RL, Rao KS, Saxena KG, Das A K (2001) Indigenous techniques of agricultural soil fertility maintenance in the central Himalaya. Ecol. Environ. Conserv, 7:15-20.

Maikhuri RK, Rawat LS, Maletha A, Phondani PC, Semwal RL, Bahuguna YM, Bisht TS (2019) Community response and adaptation to climate change in Central Himalaya, Uttarakhand, India. In: Garkoti SC, Bloem SJV, Fule, PZ, Semwal RL (eds) Tropical ecosystems: structure, functions and challenges in the face of global change, pp 213–231. https://doi.org/10.1007/978-981-13-8249-9_11

Majumder M (2007) Crop diversity, microbial biomass and soil nutrient dynamics of agroecosystems in 'Chakma' villages adjoining Namdhapa national park, Arunachal Pradesh Ph.D. thesis. Rajiv Gandhi University, Itanagar

Majumder M, Shukla AK, Arunachalam A (2011) Agricultural practices in Northeast India and options for sustainable management. Biodiversity, Biofuels, Agroforestry and Conservation Agriculture, pp 287–315

Mathur BL (1995) Kana bundi indigenous method of controlling wind erosion. Honey Bee 12(7):55–81

Men BX, Tinh TK, Preston TR, Ogle RB, Lindberg JE (1999) Use of local ducklings to control insect pests and weeds in the growing rice field. Livest Res Rural Dev 11(2):1–10

Mishra PK, Rai SC (2013) Use of indigenous soil and water conservation practices among farmers in Sikkim Himalaya. Indian J Tradit Knowl 12(3):454–464

Mishra P, Rai A, Rai S (2019) Agronomic measures in traditional soil and water conservation practices in the Sikkim Himalaya, India. Am Res J Agric 5:1–16

Mishra PK, Rai A, Rai SC (2020) Indigenous knowledge of terrace management for soil and water conservation in the Sikkim Himalaya, India. Indian J Tradit Knowl 19(3):475–485

Namtak S, Sharma RC (2018) Traditional knowledge and innovative practices for water management and conservation in Indian cold desert, Leh Ladakh. Indian J Ecol 45(4):737–744

Negi GCS, Kandpal KD (2003) Traditional methods of water management in the Central Himalayan agriculture. Indian J Tradit Knowl 2(3):256–264

Negi BK, Joshi RK, Pandey A (2018) Status of large cardamom (*Amomum subulatum* roxb.) Farming systems in the changing scenario of modern economics of Sikkim, Himalaya. Glob J Biosci Biotechnol 7:189–199

Negi VS, Pathak R, Thakur S, Joshi RK, Bhatt ID, Rawal RS (2023) Scoping the need of Mainstreaming indigenous knowledge for sustainable use of bioresources in the Indian Himalayan region. Environ Manag 72(1):135–146

Nüsser M (2017) Socio-hydrology: a new perspective on mountain waterscapes at the nexus of natural and social processes. Mt Res Dev 37(4):518–520

Nüsser M, Baghel R (2016) Local knowledge and global concerns: artificial glaciers as a focus of environmental knowledge and development interventions. In: Meusburger P, Freytag T, Suarsana L (eds) Ethnic and cultural dimensions of knowledge. Knowledge and space, vol 8. Springer, Cham. https://doi.org/10.1007/978-3-319-21900-4_9

Nüsser M, Gerwin M (2008) Diversity, complexity and dynamics: land use patterns in the Central Himalayas of Kumaon, Northern India. Divers Mountain Syst:107–119

Nüsser M, Schmidt S (2018) Nanga Parbat revisited: evolution and dynamics of sociohydrological interactions in the Northwestern Himalaya. In Mountains: physical, human-environmental, and sociocultural dynamics. Routledge, pp 166–178

Nüsser M, Dame J, Parveen S, Kraus B, Baghel R, Schmidt S (2019) Cryosphere-fed irrigation networks in the northwestern Himalaya: precarious livelihoods and adaptation strategies under the impact of climate change. Mt Res Dev 39(2):R1–R11

Pandey V, Mittal R, Sharma P (2017) Documentation and application of indigenous traditional knowledge (ITK) for sustainable agricultural development. Asian J Agric Extension Econ Sociol 15(3):1–9

Patel SK, Singh R, Sharma A, Singh GS (2018) Biodiversity utilization and human well-beings: current state and trends. In: Singh S (ed) Biodiversity: monitoring management and utilization. Daya Publishing House, New Delhi, pp 371–402

Patel SK, Singh A, Singh GS (2019a) Food production through traditional agriculture: an urgent need to improve soil health by sustaining soil microbial diversity. Int J Curr Microbiol Appl Sci 8(1):183–196

Sarkar S, Padaria RN, Vijayragavan K, Pathak H, Kumar P, Jha GK (2015) Assessing the potential of Indigenous Technological Knowledge (ITK) for adaptation to climate change in the Himalayan and arid ecosystems. Indian J Tradit Knowl 14(2):251–257

Saxena KG, Maikhuri RK, Rao KS (2005) Changes in agricultural biodiversity: implications for sustainable livelihood in the Himalaya. J Mt Sci 2:23–31

Semwal RL, Maikhuri RK (2015) Valuing traditional agrobiodiversity for sustainable development in Uttarakhand. Ecosyst Services Mainstreaming Dev Plann Process:92–114

Sharma G, Dhakal T (2011) Opportunities and challenges of the globally important traditional agriculture heritage systems of the Sikkim Himalaya. Biodiversity of Sikkim: exploring and conserving a global hotspot. IPR Department, Government of Sikkim, Gangtok, pp 411–440

Sharma N, Kanwar P (2009) Indigenous water conservation systems—a rich tradition of rural Himachal Pradesh. Indian J Tradit Knowl 8(4):510–513

Sharma G, Liang L (2006) The role of traditional ecological knowledge systems in conservation of agrobiodiversity: A case study in the Eastern Himalayas. In: Proceedings of international policy consultation for learning from grassroots initiatives and institutional interventions, 27–29 May. Indian Institute of Management, Ahmedabad, India

Sharma R, Xu J, Sharma G (2007) Traditional agroforestry in the eastern Himalayan region: land management system supporting ecosystem services. Trop Ecol 48(2):189

Sharma G, Liang L, Sharma E, Subba JR (2009) Sikkim Himalayan-agriculture: improving and scaling up of the traditionally managed agricultural systems of global significance. Resources Science, vol 31, No 1 Jan 2009 China. (ISSBN:1007-7588(2009)01-0021-10)

Sharma A, Singh R, Patel SK, Singh GS (2018) Conservation and management of biological diversity: global challenges and opportunities. In: Singh S (ed) Biodiversity: monitoring management and utilization. Daya Publishing House, New Delhi, pp 257–283

Sharma A, Patel SK, Singh GS (2021) Traditional knowledge of medicinal plants among three tribal communities of Vindhyan highlands, India: an approach for their conservation and sustainability. Environ Sustain 4:749–783

Sharma A, Patel SK, Singh GS (2023) Variation in species composition, structural diversity, and regeneration along disturbances in tropical dry forest of Northern India. J Asia-Pac Biodivers 16(1):83–95

Singh JS (2004) Sustainable development of Indian Himalayan region: linking ecological and economic concerns. G.B. Pant Memorial Lecture – X, GBPIHED, Kosi-Katarmal, Almora

Singh M (2007) Status, prospects and problems of organic farming in India. In: Munda GC, Ghosh PK, Anup Das, Ngachan SV, Bujarbaruah KM (eds) Advances in organic farming technology in India. published by the Director, ICAR Research Complex for NEH Region, Umiam-793103, Meghalaya, 73–83

Singh RD, Munda GC (2008) Status of natural resources and suitable crops & cropping systems for North Eastern Region. Agric Situat India 1(1):3–9

Singh R, Singh GS (2017) Traditional agriculture: a climate-smart approach for sustainable food production. Energy Ecol Environ 2:296–316. https://doi.org/10.1007/S40974-017-0074-7

Singh RK, Sureja AK (2008) Indigenous knowledge and sustainable agricultural resources management under rainfed agro-ecosystem. Indian J Tradit Knowl 7(4):642–654

Singh KA, Rai RN, Patiram, Bhutia DT (1989) Large cardamom (Amomum subulatum Roxb.) plantation—an age-old agroforestry system in Eastern Himalayas. Agrofor Syst 9:241–257

Singh GS, Ram SC, Kuniyal JC (1997) Changing traditional land use patterns in the Great Himalayas: a case study of Lahaul Valley. J Environ Syst 25:195–211

Singh RK, Sureja AK, Singh D (2006) Amta and Amti (Hibiscus sabdariffa L.)-cultural and agricultural dynamics of agrobiodiversity conservation. Indian J Tradit Knowl 5(1):151–157

Singh AK, Arunachalam A, Ngachan SV, Mohapatra KP, Dagar JC (2014) From shifting cultivation to integrating farming: experience of agroforestry development in the northeastern Himalayan region. Agroforestry Systems in India, Livelihood Security & Ecosystem Services, pp 57–86

Singh A, Kumar P, Kour H (2016) A case report on cow dung composting: traditional practice followed by women in rural areas of Jammu & Kashmir for maintaining agroecology. J Rural Agric Res 16:79–80

Patel SK, Verma P, Shankar Singh G (2019b) Agricultural growth and land use land cover change in peri-urban India. Environ Monit Assess 191:1–17

Patel SK, Sharma A, Singh GS (2020a) Traditional agricultural practices in India: an approach for environmental sustainability and food security. Energy Ecol Environ 5:253–271

Patel SK, Sharma A, Tiwari AK, Singh GS (2020b) A study of ethnomedicinal plant diversity of sacred natural sites of Panchkoshi pilgrimage of Varanasi, Uttar Pradesh India. Int J Pharmaceut Sci Res 11(2):710–720

Patel SK, Sharma A, Singh R, Tiwari AK, Singh GS (2022) Diversity and distribution of traditional home gardens along different disturbances in a dry tropical region, India. Front For Global Change 5:822320

Patel SK, Sharma A, Barla A, Tiwari AK, Singh R, Kumar S, Singh GS (2023a) Socio-ecological challenges and adaptation strategies of farmers towards changing climate in Vindhyan highlands, India. Environ Manage:1–17

Patel SK, Sharma A, Barla A, Singh GS (2023b) Assessing morphological and physiological crop functional traits of underutilized crops in response to different nutrient amendments in Vindhyan Highlands, India. Int J Prod. https://doi.org/10.1007/s42106-023-00272-1

Phondani PC, Maikhuri RK, Rawat LS, Negi VS (2020) Assessing farmers' perception on criteria and indicators for sustainable management of indigenous agroforestry systems in Uttarakhand, India. Environ Sustain Indicators 5:100018

Pimentel D, Kounang N (1998) Ecology of soil erosion in ecosystems. Ecosystems 1:416–246

Pulido JS, Bocco G (2003) The traditional farming system of a Mexican indigenous community: the case of Nuevo San Juan Parangaricutiro, Michoacan, Mexico. Geoderma 111:249–265

Raj S (2010) Traditional knowledge, innovation systems and democracy for sustainable agriculture: a case study on Adi tribes of eastern Himalayas of north-east India. In ISDA. Cirad-Inra-SupAgro 1–10

Ramakrishnan PS (1993) Shifting agriculture and sustainable development: an interdisciplinary study from North-Eastern India. Oxford University Press, Delhi, p 424

Ramakrishnan PS (1994) The Jhum agro-ecosystem in north-eastern India: a case study of biological management of soils in a shifting agricultural system. In: Woomer PL, Swift MJ (eds) The biological management of tropical soil fertility. Wiley-Sayce, Chichester, pp 189–207

Ramakrishnan PS (1997) Scientific basis of traditional wet rice cultivation by North-East India hills tribes. In: Behera MC, Roy NC (eds) Trends in Agrarian structure in the hills of North_ East India. Commonwealth Publishers, New Delhi, pp 233–247

Ramakrishnan PS, Patnaik S (1992) Jhum: slash and burn cultivation. India Int Centre Q 19(1/2):215–220

Rao KS, Semwal RL, Maikhuri RK, Nautiyal S, Sen KK, Singh K, Saxena KG (2003) Indigenous ecological knowledge, biodiversity and sustainable development in the central Himalayas. Trop Ecol 44(1):93–111

Rathore SS, Karunakaran K, Prakash B (2010) Alder based farming system a traditional farming practices in Nagaland for amelioration of jhum land. Indian J Tradit Knowl 9(4):677–680

Rodríguez BC, Zuazo VD, Rodríguez MS, García-Tejero IF, Ruiz BG, de Torres MRR, Tavira SC et al (2022) Legumes protect the soil erosion and ecosystem services. In: Advances in legumes for sustainable intensification. Academic Press, pp 247–266. https://doi.org/10.1016/B978-0-323-85797-0.00010-0

Roy SD (1986) Paddy cum fish culture in Apatani plateau in Arunachal Pradesh. Fishing Chimes 6:22–24

Samal PK, Fernando R, Rawat DS (2000) Influences of economy and culture in development among mountain tribes of Indian central Himalaya. Int J Sust Dev World 7:41–49

Sandor JA, Furbee L (1996) Indigenous knowledge and classification of soils in the Andes of southern Peru. Soil Sci Soc Am J 60:1502–1512

Saravanan R (2010) Traditional knowledge, innovation systems and democracy for sustainable agriculture: a case study on Adi tribes of Eastern Himalayas of North-East India. ISDA Montpellier, June 28–30

Singh R, Sharma A, Patel SK, Singh GS (2018) Global biodiversity assessment: current status and threats. In: Singh S (ed) Biodiversity: monitoring management and utilization. Daya Publishing House, New Delhi, pp 129–158

Slathia PS, Sharma BC, Paul N, Kumar R, Gupta SK (2016) Traditional water conveyance modes, their management and use in Udhampur district of Jammu and Kashmir. Indian J Tradit Knowl 15(3):503–508

Steinfeld H, Gerber P, Wassenaar T, Castel V, Rosales M, de Haan C (2006) Livestock's long shadow: environmental issues and options. Food and Agriculture Organization of the United Nations (FAO), Rome

Subba JR (2009) Indigenous knowledge on bio-resources management for livelihood of the people of Sikkim. Indian J Tradit Knowl 8(1):56–64

Tyagi S, Singh MK, Singh BD, Kumar S (2018) Conservation and management of indigenous technical knowledge for livelihood upliftment of small and marginal farmers in rural areas. Int J Inclusive Dev 4(2):53–58

Ulian T, Diazgranados M, Pironon S, Padulosi S, Liu U, Davies L, Mattana E (2020) Unlocking plant resources to support food security and promote sustainable agriculture. Plants People Planet 2(5):421–445

Upadhaya K, Barik SK, Kharbhih VM, Nongbri G, Debnath G, Gupta A, Ojha A (2020) Traditional bun shifting cultivation practice in Meghalaya, Northeast India. Energy Ecol Environ 5(1):34–46

Upadhyay A, Rai SC (2022) Climate change analysis in Rangit Basin of Sikkim Himalaya. Oriental Anthropol 23(1):103–130

Upadhyay A, Rai SC (2023) Climate change analysis in Rangit Basin of Sikkim Himalaya. Oriental Anthropol 23(1):103–130

Van der Kooij S, Zwarteveen M, Boesveld H, Kuper M (2013) The efficiency of drip irrigation unpacked. Agric Water Manag 123:103–110

Vanaja T (2013) KAIPAD–a unique, naturally organic, saline prone rice ecosystem of Kerala, India. Am J Environ Protect 2(2):42–46

Venkateswarlu B, Shanker AK (2012) Dryland agriculture: bringing resilience to crop production under changing climate. In: Venkateswarlu B, Shanker A, Shanker C, Maheswari M (eds) Crop stress and its management: perspectives and strategies. Springer, Dordrecht. https://doi.org/10.1007/978-94-007-2220-0_2

Viana CM, Freire D, Abrantes P, Rocha J, Pereira P (2022) Agricultural land systems importance for supporting food security and sustainable development goals: a systematic review. Sci Total Environ 806:150718

Vishvakarma SCR, Kuniyal JC, Singh GS (1998) Indigenous agroforestry system of north western Himalaya. Research for mountain development. Some initiatives and accomplishments, Gyanodaya Prakashan, Nainital, pp 99–118

Warren DM, Cashman K (1988) Indigenous knowledge for sustainable agriculture and rural development. International Institute for Environment and Development, Sustainable Agriculture Programme, London

Wilken GC (1987) Good farmers: traditional agricultural resource management in Mexico and Central America. University of California Press, Berkeley

Traditional Agri/Horticultural Crops, Farming Practices and Conservation in the Himalayan Region

P. Kiran Babu

1 Introduction

The Himalayan region is known as a hotspot for agro-biodiversity, and its residents primarily depend on agriculture. Over 75% of the population works in agriculture or related fields, with traditional subsistence cereal farming accounting for the majority of these activities (Gururani et al. 2021). Native aboriginal communities in the Himalayan region have long demonstrated the durability of traditional agricultural methods. The location provides an abundance of crop diversity and resilience methods and innovations to assist adaptation to climate change, particularly in light of the fact that many parts of India are experiencing an agrarian crisis, with fast declining productivity and soil quality and an increase in instances of drought. Because the populations in this Himalayan region live in places rich in biodiversity and use agricultural practices that have developed over millennia, it is particularly significant from the standpoint of bio-cultural heritage agriculture.

Indigenous and marginalized populations are present in many Himalayan areas, and because they depend on natural resources for their livelihoods, they are particularly vulnerable to the effects of climate change. These settlements are frequently found in isolated, difficult-to-reach locations with little access to infrastructure and essential services like healthcare, especially in the ecologically delicate mid-hills and mountain regions (Sujakhu et al. 2019; Subedi et al. 2023). The distinct local environmental conditions of mountain habitats, such as the Himalayas, influence cultivation practices. High altitudes, steep slopes, and severe weather are characteristics of mountain agriculture, which present considerable difficulties for farmers (Hussain et al. 2016; Dahal et al. 2023). In the event of climate change, traditional agricultural methods support energy conservation, human health safety, the

P. K. Babu (✉)
ICAR-Indian Institute of Oil Palm Research, Pedavegi, Andhra Pradesh, India

A. Borthakur, P. Singh (eds.), *The Resilience of Traditional Knowledge Systems for a Sustainable Future*, Advances in Asian Human-Environmental Research,
https://doi.org/10.1007/978-3-031-56858-9_3

management of natural resources, and socio-ecological integrity. Historically, hill farmers have used an integrated farming system, sometimes referred to as traditional farming, to maintain strong ties and balance between agriculture, animal husbandry, and forestry. Additionally, these crops from traditional agriculture offer high-quality fodder, which boosts the output of meat, milk, or eggs to raise farm profitability and give both producers and customers more freedom (Padulosi et al. 2011; Singh and Singh 2017).

2 Farming in the Himalayan Region

In traditional agricultural system in the Himalayan region, it is customary for farmers to raise animals, which in turn supply agricultural fields with manure and the cattle with crop leftovers for feeding. Indigenous bee keeping is an essential component of agricultural methods, apart from cattle. The placement of the bee hives, or Madhams, in homes or close to fields of crops allows farmers to produce healthy crops and earn a valuable byproduct in the form of honey. Conventional wood-and-iron tools are used for the ploughing with the aid of an ox (Fig. 1a, b). The other farming methods requires a lot of labour and are carried out by hand. Chemical

Fig. 1 Agricultural operations in the Himalayan region. (**a**) Dryland ploughing with animals, (**b**) Wetland terrace ploughing, (**c**) paddy cultivation and harvested bundles in the field, (**d**) Post harvest processing

fertilizers are often not used in any agricultural operations in traditional farming. After harvesting and sun-drying, the crops, mostly cereal crops and pulses, are kept in big wooden boxes that the locals referred to as "datha" for storage in many parts of the Himalayas. The farmers ground the grain in traditional water mills called "gharats" in order to use it as flour. As a result, the entire agricultural system is self-sustaining and dependent on locally available resources. This system's sole drawback is its restricted crop output, which could only satisfy domestic demand. Millions of people's need for food cannot be met by the existing agricultural system, endangering the nation's food security. Very few crops have enough surplus to be sold by farmers as a cash crop, which may increase their income.

2.1 Agricultural Operations

The major agricultural operations are done with the help of locally made implements. For instance, in parts of North-East India, ploughing is carried by Khasi spade and local plough; harrowing done by peg-tooth harrow; land leveling by the leveller; seeding/planting/sowing with the help of dibbler; weeding by Khasi spade (small); inter-culture done by hand fork; bund/furrow making with Khasi spade; harvesting/digging/uprooting with the help of sickle (Fig. 1c). Threshing is done manually and with the help of animals; milling of cereals, pulses and oilseeds is done with the help of Counter poise foot pounder (Fig. 1d). The majority of agricultural tasks are manual and labor-intensive. Even if locally developed tools work well, they produce little and require a lot of labour. Very little mechanical power is used. It is necessary to selectively adopt better farm equipment and increase its availability into the current farming systems. In steep areas, tractors and power tillers are already in high demand.

2.2 Weed Management

Under organic management production methods, weeds pose a serious challenge because they compete with crops for nutrients and soil moisture, two key constraints, particularly in the Himalayan region with heavy rainfall. Consequently, effective weed control becomes even more crucial, and weed-management techniques—such as cultural, mechanical, physical, and biological methods—when used singly or in combination to lessen the threat of weeds without the use of artificial pesticides and chemicals, help to eradicate them. In the Himalayan area, hand weeding is the most widely used way of weed removal among all the control measures (Singh et al. 2016). Intercropping creates a good ground cover that inhibits weed growth, it is also essential for managing weeds. In contrast to the corresponding solo crops, lower levels of weed biomass, weed density, weed diversity, and weed infestation are noted in intercropping systems (Choudhary and Choudhury 2018).

2.3 Plant Protection Measures

In the Himalayan region, farmers do not usually choose high yielding cultivars; rather, their field performance often led them to do so. However, because of soil, air, and water pollution, the majority of crops are grown in monoculture using high-yielding hybrid crop varieties, and there are constant use of chemical pesticides and synthetic fertilizers. This is leading to soil degradation and a decline in biodiversity (Aktar et al. 2009). Sustainable farming allows for the employment of a variety of tactics, including soil solarization, crop rotations, intercropping, cultivating resistance types, use of organic amendments, and physical crop protection techniques (Haldhar et al. 2017). Wetland paddy crops can effectively be free of insect pests and their vanishing stages in the soil by adopting the indigenous repeated ploughing approach. Many illnesses and certain insects can be managed by clean farming, crop-residue management, and the elimination of alternative hosts (Reddy 2017). When intercrops such as long-duration pigeonpea, sorghum, maize, and others are grown alongside jhum rice, pest attacks on rice are significantly decreased (Chhetry and Belbahri 2009). Application of neem oil, neem cake, neem-based products are often opted by the farmers.

Indigenous methods of controlling pests and diseases are based on organic methods and are thought to be safe, simple, easy to use, and made from locally available materials (Das et al. 2017a, b). Farmers in this region plant trap/repellent crops around the field's edges to combat pests, and seed barnyard and konri millet along the plot's edge rather than in the centre. In paddy crops, summertime field bund trimming eliminates pests' and diseases' alternate hosts. In potato and chilli crops, ash is spread to shield plants from insects that bite and suckers. In order to shield potato tubers from sunlight, earthing-up is done at the second weeding stage, up to a height of one foot. One kilogram of cow dung mixed with 2 L of water is sprayed with 100 L of cow dung solution to control onion blight (Rathore et al. 2021). The hilly aboriginal people successfully managed the rodent incidents when the holes are filled with horse dung. To divert the mouse's attention from the field, urea is kept at the opening of the mouse hole and by putting prickly kilmora (*Berberis asiatica*) shrubs and bichhu ghaas (*Urtica dioica*) plants at the opening of a mouse hole. One kilogram of wheat flour and half a kilogram of glass, crushed and kneaded with a small amount of water, is what kills rats. Rat breeding is prevented by keeping the area in and around the field clean (Rathore et al. 2021). When ploughing in the summer, ash is sprayed on garlic and onion fields to keep eating insects away from the plants. Cow urine acts as an insect repellant too, thus it is sprayed in the kitchen garden to check for signs of plant wilting. One can cure cabbage plants by using a mixture of soil, ash, and cow urine. Crop growth is also aided by the nitrogen found in urine. Cow urine has a molluscicidal function, which keeps snails and slugs from damaging rice fields (Srivastava et al. 1985). Before relocating each plant, holes are constructed and cow dung cakes are burned inside to protect the chilli crop from white grub. After the wheat crop is harvested, fire is started in the field to control white grub. Farmers spread the farmyard manure that has been properly degraded in the field and pine leaves are burned in the field after the wheat crop is harvested. It lessens the likelihood that the next paddy crop will be attacked by white grubs. White grub attacks are less likely in rice fields when there is a constant 2 cm depth of water in the area. Spreading of the common salt uniformly across the field is effective, by exosmosis, as salt may kill the grubs (Rathore et al. 2021).

One of the current problems has been the loss in horticulture crops' productivity and production due to biotic and abiotic stressors. Apple output has decreased mostly in lower altitudes of the Himalayan region due to insufficient chilling as the temperature at lower altitudes is rising due to climate change, and apple production has drastically decreased as a result of apple farming shifting to higher altitudes. The timing of bud-break is impacted by the reduction in chilling hours for apple trees caused by the shift in snowfall. The entire socio-economic structure of some Himalayan areas is being thrown off balance by the frequency of various catastrophic events including landslides, heavy rains, floods, and temperature changes.

Compared to the plain regions, the Himalayan region has a distinct agricultural environment that is older and less affected by abrupt changes in farming practices and cultural norms. The mountainous and sub-mountainous areas are distinct civilization hotspots where crop adaptations and environmental variances impact agricultural and food culture systems. The traditional agro-ecosystems of the Himalayan Mountains, which are rich in crop diversity and traditional agriculture, rely heavily on locally accessible resources and indigenous technologies for its farmers (Kuniyal et al. 2004; Nautiyal et al. 2007).

2.4 Farm Mechanization

Traditional farmers in many Himalayan regions have created intricate farming systems that are tailored to the unique requirements of the region, enabling them to manage the harsh settings in a sustainable manner and meet their subsistence needs without relying on outside inputs or modern agricultural technologies. Indigenous farmers have created integrated and varied polyculture production systems that are tailored to the various ecological systems by utilizing the knowledge they have acquired through experience and locally accessible resources. The Himalayan region has vast stretches of hilly and mountainous arable land, with little to no agriculture mechanization. Farmers should have access to inexpensive farming equipment and tools for the difficult terrain relative to other regions of the Himalayan countries like India.

The majority of people in the Himalayan region still work mostly in agriculture and animal husbandry as their primary source of income. The Himalayan region is rich in animal diversity which is essential to several hilly state's agricultural economy in India and are mostly dependent on various marginal, sub-marginal, and landless farmers. Nearly all of a state's arable farming, which is nearly entirely rainfed, is done on small terraced pieces of land, frequently measuring less than 100 m^2, found on the state's high slopes. Mechanizing agriculture in the hills seems not feasible in the near future, as arable farming relies solely on the modest, unremarkable working bullocks for farm power.

3 Agri and Horticultural Diversity in the Himalayan Region

Rice, wheat and maize are the major crops in the most Himalayan region. These crops are having more diversity in nature and are grown in a traditional way. "Do no harm and use natural resources wisely" has been the custom. Local seed saving was common in the past and is still used to some extent in mixed cropping systems to preserve diversity. Many species and kinds are specifically kept to be used in rituals and sacrifices; in the Himalayan cultures, ancestors are often offered crops and animals as offerings. In the case of regular food crops, community views of conservation are becoming more utilitarian, and communities are increasingly restricting their conservation efforts to species or varieties that have a commercial benefit. It is often felt that the unexplored areas in the region need to be surveyed vigorously, as they can provide very valuable germplasm material, and that the explored areas need to be revisited to check on the genetic erosion of the local landraces.

With its exceptional biodiversity and habitat richness in the Himalayan region, this biodiversity hotspot has the potential to act as a gene reserve for the in-situ conservation of resilient landraces and agricultural biodiversity, as well as a sanctuary for other wildlife and cultural traditions. It must be acknowledged and protected from unforeseen development interventions like highways, canals, or the repurposing of land for other uses. Simultaneously, local residents' means of subsistence must be improved in a way that supports the region's growth as a hotspot for biological and cultural diversity. The national biodiversity legislation of India ought to designate the biocultural landscape as a Biodiversity Heritage Site (Biological Diversity Act, 2002).

3.1 Millets

Because of the nutritional importance of millet, it is necessary to investigate the functional qualities and nutritional traits of various millet cultivars and to create value-added millet products. In addition to offering protein, fatty acids, minerals, vitamins, dietary fibre, and polyphenols, millets are a good source of energy. A high concentration of essential amino acids, particularly those containing sulphur, are present in typical millet protein. When millet is processed, the fibre- and phytochemical-rich bran and germ layers are removed, resulting in a large loss. Antioxidants such as phenolic acids and glycated flavonoids abound in the millets. Foods containing millet are thought to have probiotic potential and can improve the viability or effectiveness of probiotics with important health advantages.

A vital component of the subsistence farming practiced by small farmers in the Himalayan region is millet. While millets are utilised for food and drink, their use as animal fodder and bird feed is minimal. Major millet crops grown in this area are sorghum (jowar), proso millet (cheena/common millet), finger millet (ragi/mandua), foxtail millet (kangni/Italian millet), or little millet (kutki), kodo millet, barnyard millet (sawan/jhangora), and brown top millet (korale). Every state in this

Indian Himalayan region is further culturally divided into numerous diverse tribes, each with unique eating customs, and millets play a variety of roles in these traditions. This area uses every kind of farming, including hybrids, mixed crops, and shifting cultivation or *jhum* systems. Similar to how different cultures around the Himalayan region use millet as a primary grain, fermented foods and beverages are also popular to varied degrees. Indian Himalayan states have different patterns of millet consumption, which have either decreased or stayed constant over time.

To sustain the food security of India, several traditional crops—millets, for example—are being promoted at the national level due to their increased productivity. Since most crops in the country, including those in the Himalayan region, are rain-fed and dependent on seasonal rainfall to thrive, millets are an important part of the cropping system used in traditional farming. There are two landraces of millet: *dolley* and *nangrey*, however *dolley* is becoming rare. Both are robust, but dolley yields more. Even though millet is now considerably less widespread, some homes in the Himalayan region in India still grow it for food or religious reasons. In Sikkim's steep regions, for instance, millet is a labor-intensive crop whose yield has also declined. In Uttarakhand, broomcorn—a domesticated variety of sorghum—is farmed for its panicles and utilized as food, feed, and a raw material for broom production (Elangovan et al. 2015).

3.2 Pseudo Cereals

Job's tear *(Coix lacryma—jobi* (L)) is also referred to as tear grass, adlay, adley, pingpih (Mizo), gurgur (Bengali), samkru (Hindi), and coix millet. The hard, glossy fruit of this crop is what is most associated with it and is the source of the word "tear." It may serve as a staple meal to ensure food security and improve the community's overall health (Laxmisha et al. 2022) in the Himalayan region. The main reasons coix are underutilized in the region are because of the fact that some other local varieties are more common and proper agronomic practices are not followed. Despite coix's high nutritional value, resilience to the climate, resistance to insects and diseases, and significant medicinal value, these factors result in a lower crop yield (Devaraj et al. 2020).

3.3 Legumes

Legume crops are valuable for many purposes and are crucial to the hill farmers' ability to secure a living through agriculture, food, and for nutrition. Legume crops are very nutritious and have a major role in preventing malnutrition in the Himalayan region. In addition to providing energy in the form of carbohydrates, legume crops also contain minerals, vitamins, antioxidants, biologically significant proteins, and anti-cancerous compounds that support good health and help fend off disease (Sarita and Singh 2016). They have a tight relationship with cereals; in agriculture, legumes complement cereals in terms of crop cycle and cropping pattern. Pulses' high protein content has earned them the name "poor men's meat." Legume species are important for more reasons than only fuel, fodder, and other practical uses; they are also used in customary rites and rituals. Legumes can also be grown to greatly increase productivity and improve the sustainability of cropping systems based on cereals. Legumes are an excellent way to restore degraded soils (Das et al. 2018).

More varieties of legumes are grown in the Himalayan region than anywhere else in the world, including broad beans (*Vicia faba* L), lablab beans (*Lablab purpureus* (L.), cowpea (*Vigna unguiculata* (L.) pigeon pea (*Cajanus cajan* (L.), pea (*Pisum sativum* L.), rice bean (*Vigna umbellata* Thunb), French bean/rajmash (*Phaseolus vulgaris* L), soybean (Glycine max), green gramme (*Vigna radiata* L), black gramme (*Vigna mungo* L), lentil (*Lens culinaris*), adzuki bean (*Vigna angularis*), and horse gramme (*Macrotyloma uniflorum*) (Fig. 2).

3.4 Indigenous Legumes of Himalayan Region

The black seed coat soybean, also referred to locally as "bhat", "bhatmash", and "kala bhat" is grown in the Indian Himalaya's Uttarakhand state's borders as well as in the Kumaon and Garhwal regions. Kala bhat is regarded as a treasure trove of many therapeutic virtues, high concentration of iso-flavones which can be found in its products and it is highly rich in protein which is recommended to jaundice

Fig. 2 Pulses diversity in the Himalayan region

patients and also effective for stomach infection (Shah 2006; Hipparagi et al. 2017) (Fig. 3). Owing to their capacity to adapt to the region's varied agroclimatic conditions, some unique pulses have been farmed here for ages. The adzuki bean, wide bean, moth bean, native khesari, winged bean (*Psophocarpus tetragonolobus*), tree beans (*Parkia roxburghii*), jack bean (*Canavalia ensiformis*), sword bean (*Canavalia gladiata*), and other legumes are among the native variety. Not as well-known as cowpea, *Vigna vexillata* is a possible pulse/tuber crop in this area that yields tasty

Fig. 3 Soybean diversity in the Himalayan region

seeds and tubers. In Eastern Himalayas in India, Nagaland is home to the famed kholar bean, a rajmash variation. Growing primarily in the Eastern Himalayan region, the Bakthul bean (*Sesbania grandiflora*) is another native, underutilized potential leguminous crop. Native legumes have more nutritional content than many of the more widely cultivated pulses. Further, Sikkim's most widely farmed pulses are Pahenlo dal, a rare green-seeded urd bean, and Kalo dal, a black-seeded urd bean. Compared to Kalo dal (black-seeded urd beans), pahenlo dal has more protein and has a superior flavour (Das and Ghosh 2012; Das et al. 2018).

3.4.1 Rice Bean (*Vigna umbellata* (Thunb.) Ohwi and Ohashi)

Other names for rice beans are ascending mountain bean, Mambi bean, Oriental bean, and haricot bean (*Vigna umbellata* (Thunb) Ohwi and Ohashi). Due to its underutilization, little information about the many qualities and anti-qualities of rice beans is now available. In order to promote health and fight malnutrition in humans, seeds are a well-balanced source of essential nutrients like protein, carbohydrates, minerals, vitamins, polyunsaturated fatty acids (PUFAs), and antioxidants. Underutilized legumes are a great source of phyto-chemicals, antioxidants, and phenols that can help prevent a variety of human illnesses and improve overall health. Thus, further research into these possible food sources is necessary in order to reduce the world's food and nutritional insecurity. Because of rice beans' greater genetic diversity, nutritional value, resistance to yellow mosaic virus, and storage quality, they have drawn interest as a possible multipurpose grain legume crop.

3.4.2 Adzukibean (*Vigna angularis* (Willd.) Ohwi and Ohashi)

Because of its distinct flavour, adzuki beans are used to make traditional desserts in addition to being high in nutrients and taste sweet. In the present era, its low calorie and fat content makes it more and more valuable for creating new desserts. The dried seeds are used for human consumption in soups, cakes, and confections, either cooked whole or ground into meal. Beans can be eaten as sugar candies, popped like corn or substituted for coffee. Face creams and shampoos are made with its flour (Kim et al. 2020; Bi et al. 2021). Adzuki beans also include a lot of protein, minerals, vitamins, and flavonoids. Since ancient times, they have been utilised in medicine to treat immune system regulation, inflammation, blood sugar and blood lipid regulation, as antioxidants, and more (Lim et al. 2020; Wu et al. 2020; Wang et al. 2023).

3.4.3 Jack Bean (*Canavalia ensiformis*)

The nutritional relevance of the *Canavalia ensiformis* seed material appears promising, as a moderate intake of these seeds significantly improves the consumers' total dietary protein intake. Overall, these seeds are high in protein, carbs, and vitamin C (Bressani et al. 1987; Abitogun and Olasehinde 2012; Patel et al. 2016.)

3.5 Vegetables

There is a significant discrepancy between the supply and demand for plant-based foods, especially vegetables, and this gap will only widen as the population grows. Vegetable productivity needs to be raised while resources are used effectively and production costs are kept to a minimum. Vegetable crops require the identification, design, and introduction of mechanization equipment for tasks including picking and digging, watering, inter-cultural activities, planting, transplanting of seedlings, and seed bed preparation. Vegetables are basically any parts of herbaceous plants that can be utilized in cooking. Vegetables are a vast group of plants, predominantly annuals with a few biennials and perennials mixed in. Different parts of these plants, including seeds, roots, and fruits, as well as their immature sections, like stems, leaves, flowers, and succulent bulbs, are consumed. The income and well-being of the people living in mountainous regions depend heavily on vegetables. In addition to being a natural supply of fibre, vitamins, minerals, proteins, carbohydrates, and lipids, vegetables also serve as a protective food. Vegetables improve meal palatability, reduce acidity, offer beneficial roughages, and are the least expensive source of antioxidants. At least 300 g of various veggies should be included in any diet to be deemed balanced. They are a significant natural nutrient source. Since most vegetables grow quickly and have short growing seasons, more crops can be produced on the same amount of land. They can be grown on tiny and marginal lands. They can also be cultivated in neglected areas including backyards, roofs, hallways, verandahs, windows, and partially shaded spots, where other crops might not grow well, contributing significantly to fulfilling daily needs and raising aesthetic standards in the process. When compared to grains, veggies have a relatively high yield potential and productivity. Vegetable growing in the kitchen has been shown to be the most effective way to recycle household wastes like organic waste and spent water. Given that the majority of the processes are labor-intensive and time-sensitive, it offers rural impoverished people the opportunity for employment.

One of the traditional foods of the Naga community in Nagaland in India's Eastern Himalayas is the Naga King chilli (*Capsicum chinense sacq*), also referred to locally as "Raja Merja" or "Raja Mircha." In Assam, it is referred to as Bhut jolokia, while in Manipur, it is called U-morok or Go-morok. India's northeast is known as a hotspot for a wide variety of chillies (Meetei et al. 2016). The Naga King chilli, one of the many landraces of chilli grown in the northeast, is the most well-known worldwide for its unique flavour, aroma, and spiciness. It is regarded as one of the hottest chilies in the world and was recorded in the Guinness Book of World Records with a measurement of 855,000 Scoville units, surpassing the record held by the Mexican red savana habaneros (5,77,000 Scoville units) (Malangmeih and Rahaman 2016). The Naga King chilli's geographical designation and patent rights were also acquired by the state government of Nagaland in compliance with the Registration and Protection Act of 1999 (GoN 2017). Similarly, one of the hottest chillies in the world is the Himalayan fireball, or Dalle Khursani. The two Himalayan locations, where it is mainly grown, are Darjeeling and Sikkim. When completely ripe, Dalle Khursani's spherical appearance almost resembles a cherry, and it turns a bright red in colour. It is one of the hottest chilies found in the Himalayan region. The major vegetables cultivated in the Indian Himalayan region are, Amaranthus, Basella, Bhindi/Okra, Bottle gourd, Bitter gourd, Brinjal, Broccoli, Cauliflower, Cabbage, Colocasia, Coriander, Chilli, Capsicum/Sweet Pepper, Ginger, Cucumber, Muskmelon, Watermelon, Round gourd/Indian Squash (Tinda), Pumpkin, Sponge gourd and Ridge gourd, Pointed gourd, and Ivy gourd, Knol-khol, Radish,Carrot, Potato, Onion, Garlic, Pea, French Bean, Broad Bean, Cowpea, Beet Leaf, Spinach, Kasuri Methi,Common Methi, Asparagus, Lettuce, Swiss Chard, Tomato, Turmeric and Turnip.

3.6 Fruit Crops

The most significant fresh fruit markets in the Indian Himalayas include those for apple, peach citrus, kiwi (Fig. 4a, b), litchi, apricot, walnut, pistachio, pear and mango. Gooseberries, raspberries, and blackberries are among the fruits that are rich in phenols, anthocyanins, and ascorbic acid, all of which are vital to human health (Nile and Park 2014; Pillai 2020). Mango, guava, citrus, litchi are subtropical fruits; while apple, pear, peach, plum, apricot, almonds, cherry are examples of temperate fruits found in the Indian Himalaya. With apples in particular, both producers and consumers are quite particular about their quality and flavor in the region (Moyer et al. 2002; Pantelidis et al. 2007).

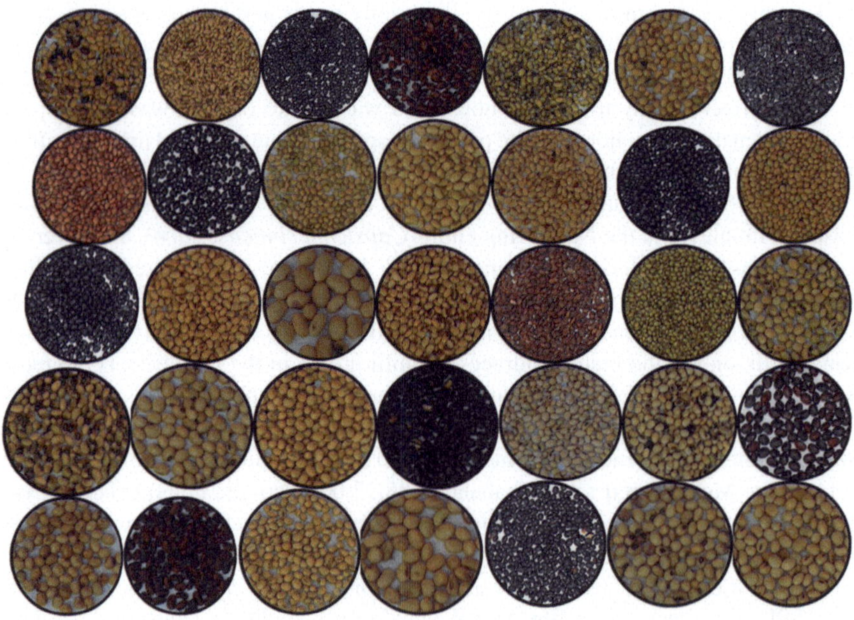

Fig. 4 Commercial crops in the Himalayan region; (**a**) Sikkim Mandarin, (**b**) Kiwi fruits, (**c**) Jerusalem Artichock, (**d**) Large cardamom

3.7 Underutilized Minor Fruit Crops

In addition to cultivated fruits, the Indian Himalayas are home to a range of underutilized wild fruit crops. Although local societies have traditionally used the underutilized fruits for food, fibre, animal feed, oil, or medicine, there are still unrealized potential applications of them. Eating fruits that grow naturally have a higher nutritional value than fruits that are produced, and they are essential for enhancing socioeconomic conditions (Maikhuri and Ramakrishnan 1992; Maikhuri et al. 1994). Minor fruit crops are typically believed to be grown outside of organized plantations, consumed in smaller quantities, and have lower market demand. However, there is little question that the important component enabling minor fruit crops to spread widely across a vast area in a single region is their edapho-climatic flexibility (Maikhuri et al. 2004; Kayang 2007).

With the exception of a few recently developed fruits like kiwis and persimmons, most minor fruit crops do not require organized orcharding because they have many desirable qualities, such as hardiness, the ability to yield without the need for external agro-inputs, the ability to store genes for a variety of uses, the ability to be processed, and—above all—nutritional and medicinal qualities that outweigh those of several cultivated fruits (Dhyani 2013; Lata et al. 2023). For this reason, extra consideration must be given to the conservation and sustainable use of these species. Appropriate handling and use of small fruits are crucial to boosting food security, eradicating poverty, and preserving the environment. Research attempts have been made to gather information on several novel fruits that National Bureau of Plant Genetic Resources (NBPGR) has introduced and popularized, as well as minor temperate fruits that are found in the North-Western Himalayas (Rana and Verma 2011). Underutilized fruits of the Himalayan region are listed in the Table 1. Many of the fruits that are currently popular in the Eastern Indian Himalayan region are: Sikkim mandarin, Sikkim passion fruit for its concentrated syrup, citrus fruits from Garo Hills of Meghalaya (First gene sanctuary in India for citrus).

3.8 Landraces

The landrace population and traditional varieties exhibit significant variation in appearance; yet, they are all recognizable because to their unique attributes, which may include early or late maturation, soil type adaptability etc. Any cultivated species' core centre of genetic diversity contains many important genes, including those that confer resistance to illnesses, insect pests, physiological stress, and qualitative attributes. It might offer genetic material that is useful for any plant breeding initiative. And the farmers' decades-long efforts in breeding, selection, and conservation—all carried out in an environment that supports the ecosystems—are directly responsible for this blessing for our future food security. Therefore, a larger genetic base aid in the production of enhanced crop varieties by the breeders, which are intended to combine greater quality, resistance to pests and diseases, and better tolerance to abiotic stress settings with a high yield potential.

Table 1 Underutilized fruits of the Himalayan region

Name of the fruit	Local names and other properties
Apricot Wild *(Prunus armeniaca* L.)	*Chufi, Chuaru, Chola, Chult, Chuari, Gurdalu, Zardalu, Sarha, Chufi, and Chuari.* Chuli oil, which is extracted from kernels, is frequently used in cooking and medicine, particularly for body massages in cases of arthritis. Fruits are used to make an alcoholic beverage in the western Himalayan region
Asiatic pear (P*yrus pyrifolia* Burn. f.)	*Nashpati,* Nashipat, *Pathernakh* and *Tang*
Assam apple *(Docynio indica* Decne.)	Indian crab apple, Soh-phoh, Tong, Thunghurpu, Mel, Ukung, and Mehul. The Assam apple originates from the Eastern Himalayas
Behmi *(Prunus mira Koehne.)*	
Chestnut *(Castanea sativa)*	This particular type of chestnut tree is the only one that was successfully brought to the Himalayas and other temperate regions of India during the colonial era. It is typically planted as an ornamental tree around daak-bungalows, botanical gardens, and official government residences
Fig *(Ficus carica* L.)	
Grape *(Vitis* species)	*Bhambti* and *purain*
Hazelnut *(Cory/us avellana* L.)	Bhutia badam in Garhwal, Uttarakhand; Winri, Wiri, Warrawi, and Wuriya in Kashmir; Thangi, Thangoliin in Pangi, HP; and Kapasi in Kumaon, Uttarakhand
Mountain ash *(Sorbus* species)	
Himalayan strawberry *(Fragaria nubicola)*	
Kaphal *(Myrica nagi* Thumb.)	Bay-berry, Box myrtle, *Kaya-phal, Katphala, Kariphal, Soh-phie, Kaidar-yamu, Marudam, Maruta* in north-western Himalayas
Kiwifruit *(Actinidia deliciosa* var. *deliciosa)*	
Loquat *(Eriobotrya japonica* lindl.)	Lokat, loquat phal, Lakota, Lakkotta, Lakkote, Lataku and Japanese medlar
Pine nut *(Pinus gerardiana* Wall. ex. D. Don)	*Ri, Rhi, Shangti, Neja, Neje, Neojo, Noosa*
Raspberries *(Rubus* species)	*Akhain, heer, hinsalu, hisalu* and *kala hinsalu*
Sea buckthorn (*Hippophae rhamnoides* L)	*Chharma, Ames, Dhurchuk, Gartsak, Soria, Treskas, Soma, Tasru* and *Dhurchule*

The rice landraces such as Malkangiri-Maharajpali-Assam Chudi, Nbr-Assamchudi, Assamkoti Kirti are from Assam, Tilakchandan, Hansraj, Indrasan from Uttarakhand, Chhohartu from Himachal Pradesh were successfully registered with Protection of Plant Varieties and Farmers' Rights Authority (PPV&FRA). Whereas, distinct agricultural products and crops are identified according to their place of origin and are assigned a Geographical Indication Tag like: Arunachal Pradesh's orange, Assam's Karbi Anglong ginger, Boka Chaul, Judima, Tezpur litchi, Joha Rice of Assam, Basmati (Himachal Pradesh), Ladakh Raktsey Karpo apricot (Ladakh), Chak—Hao (Manipuri black rice), Tamenglong Orange, Kachai Lemon, Hathei Chilli (Manipur), Khasi Mandarin, Memong narang (Meghalaya), Naga cucumber, Nagar mircha, Naga tree tomato (Nagaland), Dalle khursani (Sikkim), Tripura queen pineapple (Tripura), Kumaon Chyura Oil, Mansyari Razma (Uttarakhand), Tulai panji rice, Gobindbhog rice, Fazil mango, Khirsapati (Himsagar) mango (Babu and Rampal 2024).

Strong resistance to climatic changes is seen by many traditional landraces. As a result, they present excellent chances for collaborative plant breeding to create new, more productive and robust types. The local varieties and their wild counterparts are the richest sources of agricultural genetic variation—genes associated to adaptation, and genes related to stress. Conventionally cultivated plant genetic resources are unique global legacies that require immediate preservation and utilization. In addition to being important for conservation, these crop plants also provide information on their genetic backgrounds and important genes or alleles that enable their survival in harsh settings (Goron and Raizada 2015; Sood et al. 2016). The "centres of diversity"—mostly found in emerging nations—are where crop genetic variation is preserved. One of India's hotspots for biodiversity is the Himalayan region, where farming communities continue to practice ancient farming methods like Barahnaja by preserving their own seeds (Gururani et al. 2021). In the steep regions of Uttarakhand state in India's Western Himalaya, broomcorn sorghum and seasonal fodder grasses are grown due to the small number of agricultural holdings and lack of irrigation facilities. Sorghum cultivation ranged from 813 to 1975 m mean above sea level (msl) based on altitudinal variables. It appears that genotypes that have been shown to be cold tolerant are grown close to the Himalayan mountains' higher altitudes (Elangovan et al. 2015).

3.9 Underutilized/Neglected Crops

Only 120 plant species, out of the millions that are known to exist, are farmed for human consumption, and among them, maize, rice, and wheat provide more than half of the energy that comes from plant sources in the diet. Due to the significant influence that a threat to any of these important species has on global food availability and prices, this reliance on a small number of species seriously jeopardizes food security. For example, tree tomatoes are found from 1000 feet to 10,000 feet elevation in its native environment in the Indian Himalayan region and it is quite adaptable in the kitchen (Fig. 5d). They can be added to hot sauces, chopped fresh and used in salads, or served to sweeten desserts. They can also be used as a tomato alternative. In many other countries, chutney produced from this fruit is highly prized and frequently served in place of tomato ketchup.

Supporting a few neglected and underutilized crops can help mitigate the negative effects of climate change on food security in the highlands by boosting nutritional diversity and creating economic opportunities. Many of these crops have adapted to live in challenging environments. They don't need a lot of resources and can be successfully grown on marginal or degraded waste sites with little input. They also contribute to increased agricultural productivity, crop diversification, and environmental improvement. They provide valuable genes that can be employed to maintain productivity in the face of climate change and improve the variety.

Fig. 5 Mixed cropping in the Himalayan region. (**a**) Broom corn sorghum mixed with finger millet, (**b**) Broom grass, (**c**) Mixed leafy vegetable cultivation (**d**) Tree tomato

4 Cropping System in Himalayan Region

One example of a system, that has been endured for ages in the Central Himalayan Region, is the traditional mixed cropping system, or barahnaja. Research by Gururani et al. (2021) outline the Barahnaja system's elements, the crops' spatio-temporal distribution within it, its advantages, and upcoming research prospects related to different system components (Gururani et al. 2021). In order to mitigate the primary issues related to climate change—crop failure, pests, surface runoff, and disease—the multiple or mixed cropping system is highly helpful. It also helps to increase soil fertility (Lithourgidis et al. 2011). Farmers in hill states own small, dispersed plots of land. Farmers practiced crop diversification through the traditional mixed cropping strategy of agriculture. Insufficient infrastructures, including inadequate road networks and connection, force farmers to cultivate a variety of crops on their small plots of land in order to provide enough food for their families. The practice of traditional mixed cropping farming makes it easier to seed vegetables, pulses, grains and oil seeds on the same areas (Gururani et al. 2021).

Legume intercropping, with its sustained intensification under changing climatic conditions, offers immense potential to feed the world's growing population. Raising two or more crops concurrently on the same plot of land at the same time is a significant agronomic practice (Kumar et al. 2020; Jhariya et al. 2021). Temperate locations have also shown compatibility between winter faba bean and winter common vetch (*Vicia sativa*), spring faba bean and spring grass pea (*Lathyrus sativus* L.), and pigeon pea and lablab bean (Mikić et al. 2015). Intercropping winter oilseed rape with legumes and beans increases growth and above-ground nitrogen content, improves nitrogen utilization efficiency, decreases weed problems, and decreases incidence of stem weevil (Cadoux et al. 2015).

While Sikkim and Nagaland in the Eastern Himalaya practice terrace or wet rice growing, Arunachal Pradesh also uses the traditional rice-based agricultural method known as "*Apatani wet rice cultivation*". As a result, all of these methods are naturally organic. Farmers utilise composts, farmyard manure, green manure, and locally produced, self-prepared goods and techniques to manage pests in both crops and vegetables. The Eastern Himalayan region is notable for its "*Jhum*" cultivation, or "Slash and Burn" agricultural methods. While farmers in Assam, Manipur, and Tripura use an integrated farming system centred around ponds, "*Zaboo*" farming—an indigenous farming method carried out on sloping terrain with water harvesting tanks on the higher slope and farming techniques on the mid- and lower slopes—is performed in Nagaland (Das et al. 2018).

4.1 Organic Farming

Genetically modified organisms (GMOs), artificial additives, and synthetic chemicals are either completely eliminated or used very sparingly in organic farming. Rather, it depends on organic methods and procedures to guarantee wholesome agricultural development and animal welfare. In addition to providing high-quality, chemical-free food, organic farming places a priority on soil health, biodiversity, and ecosystem preservation (Singh et al. 2021). The use of artificially manufactured agro-inputs is forbidden in the production method of organic farming. Rather, in order to maintain soil productivity and fertility and manage pests under conditions of sustainable natural resources and a healthy environment, it depends on organic material. But removing artificial or chemical-based inputs is just one part of an organic production system. More significantly, it promotes a whole-system approach in which every element—soil, minerals, organic matter, microorganisms, insects, plants, animals, and humans—works together to form an ecosystem that is self-regulating and sustainable. It is an ecological approach of production management that strengthens biological cycles, biodiversity, and biological activity in the soil. Its foundation is the limited application of inputs from off-farm sources and management techniques that improve and maintain ecological balance. Many indigenous farming practices, the majority of which are organic in nature and based on long-standing indigenous technical knowledge systems, can be found in the Himalayan region. The actual epicenter for the production and development of organic farming products, including cashew, oranges, tea, pineapples, spices, and condiments, as well as numerous other regional land races of food grain crops and vegetables rich in nutraceuticals, may be found in the Himalayan region (Singh et al. 2021).

Compared to other regions of the country, Sikkim farmers deliberately work to maintain high levels of organic matter in their fields by using organic manure consistently to replace nutrients lost due to crop removal and erosion (Avasthe et al. 2020). Plant nutrients are obtained from plant and animal wastes. They are an excellent source of macro and micronutrients, enhance the physical characteristics of the soil, increase its ability to hold water, and make more nutrients available to plants. For the purpose of enhancing soil fertility, a variety of organic manures, such as compost, phospho-compost, night soil, sewage, green manure, sheep-, goat-, and chicken-manure, can be prepared on the farm (Avasthe et al. 2012; Yadav et al. 2018). Finding enormous amounts of organic inputs to meet farm demand is the main obstacle in organic agriculture. It entails the appropriate blending of all accessible nutrition sources which is appropriate to the land use system and ecological, social, and economic circumstances (Singh et al. 2021).

4.2 Crop Diversification

Crop diversification by involving several crops is the need of the hour for the development of sustainable farming systems in the Himalayan region. Inclusion of commercial crops in cropping system either as intercrop or in sequence of crop rotation is one of the most promising options for diversified sustainability of the system and it creates an opportunity to enhance the cropping intensity manifolds (Saha et al. 2011; Meena et al. 2015). The Himalayan region differs from other mountainous regions in terms of climate, edaphic characteristics, vegetation, cropping patterns, crop rotations, and other traits because of the intricate variation of agro-ecosystems. The Himalayas produce a picture of a variety of agro-ecological zones that supply the numerous microhabitats. The hill farmers have been cultivating an incredible array of horticultural and agricultural crops for thousands of years, thanks to this varied ecosystem, which has made this area the wealthy hub for crop diversification. On the other hand, the Himalayan hill and mountain regions are severely limited in their ability to produce resources due to a number of issues, including their poor economic development and ecological fragility (Bhatta and Vetaas 2016). Farmers may find it easier to adopt soil conservation and management strategies like conservation tillage, cover crops, and integrated pest management, as well as water conservation and management strategies like rainwater harvesting, irrigation, and water-efficient farming practices, with the support of training and technical assistance (Ogunsola et al. 2021; Zhao and Boll 2022). Rainfed, monocropping, and subsistence farming are the three main agricultural practices found in the Himalayan region. The primary crop farmed in the area is paddy, which is followed by maize.

5 Transforming Agriculture by the Himalayan Farmers

The Himalayan farmers often have little chance of escaping poverty with cereals produced by themselves. As a result, new cropping systems, such as cash farming of fruits and vegetables, have emerged. Payrolls from cash crops are creating jobs and much greater family incomes. Currently, the Himalayan states in India grow over 175 distinct varieties of vegetables. Generally, this information is disregarded due to the belief that native agro-biodiversity must be preserved. In recent decades, cropping patterns have shifted from agriculture to fruit crops due to the hills and mountains' unique appropriateness for horticulture. These data unequivocally show that an apparently irreversible process of transformation has begun in the Himalayan nations, leading farmers to repurpose their agricultural areas for the production of commercial commodities. The degree to which states in the Western and Eastern Himalayan area in India have adopted diversification varies greatly. There are significant attempts that are being made in the Eastern Himalayan region to diversify the crops. Through this method, farmers have been able to fully utilize the local niches created by the socioeconomic and agro-climatic conditions. Therefore, it appears that the process of changing the cropping pattern to cash crops has pushed food grain cultivation in the Himalayas to a smaller extent. Looking ahead, trends point to cash cropping of fruits, vegetables, floriculture, medicinal, and aromatic plants becoming dominant throughout the Himalayan states. For example, Tiger grass, often known as broom grass (*Thysanolaena maxima*), is a tall, valuable perennial plant with several uses. This grass grows in tussocks or hills, and its inflorescence is used to make environmentally friendly, biodegradable broom. In addition to being used to make brooms, the sticks can be used as fuel and to make paper in the pulp and paper industries. Again, the plant known as Jerusalem artichoke (*Helianthus tuberosus* L.) is native to temperate North America and has been used for food and animal feed for a long time. In recent years, it has also been used to produce bioactive food ingredients like oligo-fructans and inulin, and most importantly, as a raw material for the production of bioethanol. It was successfully introduced and well adopted in the Sikkim Himalayan region (Fig. 4c). The primary cash crop grown in the sub-Himalayan states of Sikkim and the Darjeeling district of West Bengal is large cardamom (*Amomum subulatum* Roxb.). It should be noted that one of the earliest spices that humans have utilised is large cardamom. The majority of the world's and India's supply of large cardamom comes from Sikkim, the world's largest producer. Large-scale cardamom cultivation has just begun in Arunachal Pradesh, Nagaland, and Meghalaya too (Fig. 4d).

Intensive agriculture offers the key to making agriculture profitable for small and marginal hill farms. The commercialization of Himalayan agriculture has resulted in a greater reliance on outside resources for farmers, including manpower, seeds, fertilizer, and machinery. Farmers are experiencing a general labour shortage due to the growing demand for cash crops, their own affordability, and the trend of locals choosing not to work as agricultural labourers on other people's farms because there are other options available to them for employment under government programmes. Currently, labour makes up almost half of all the variable costs associated with farming. Because of this, migrant labourer are available on a seasonal and permanent basis, and small and medium-sized farmers alike are increasingly using them. Farmers' interaction with market functions has greatly risen as a result of farming's market orientation.

New generation farmers in the Himalayan region are often also familiar with online marketing, which is particularly useful for promoting produce, fruits, and flowers. Nevertheless, the degree to which agriculture is commercialized varies greatly throughout the Himalayan states in India. Small-scale farmers in the Himalayan region are still involved in the production of traditional food crops for livelihood. Yet the reality is that commercial crops are dominating hill farming and are expanding both in terms of acreage and productivity. However, the acreage and productivity of the indigenous crops have significantly decreased. Although traditional methods of commercializing hill farming have increased economic well-being, they have also reduced native agro-biodiversity and eroded the traditional resource base. A new breed of farmers, known as progressive farmers, agribusiness entrepreneurs, or new age farmers, has emerged—thanks to the transformation process. They frequently have education and are starting to serve as role models for most other farmers.

It must be acknowledged that traditional agro-ecosystems are the tried-and-true prototypes of contemporary sustainable agricultural systems. There are potential to increase crop yield and reduce climate change impacts using traditional methods like agro-forestry, intercropping, crop rotation, cover crops, traditional organic composting, and integrated crop-animal husbandry. Locals and indigenous farmers have different perspectives on climate change, and they have different strategies for preparing for it (Tripathi and Singh 2013). Further, landscapes that maintain biodiversity and traditional sustainable agriculture methods are referred to as traditional agricultural landscapes. Their aesthetic, natural, cultural, historical, and socioeconomic values are respected (Barankova et al. 2011; Lieskovský et al. 2015). Traditional farming landscapes are found in areas where agricultural techniques have either remained mostly unchanged for a long time or have changed very little (Fischer et al. 2012).

6 Seed System and Community Seed Banks

In the Himalayan region, hybrid and enhanced seeds are not as popular as traditional seeds since people feel they are not well adapted to their ecosystem and are needed to be bought every season. In the region, the main ways to obtain improved seeds are through regional extension offices and local marketplaces. Farmers in this region have continued to use selection to improve conventional seeds. For hybrid or enhanced kinds of vegetables and grains, they only rely on seeds provided by agricultural extension agencies; however, people only plant these when they are given away at no cost. Except for potatoes, farming communities are largely self-sufficient with their crops and do not have a systematic need to buy seeds. When yield starts to fall, usually after 7–8 years of continuous cultivation, farmers replace their seeds. Before swapping or buying the crop, farmers discuss the crop's longevity, yield, and byproducts with the owner of the crop. They then determine whether or not to accept it. Soon after the crop is harvested, if the crop owners are friends or family, they often mutually agree to trade seeds for the same quantity or 20% more. Due to storage losses and holding onto wastages during sowing, farmers receive a quarter more than they were supposed to during seed exchange or purchase (Samberg et al. 2013). The majority of vegetables grown using community-improved seeds are native ones, for which the villages consistently preserve their own landrace seeds through customary selection procedures. Few farmers rely on local markets for the seeds and seedlings of crops like vegetables and commercial crops; instead, they rely on their own conserved seeds to acquire landraces. The need for seeds of contemporary kinds from agricultural departments is minuscule. Additional funding sources from non-governmental organizations, presents and remittances or food assistance grains are unimportant. New or better varieties are not freely exchanged between family members and neighbours. It seems that people prefer not to test these kinds before exchanging them with friends and neighbours because relationships may suffer if harvests don't work out.

In the Himalayan region, community seed banks have been developed to address the pressing needs of crop diversity conservation and seed access. Many seeds that had become extinct in an area have been brought in from other parts of the region, and some of them have gained enormous popularity. Communities need some direction and processes to be developed because they are unsure about the collaboration and safeguards required in this seed exchange. It is necessary to support new community entrepreneurial options like growing and selling bio-cultural goods and providing wholesome grains and cereals.

7 Innovations in Himalayan Farming System

7.1 Technological Innovations

In the Himalayan region, the majority of blacksmiths who create hand tools and implements for agriculture use traditional methods. The only fuel used in the furnaces is wooden charcoal. Government initiatives, mechanization, greater chemical use, specialization, and new technology—all contribute to the increased growth in food productivity in India More food is produced by fewer farmers, but at the expense of increased risks to human health, air and water pollution, soil erosion, and many other problems. The plains of India show these changes more than the hills. Agriculture in the plains is very different from that of the Himalayas, where traditional crops and farming methods still predominate. In the Himalayan region, the idea behind this kind of farming was to attain year-round self-sustainability free from any commercial interest (Pandey et al. 2016).

Agriculture's annual growth is very unpredictable, but its decadal growth rate paints a clear picture of the economic achievements made possible by the sector. The majority of agricultural experts view the use of contemporary technology as a way to reduce the natural control over agricultural growth, which has been attained to some extent through the construction of irrigation infrastructure and the use of pesticides and fertilizers, among other things. An important issue that has forced farmers to come up with creative solutions recently is the rise in crop raiding by untamed animals. Economic and market demands, cooperation with scientists, and experimentation are all significant drivers of innovation. Due to economic pressures, several crops that had vanished but were in high demand are brought back. Through their connections with scientists, some farmers have learned new and improved farming practices to increase productivity and improve soil quality and land productivity. They have since experimented to modify them to fit local requirements. Considering the patterns and trends of growth in the agricultural sector in India, Bhalla and Singh (1997) state, "the growth of land productivity and production, at the country and at the state levels, are positively associated with the use of modern inputs" (Bhalla and Singh 1997).

Modern agriculture helps to stabilize agricultural output and offers some protection from unusual monsoon conditions in India, but it does not guarantee that agriculture will continue to expand. The use of contemporary agricultural implements has its limitations. Due to greater potential, the adoption of contemporary technology initially causes stronger growth; but, as soon as potential is realized, agricultural expansion tends to slow down. There are significant regional disparities as a result of the states' differing approaches to the use of contemporary agricultural technologies and the development of their agricultural infrastructure throughout time and space (Bhalla and Singh 1997; Chand et al. 2007).

7.2 Market Innovations

It is challenging for farmers in mountainous regions to market their harvest, particularly perishable goods like vegetables. They perform mixed farming, have small landholdings, produce in modest quantities, and deal with unreliable transportation systems and rugged terrain. Amounts of agricultural produce cannot be transported to the market by one household. A group of creative, industrious young people from a hilly area have to take the lead in combining farmers' produce to make quantities that can be sold. Small farmers are encouraged to increase their revenue by growing agricultural items like honey, mushrooms, and vegetables, especially if they are women. The community's trust and cooperation with these young people, as well as their aptitude at organizing this group activity, are the main driving forces behind many innovative approaches.

7.3 Economic Growth Through Farming

The Himalayan region's economic situation is influenced in part by the resources that are distributed among its numerous ecological zones. Animal husbandry is a major industry, but forestry, trade, and tourism are also significant. In this region, agriculture is a significant source of revenue and essentially a subsistence effort. In the Himalayas, a shortage of arable land, land degradation, and unsustainable farming practices have a detrimental effect on livelihoods and food security (Das et al. 2014). Economic resources abound in the Himalayas. These consist of vast grasslands and forests, abundant agricultural land areas, easily accessible mineral resources, easily harvested waterpower, and breathtakingly beautiful natural surroundings. Rice, maize, wheat, potatoes and millets are produced in the most productive arable land areas. It could be predicted that the demand for rice could decline as a result of lifestyle changes and income growth, but consumption of fruits, vegetables, and pulses would rise. Therefore, it is imperative that farming systems based on rice and maize be diversified to include pulses and vegetables. Additionally, a lot of farmers and business owners are interested in raising animals and poultry for profit. Fruit-based alcoholic beverages, ginger processing, apple cultivation and processed products, processed citrus products, tapioca-based starch and sago, cold storages for horticultural produce, and multipurpose fruit and vegetable processing products are just a few of the horticulture-based industries that have been developed on the product lines in the region. Such initiatives have the power to transform the area and elevate the status of its populace as economically sound (Singh et al. 2021).

8 Degree of Climatic Variations and Related Phenomena in Himalayan Agro-ecosystem

The Himalayan region has been experiencing irregular precipitation due to climate change, which has also affected its crops. When it comes to rice for instance, hybrid grains are being developed, whereas conventional rice varieties are dwindling and are now nonexistent in some regions. Along with climate change, other primary causes of the decline in agricultural activity are labour shortages, farmer exodus, shifting land uses, globalization, and dwindling traditional knowledge. Farmers in the Indian Himalayan region have also stated that the primary causes of their switch from millets to other crops, for instance, are severe hailstorms as well as changes in food habits. Although the introduction of new vegetable and crop varieties has been made possible by rising temperatures, the current diversity has been lost as a result.

Crop productivity, water availability, and agricultural techniques in the Himalayas have changed as a result of climate change. Contrary to the detrimental effects, it has also been suggested that several crops, including tomato, cauliflower, wheat, maize, and rice, may benefit from the influence of increasing temperature in the hilly regions (Dahal et al. 2023). The primary climatic shifts that the Himalayan region experience include erratic rainfall, protracted dry spells in the winter and summer, lower annual rainfall, rising temperatures, and an increase in pests and diseases. Impacts of climate change on winter crops and apple farming can be detected due to the decreasing snowfall and rising temperature. Additionally, the lack of water for irrigation brought on by the winter drought has a detrimental effect on people's livelihoods (Bhandari 2018). The mountainous and sub-mountainous areas are distinct civilization hotspots where crop adaptations and environmental variances impact agricultural and food culture systems. This region is unique in that—It cultivates a wide variety of crops, including fruits-bearing plants, vegetables, oilseeds, legumes, grains, and spices. However, a key barrier to the region's food security is the severe circumstances of the hill agriculture system and the low yield of cereals, oilseeds, and pulses.

In the Himalayan region, flooding and landslides have become more common as a result of deforestation, extensive rural road development, and high-intensity rainfall over shorter time intervals. Deforestation is the primary cause of the climate shifts, according to the locals in the region. They also attributed the rise in illnesses and pests in plants and animals to the high humidity levels in July and August. The occurrence of biotic and abiotic stresses is common, such as rainfall, floods, wind strength at specific times, insects/pests, diseases, and droughts during the summer. Due to longer summers, a shorter rainy season, an increase in pest and disease prevalence, and strong winds in February and March, farmers are experiencing extended dry spells. Major effects of these changes are noted by them, including the removal of wetlands, modifications to tree seeding patterns, extinction of local species, and adjustments to the duration of seasons and agricultural growth period.

The Himalayan region's small farmers have been using a variety of tactics to address the problems brought on by climate change. Their transition to a cash economy has resulted in increased agricultural activity. They have adopted and domesticated drought-tolerant paddy varieties like addey, altered cropping patterns, and chosen to plant cash crops like broomstick grass and fodder trees for cattle farming. They have also adjusted the planting and harvesting times of crops (paddy, millet, and maize). Planting conditions and timing adjustments are crucial. This could be as a result of its location in the ecotone zone, which separates higher and lower altitudes. They typically cultivate both average-yielding low-altitude crops and high-altitude ones. Their growing circumstances and time are greatly impacted by even a slight change in climate. Overall though, there was minimal variation in the location of staple crops, their resistance to pests and diseases, and their production.

Some temperate fruit crops are becoming harder to continue farming as a result of localized temperature increases brought on by global warming. One such fruit crop that is being challenged in many parts of Himalaya is the apple; plums, pomegranates, and persimmons are the other fruit crops. Himalayan farmers are afraid of the their futures because of the unpredictable rainfall patterns, local weather conditions, and warmer, sometimes drier, sometimes excessively wet, sometimes unexpectedly cold, conditions that have been extensively experienced over the previous 10 years. It is believed that the supply of irrigation water, which is essential for cash crops, has too been impacted by climate change.

9 Factors Effecting the Degradation of Traditional Farming

The technology used by farmers with limited resources and the technology created by experts at research farms equipped with all the amenities are very different. Appropriate agriculture equipment and its expensive acquisition are difficult for the local people in the Himalayan region. Poor performance of multifaceted intercropping systems is caused by a lack of scientific knowledge about cropping patterns in the Himalayan region, limited and non-availability of quality seeds, synthetic fertilizers, bio-fertilizers, and other farm inputs, as well as a lack of technical support. A change in the growing season, crop duration, and soil moisture regime due to rising temperatures and shifting precipitation patterns may cause drought in crop plants, which would have a detrimental effect on crop productivity (Jasper et al. 2006; David and Naylor 2009; Calzadilla et al. 2013). The lack of infrastructure, government policies, financial assistance, and adequate market facilities limit the spread of the native crops. Farmers switched from traditional to modern agriculture as a result of their tenacious labour and meagre profits. While modern agriculture has increased food production, it has also accelerated the rise of a number of environmental issues, including pollution, soil degradation, biodiversity loss, climate change, and hazardous food (Zhang et al. 2017).

Due to the adoption of high yield varieties and the planting of genetically homogeneous cultivars across broad fields, modern agriculture is a major contributor to the loss of crop genetic resources in the Third World. Small farmers are typically the only ones who utilize traditional farming methods. The loss of habitat, the introduction of new kinds, past policies, the stigma associated with using traditional crops, and changing lifestyles have put traditional farming expertise in grave danger (Dweba and Mearns 2011). There is an immediate need for climate-smart traditional practices and sustainable food production amid the changing environment. The moment has come to resurrect and use old methods in order to enhance the socio-ecological integrity of agro-ecosystems. The current situation demands that traditional and modern agriculture be integrated. This unification would close the vast divide between traditional and contemporary peasants. Furthermore, it would strengthen the bonds between humans and nature. Traditional agriculture's future outlook calls for a climate-smart strategy for the production of sustainable food. Conventional farming methods made a substantial contribution to the advancement of agricultural science. They continue to feed humans in various parts of the world and have sustained a sizable population for ages (Koohafkan and Altieri 2010).

10 Conservation of Traditional Crops

Due to the limited area under cultivation, many crop species in India's northwest Himalayan region are hardly distributed and have limited potential to rise with the changing climate (Kumari et al. 2017). They play a big part in local production and consumption systems and are adaptive to agro-ecological niches. Landraces and ecotypes are given priority, and these crops have not gotten much attention in the research studies and in national agricultural and biodiversity conservation programmes (Williams and Haq 2002; Padulosi et al. 2002). Protecting our traditional crop landraces, farmer's varieties, and the agricultural knowledge that goes along with them is crucial under the new intellectual property rights regime. This allows disputes pertaining to benefit sharing to be settled and infringement and biopiracy issues to be handled with care (Barrueto et al. 2018; Malhotra et al. 2022). Furthermore, by offering a wealth of opportunities for any research related to the genetic enhancement of crop varieties—namely, yield, quality, nutrition, and resistance to significant biotic and abiotic stresses—these aspects also significantly contribute to the long-term sustainability of any particular agroecosystem (Ebert 2014; Tontisirin 2014; Roy et al. 2016; Ba et al. 2018).

The collection and conservation of inter- and intra-specific racial diversity in gene banks and agricultural farms, along with the protection of cold tolerance genotypes, disease resistant genotypes, plant populations, and traditional crop varieties where they have evolved (in-situ), are examples of complementary conservation strategies. They also utilize resources that are accessible outside of their fields. PGR Coordination between scientists and farmers is required to save the diversity of landraces from both studied and un-surveyed regions, for instance. With the help of the natural richness, trait-specific collections and conservations must be strengthened in order to address climate change. For instance, crop diversification is emphasized in traditional farming practices as a mean of mitigating risks associated with market volatility and climate susceptibility. While biodiversity is in danger of extinction due to the increasing effects of climate change, agro-biodiversity also suffers from unpredictable weather patterns. Promoting environmentally friendly policy and technology as soon as possible is therefore essential. These initiatives are still in their infancy, and for them to succeed in the long run, farmers and the government will need to put in more work and participate together in problem solving. For instance, traditional agriculture could be promoted. The knowledge gained through thousands of years of local farming traditions has resulted in traditional agriculture. Among the key characteristics of traditional agricultural systems are their high production, preservation of biodiversity, low energy inputs, and ability to mitigate climate change (Pulido and Bocco 2003; Srivastava et al. 2016).

Our ancestors' invaluable knowledge in traditional agriculture, which enables us cultivate food for ourselves while also preserving the natural balance, has left us with an invaluable legacy. The creation of new agricultural systems has been ongoing in response to the changing needs of humanity. Different places have different environmental conditions. This is true particularly for eco-sensitive areas like the Himalayas, where building substantial irrigation and farming infrastructure is difficult for contemporary agriculture. In order to attain food security, modern inputs must be incorporated into traditional agricultural systems to boost crop yields and mitigate the negative impacts of chemical fertilizers used in contemporary farming. Farmers utilize local, traditional, or landrace varieties of both main and minor crops all across the world, but especially in developing nations. Smallholder farmers use their traditional knowledge and skills to adapt to environmental changes by growing crops that are suited to the new environment and altering their farming methods (Jackson et al. 2007; Lasco et al. 2014). Such factors and measures may be especially important for the Himalayan Region's agriculture system.

11 Conclusion

The loss of biodiversity will not stop accelerating unless we alter the manner in which our food is produced. Our capacity to support human populations are threatened by additional ecosystem and habitat loss. Several research results indicate that agro-biodiversity management systems are needed to be urgently transformed. Three interrelated steps are recommended: altering the global agricultural pattern, preserving and allocating land for nature, and farming in a way that is more biodiversity-supporting and friendly to the environment. It is recommended that policy makers adopt a comprehensive strategy to consider the effects of agro-ecosystems, provide worldwide guidelines for reform, and translate them into national objectives. The decline in habitats and biodiversity in the Himalayas has resulted in a major loss of ecosystem services. A significant percentage of crop varieties are already vanished from the Himalayas. Because industrial patenting and other intellectual property laws are now applied to living things, fewer types and breeds are being widely cultivated and raised. As a result, the global market becomes more homogeneous, less varied, and more competitive. Therefore, there have been. Shifts in the ways that farmers and customers view and value domestic animal and crop breeds as well as their living situations. Small-scale, varied food production systems that protect farmers' crop variety are marginalized worldwide and it is the same in the Himalaya.

References

Abitogun AS, Olasehinde EF (2012) Nutritional evaluation of seed and characterization of crude Jack bean (Canavalia ensiformis) oil. IOSR. J Appl Chem 6:36–40

Aktar MW, Sengupta D, Chowdhury A (2009) Impact of pesticides use in agriculture: their benefits and hazards. Interdiscip Toxicol 2(1):1–12. https://doi.org/10.2478/v10102-009-0001-7

Avasthe R, Pradhan Y, Bhutia K (2012) Handbook of organic crop production in Sikkim. Published by Sikkim organic mission, GoI of Sikkim, Gangtok, Sikkim, India in collaboration with ICAR Research complex for NEH region, Sikkim Centre, Tadong, Gangtok, Sikkim, pp 59–72

Avasthe RK, Babu S, Singh R, Yadav GS, Kumar A (2020) Productivity and profitability assessment of organically grown vegetables embedded in rice–based cropping sequences in Sikkim Himalayas, North East India. J Environ Biol 41:111–117

Ba Q-X, Lu D-J, Kuo WH-J, Lai P-H (2018) Traditional farming and sustainable development of an indigenous community in the mountain area – a case study of Wutai Village in Taiwan. Sustainability 10:3370

Babu PK, Rampal (2024) Traditional ecological knowledge in sustainable conservation of seeds and food grains in the Himalayas. In: Borthakur A, Singh P (eds) Addressing the climate crisis in the Indian Himalayas. https://doi.org/10.1007/978-3-031-50097-8_3

Barankova Z, Dobrovodska M, Stefunkova D, Babicova D, Moyzeova M, Petrovic F (2011) Participation of local people on identifying the landscape values and future development in historical agricultural landscapes. Ekologia 30:216–228. https://doi.org/10.4149/ekol_2011_02_216

Barrueto AK, Merz J, Kohler T, Hammer T (2018) What prompts agricultural innovation in Rural Nepal: a study using the example of Macadamia and Walnut Trees as novel cash crops. Agriculture 8:21

Bhalla GS, Singh G (1997) Recent developments in Indian agriculture: A state level analysis. Econ Polit Wkly 32(13):A-2-A-18

Bhandari MP (2018) Climate change impacts on food security, a brief comparative case study of Bangladesh, India, Nepal and Pakistan. Acta Sci Agric 2(8):136–140

Bhatta KP, Vetaas OR (2016) Does tree canopy closure moderate the effect of climate warming on plant species composition of temperate Himalayan oak forest? J Vegetation Sci 27(5):948–957

Bi S, Wang A, Lao F, Shen Q, Liao X, Zhang P, Wu J (2021) Effects of frying, roasting and boiling on aroma profiles of adzuki beans (Vigna angularis) and potential of adzuki bean and millet flours to improve flavor and sensory characteristics of biscuits. Food Chem 339:127878

Bressani R, Brenes RS, Garcia A, Elias LG (1987) Chemical composition, amino acid content and protein quality of canavalia spp. seeds. J Sci Food Agric 40:17–23

Cadoux S, Sauzet G, Morison MV, Pontet C, Champolivier L, Robert C, Lieven J, Flénet F, Mangenot O, Fauvin P, Landé N (2015) Intercropping frost-sensitive legume crops with winter oilseed rape reduces weed competition, insect damage, and improves nitrogen use efficiency. OCL 22(3):D302

Calzadilla A, Rehdanz K, Betts R, Falloon P, Wiltshire A, Tol RSJ (2013) Climate change impacts on global agriculture. Clim Chang 120:357–374

Chand R, Raju SS, Pandey LM (2007) Growth crisis in agriculture: Severity and options at national and state levels. Econ Polit Wkly, 42(26):2528–33.

Chhetry GKN, Belbahri L (2009) Indigenous pest and disease management practices in traditional farming systems in north east India: a review. J Plant Breeding Crop Sci 1:28–38

Choudhary VK, Choudhury BU (2018) A staggered maize-legume intercrop arrangement influences yield, weed smothering and nutrient balance in the Eastern Himalayan region of India. Exp Agric 54(2):181–200

Dahal KR, Dahal P, Adhikari RK, Naukkarinen V, Panday D, Bista N, Helenius J, Marambe B (2023) Climate change impacts and adaptation in a hill farming system of the Himalayan region: climatic trends, farmers' perceptions and practices. Climate 11(1):11. https://doi.org/10.3390/cli11010011

Das A, Ghosh PK (2012) Role of legumes in sustainable agriculture and food security: an Indian perspective. Outlook Agric 41:4. https://doi.org/10.5367/oa.2012.0109

Das A, Patel DP, Munda GC, Ramkrushna GI, Kumar M, Ngachan SV (2014) Improving productivity, water and energy use efficiency in lowland rice (Oryza sativa) through appropriate establishment methods and nutrient management practices in the mid-altitude of north-east India. Expl Agric 50(3):353–375

Das A, Ramakrishna GI, Layek J, Babu S, Ngangom B, Suiting D (2017a) Package of practices of Meghalaya. In: Ravisankar N, Panwar AS, Prasad Kamta, Kumar Vipin, Bhaskar S (eds) Organic farming crop production guide. Network project on organic farming. ICAR Indian Institute of Farming Systems Research, Modipuram, Meerut, pp 1–592

Das A, Patel DP, Kumar M, Ramkrushna GI, Mukherjee A, Layek J, Ngachan SV, Buragohain J (2017b) Impact of seven years of organic farming on soil and produce quality and crop yields in eastern Himalayas, India. Agric Ecosyst Environ 236:142–153

Das A, Devi TM, Babu S, Ansari M, Layek J, Bhowmick SN, Yadav GS, Singh R (2018) Cereal-legume cropping system in Indian Himalayan region for food and environmental sustainability. In: Meena RS et al (eds) Legumes for soil health and sustainable management. https://doi.org/10.1007/978-981-13-0253-4_2

David BS, Naylor RL (2009) Historical warnings of future food insecurity with unprecedented seasonal heat. Science 323:240–244

Devaraj RD, Jeepipalli SP, Xu B (2020) Phytochemistry and health promoting effects of Job's tears (Coix lacrymajobi) – a critical review. Food Biosci 34:100537. https://doi.org/10.1016/j.fbio.2020.100537

Dweba TP, Mearns MA (2011) Conserving indigenous knowledge as the key to the current and future use of traditional vegetables. Int J Inf Manag 31(6):564–571

Ebert AW (2014) Potential of underutilized traditional vegetables and legume crops to contribute to food and nutritional security, income and more sustainable production systems. Sustainability 6:319–335

Elangovan M, Shrotria PK, Rao RKV, Khulbe RK, Babu PK (2015) Collection, GIS mapping and Conservation of Broomcorn sorghum from Uttarakhand, India. Int J Adv Res 3(8):57–69

Fischer J, Hartel T, Kuemmerle T (2012) Conservation policy in traditional farming landscapes. Conserv Lett 5(3):167–175

GoN (2017) Statistical handbook of Nagaland 2017. Directorate of Nagaland and Statistic Nagaland, Government of Nagaland, Nagaland

Goron TL, Raizada MN (2015) Genetic diversity and genomic resources available for the small millet crops to accelerate a New Green Revolution. Front Plant Sci 6:157. https://doi.org/10.3389/fpls.2015.00157

Gururani K, Sood S, Kumar A, Joshi DC, Pandey D, Sharma AR (2021) Mainstreaming Barahnaja cultivation for food and nutritional security in the Himalayan region. Biodivers Conserv 30:551–574. https://doi.org/10.1007/s10531-021-02123-9

Haldhar SM, Jat GC, Deshwal HL, Gora JS, Singh D (2017) Insect pest and disease management in organic farming. Towards organic agriculture. In: Gangwar B, Jat NK (eds) Today & tomorrow's Printers and Publishers, New Delhi, pp 359–390

Hipparagi Y, Singh R, Choudhury DR, Gupta V (2017) Genetic diversity and population structure analysis of Kala bhat (*Glycine max* (L.) Merrill) genotypes using SSR markers. Hereditas 154:9. https://doi.org/10.1186/s41065-017-0030-8

Hussain A, Rasul G, Mahapatra B, Thuladhar S (2016) Household food security in the face of climate change in the Hindu-Kush Himalayan region. Food Secur 8(5):921–937

Jackson LE, Pascual U, Hodgkin T (2007) Utilizing and conserving agrobiodiversity in agricultural landscapes. Agric Ecosyst Environ 121(3):196–210

Jasper K, Calanca P, Fuhrer J (2006) Changes in summertime soil water patterns in complex terrain due to climatic change. J Hydrol 327:550–563

Jhariya MK, Banerjee A, Meena RS, Kumar S, Raj A (2021) Sustainable intensification for agroecosystem services and management: an overview. In: Jhariya MK, Banerjee A, Meena RS, Kumar S, Raj A (eds) Sustainable intensification for agroecosystem services and management. Springer, Singapore. https://doi.org/10.1007/978-981-16-3207-5_1

Kayang H (2007) Tribal knowledge on wild edible plants of Meghalaya, Northeast India. Indian J Tradit Knowl 6:177–181

Kim TJ, Park JG, Ahn SK, Kim KW, Choi J, Kim HY, Ha S-H, Seo WD, Kim JK (2020) Discrimination of adzuki bean (Vigna angularis) geographical origin by targeted and non-targeted metabolite profiling with gas chromatography time-of-flight mass spectrometry. Metabolites 10:112

Koohafkan P, Altieri MA (2010) Globally important agricultural heritage systems: a legacy for the future. UN-FAO, Rome

Kumar S, Meena RS, Datta R, Verma SK, Yadav GS, Pradhan G, Molaei A, Rahman GM, Mashuk HA (2020) Legumes for carbon and nitrogen cycling: an organic approach. Carbon and nitrogen cycling in soil. Springer, Singapore, pp 337–375

Kumari J, Kumar A, Singh TP, Bhatt KC, Mishra AK, Semwal DP et al (2017) Collection, evaluation and phenotypic diversity assessment of maize (Zea mays) germplasm from North Eastern Himalayan region. Indian J Agric Sci 87:727–733

Lasco RD, Delfino RJP, Catacutan DC, Simelton ES, Wilson DM (2014) Climate risk adaptation by smallholder farmers: the roles of trees and agroforestry. Curr Opin Environ Sustain 6:83–88

Lata S, Yadav S, Kumar P, Yadav M (2023) Biodiversity of underutilized fruits of the Himalayas. Genet Resour Crop Evol 70:71–94. https://doi.org/10.1007/s10722-022-01476-6

Laxmisha KM, Semwal DP, Gupta V, Katral A, Bisht IS, Mehta PS, Arya M, Bhardwaj R, Bhatt KC (2022) Nutritional profiling and GIS-based grid mapping of Job's tears (Coix lacryma-jobi L) germplasm. Appl Food Res 2(2):100169. https://doi.org/10.1016/j.afres.2022.100169

Lieskovský J, Bezák P, Špulerová J, Lieskovský T, Koleda P, Dobrovodská M, Gimmi U (2015) The abandonment of traditional agricultural landscape in Slovakia-analysis of extent and driving forces. J Rural Stud 37:75–84

Lim HJ, Park SI, Bak SG, Cheong SH, Lee S, Baek YB, Lee CM, Lee KM, Lee SW, Lee SJ (2020) Beneficial effects of Vigna angularis extract in osteoporosis and osteoarthritis. Food Sci Nutr 8:6550–6556

Lithourgidis AS, Dordas CA, Damalas CA, Vlachostergios D (2011) Annual intercrops: an alternative pathway for sustainable agriculture. Aust J Crop Sci 5(4):396–410

Maikhuri RK, Ramakrishnan PS (1992) Ethnobiology of some tribal societies of Arunachal Pradesh, Northeastern India. J Econ Taxon Bot 10:61–78

Maikhuri RK, Semwal RL, Singh A, Nautiyal MC (1994) Wild fruit as a contribution to sustainable rural development: a case study from the Garhwal Himalaya. Inter J Sustain Dev World Ecol 1:56–68

Maikhuri RK, Rao KS, Saxena KG (2004) Bioprospecting of wild edibles for rural development in the central Himalayan mountain of India. Mt Res Dev 24:110–113

Malangmeih L, Rahaman SM (2016) Economics of fresh Naga king chilli in Manipur, India-a case study. Int J Environ Ecol Family Urban Stud (IJEEFUS) 6(1):151–162

Malhotra N, Sharma P, Sood H, Chandora R, Arya M, Rana JC, Singh M (2022) Agro-morphological characterization and nutritional profiling of traditional Himalayan CropLandraces for their promotion toward mainstream agriculture. Front Plant Sci 13:898220. https://doi.org/10.3389/fpls.2022.898220

Meena RS, Yadav RS, Meena H, Kumar S, Meena YK, Singh A (2015) Towards the current need to enhance legume productivity and soil sustainability worldwide: a book review. J Clean Prod 104:513–515

Meetei NT, Singh AK, Singh BK, Mandal N (2016) Recent advances in naga king chilli (capsicum chinense JACQ.) research. Int J Agric Environ Biotech 9(3):421–428

Mikić A, Ćupina B, Rubiales D, Mihailović V, Šarūnaitė L, Fustec J, Antanasović S, Krstić Đ, Bedoussac L, Zorić L, Đorđević V (2015) Models, developments, and perspectives of mutual legume inter-cropping. Adv Agron 130:337–419

Moyer RA, Hummer KE, Finn CE, Frei B, Wrolstad RE (2002) Anthocyanins, phenolics, and antioxidant capacity in diverse small fruits: Vaccinium, Rubus, and Ribes. J Agric Food Chem 50:519–525

Nile SH, Park SW (2014) Edible berries: bioactive components and their effect on human health. Nutrition 30(2):134–144

Ogunsola OA, Adeniyi OD, Adedokun VA (2021) Soil management and conservation: an approach to mitigate and ameliorate soil contamination. In: Marcelo LL, Sonia S (eds) Soil contamination-threats and sustainable solutions. Intech Open, London, pp 1–7

Padulosi S, Heywood V, Hunter D, Jarvis A (2011) Underutilized species and climate change: current status and outlook. In: Yadav SS, Redden RJ, Hatfield JL (eds) Crop adaptation to climate change. Blackwell Publishing Ltd, UK, pp 507–521

Padulosi S, Hodgkin T, Williams JT, Haq N (2002) Underutilized crops: trends, challenges and opportunities in the 21st century. In: Engels JMM et al (eds) Managing plant genetic resources, pp 323–338. CABI, Wallingford, UK, and IPGRI, Rome, Italy

Pandey PC, Vibhuti AP, Bargali K, Bargali SS (2016) Agro-biodiversity of Kumaun Himalaya, India: a review. Curr Agric Res 4(1):89. https://doi.org/10.12944/CARJ.4.1.02

Pantelidis G, Vasilakakis M, Manganaris G, Diamantadis G (2007) Antioxidant capacity, phenol, anthocyanin and ascorbic acid contents in raspberries, blackberries, red currants, gooseberries and cornelian cherries. Food Chem 102:777–783

Patel R, Singh RKR, Tyagi V, Mallesha, Raju PS (2016) Nutritional evaluation of Canavalia ensiformis (Jack bean) cultivated in north east region of India. Int J Bot Stud 6:18–21

Pillai RG (2020) Antiproliferative actions of Ziziphus jujube fruit extract is mediated through alterations in Bcl2-Bax ratio and through the activation of caspases. Chem Biol Lett 7:41–46

Pulido JS, Bocco G (2003) The traditional farming system of a Mexican indigenous community: the case of Nuevo San Juan Parangaricutiro, Michoacán, Mexico. Geoderma 111:249–265

Rana JC, Verma VD (2011) Genetic resources of temperate minor fruits (indigenous and exotic). National Bureau of Plant Genetic Resources, New Delhi, pp 1–61

Rathore S, Chandola M, Raghuvanshi R, Kaur M, Singh KV (2021) Indigenous pest management practices of Indian Hill farmers: introspecting their rationale and communication pattern for secure ecosystems. Sustainability 13:11608. https://doi.org/10.3390/su132111608

Reddy PP (2017) Crop residue management and organic amendments. (In) agro–ecological approaches to pest management for sustainable agriculture. Springer, Singapore. https://doi.org/10.1007/978-981-10-4325-3_3

Roy S, Marndi BC, Mawkhlieng B, Banerjee A, Yadav RM, Mishra AK, Bansal KC (2016) Genetic diversity and structure in hill rice (Oryza sativa L.) landraces from the north-eastern Himalayas of India. BMC Genet 17:107. https://doi.org/10.1186/s12863-016-0414-1

Saha S, Gupta A, Singh SRK, Bharti N, Singh KP, Mahajan V, Gupta HS (2011) Compositional and varietal influence of finger millet flour on rheological properties of dough and quality of biscuit. Food Sci Technol LWT 44:616–621

Samberg HL, Shennan C, Zavaleta E (2013) Farmer seed exchange and crop diversity in a changing agricultural landscape in the southern highlands of Ethiopia. Hum Ecol 41(3):477–485

Sarita ES, Singh E (2016) Potential of millets: nutrients composition and health benefits. J Sci Innov Res 5(2):46–50

Shah NC (2006) Black soybean: an ignored nutritious and medicinal food crop from the Kumaon region of India. Asian Agri-Hist 10:33–42

Singh R, Babu S, Avasthe RK, Yadav GS, Rajkhowa DJ (2016) Productivity, economic profitability and energy dynamics of rice (*Oryza sativa* L.) under diverse tillage and nutrient–management practices in rice–vegetable pea cropping system of Sikkim Himalayas. Indian J Agric Sci 86(3):326–330

Singh R, Babu S, Avasthe RK, Das A, Praharaj CS, Jayantalayek, Kumar A, Rathore SS, Kancheti M, Kumar S, Yadav SK and Pashte V (2021) Organic farming in North–East India: status and strategies. Indian J Agronomy 66:163–179

Singh R, Singh GS (2017) Traditional agriculture: a climate-smart approach for sustainable food production. Energy Ecol Environ 2(5):296–316. https://doi.org/10.1007/s40974-017-0074-7

Sood S, Kumar A, Babu BK, Gaur VS, Pandey D, Kant L, Pattnayak A (2016) Gene discovery and advances in finger millet [Eleusine coracana (L.) Gaertn.] genomics-an important nutri-cereal of future. Front Plant Sci 7:1634. https://doi.org/10.3389/fpls.2016.01634

Srivastava AS, Mathur YK, Srivastava SK, Upadhayay KD (1985) Survey and distribution of white grub beetles in U.P. Pestology 9:34–36

Srivastava P, Singh R, Tripathi S, Raghubanshi AS (2016) An urgent need for sustainable thinking in agriculture: an Indian scenario. Ecol Indic 67:611–622

Subedi A, Raut N, Gurung S (2023) How Himalayan communities are changing cultivation practices in the context of climate change. Reg Sustain 4(2023):378–389. https://doi.org/10.1016/j.regsus.2023.11.001

Sujakhu NM, Ranjitkar S, He J, Schmidt-Vogt D, Su Y, Xu J (2019) Assessing the livelihood vulnerability of rural indigenous households to climate changes in Central Nepal, Himalaya. Sustainability 11(10):2977. https://doi.org/10.3390/su11102977

Tontisirin K (2014) Promotion of underutilized indigenous food resources for food security and nutrition in Asia and the Pacific. In: Durst P, Bayasgalanbat N (eds) Promotion of underutilized indigenous food resources for food security and nutrition in Asia and the Pacific. Food and Agriculture Organization of the United Nations Regional Office for Asia and the Pacific, Bangkok, pp 21–25

Tripathi A, Singh GS (2013) Perception, anticipation and responses of people to changing climate in the Gangetic plain of India. Curr Sci 105:1673–1684

Wang X, Choi Y-M, Jeon Y-A, Yi J, Shin M-J, Desta KT, Yoon H (2023) Analysis of genetic diversity in adzuki beans (Vigna angularis): insights into environmental adaptation and early breeding strategies for yield improvement. Plan Theory 12:4154. https://doi.org/10.3390/plants12244154

Williams J, Haq N (2002) Global research on underutilized crops: an assessment of current activities and proposals for enhanced cooperation. ICUC

Wu G, Bai Z, Wan Y, Shi H, Huang X, Nie S (2020) Antidiabetic effects of polysaccharide from azuki bean (Vigna angularis) in type 2 diabetic rats via insulin/PI3K/AKT signaling pathway. Food Hydrocoll 101:105456

Yadav GS, Das A, Lal R, Babu S, Meena RP, Saha P, Singh R, Datta M (2018) Energy budget and carbon footprint in a no–till and mulch based rice–mustard cropping system. J Clean Prod 191:144–157

Zhang Y, Min Q, Li H, He L, Zhang C, Yang L (2017) A conservation approach of globally important agricultural heritage systems (GIAHS): improving traditional agricultural patterns and promoting scale-production. Sustainability 9(2):295

Zhao MQ, Boll J (2022) Adaptation of water resources management under climate change. Front Water 4. https://doi.org/10.3389/frwa.2022.983228

Shifting Cultivation: An Analysis of Jhum Cultivation in Karbi Society of Assam

Purabi Bhagawati

1 Introduction

Shifting cultivation or Jhum cultivation is a type of shifting farming practised in the hilly regions of North-East (NE) India: Assam, Arunachal Pradesh, Manipur, Meghalaya, Nagaland, and Tripura. Shifting cultivation is a chief mode of production and a key source of economic stability in these regions (Ninan 1992). The tribes that live in hilly regions in these states have been practising shifting agriculture for several years. The allotment of plots for practising shifting cultivation within these communities is frequently based on egalitarian ideas such as shared responsibility, ownership, and partnership. The process is usually practised in areas near ills or forest. Jhum cultivation usually starts during the months of October–December and is primarily practised in hills and near forests. As the cultivation methods involve tribal modes of production, each tribal community follows a different method in selecting plots. In certain tribes, the entire community is in charge of maintaining Jhum plots, but in other tribes, the family that is given the plot is responsible for cleaning and maintaining it. Ensuring equitable and universal access to land is a prime concern in shifting cultivation. During the allotment of the plot, the size and workforce of the family is taken into consideration. Thus, the distribution of plots is based on an equitable system, where families receive plots as per the numbers of family. Availability, accessibility, and affordability for the family responsible is prioritized based on the principle of 'mouths of feed'. The process of cleaning the plot requires over a month where intensive labour is needed. Shifting agriculture plays a significant role in the tribal economy and it becomes an essential to the way of life and socioeconomic structure of hill and tribal areas. Shifting cultivation is widely practiced in West and Central Africa, Southeast Asia and South/Central America. In

P. Bhagawati (✉)
Department of Sociology, Government Model College Kaziranga, Kohora, Assam, India

A. Borthakur, P. Singh (eds.), *The Resilience of Traditional Knowledge Systems for a Sustainable Future*, Advances in Asian Human-Environmental Research,
https://doi.org/10.1007/978-3-031-56858-9_4

India, this type of cultivation is practiced among the tribes of Hilly areas of North-Eastern Region. In the seven North-Eastern states, 4.92 lakh indigenous families practiced Jhum on 4.53 lakh hectares in a year, according to the National Commission on Agriculture (1976). In these areas the tribal population's primary means of subsistence in low-density areas is shifting cultivation; nevertheless, in other states with higher densities of population, there are many forms of economic diversification among the populace. Jhum cultivation is typified by the cutting and clearance of productive forest areas, usually measuring less than 1.5 to 2.0 hectares, followed by tillage and burning of the phytomass prior to the cultivation phase (Haokip 2023). The socioeconomic, cultural and religious beliefs of these tribes are closely associated with shifting cultivation, which accounts for around 86% of the total farmed area. Again, almost all tribes believe in different religious practices before starting the cultivation, especially harvesting. It is a common belief that without worship the evil spirit or deities can affect the healthy output of their crops. Therefore, people do worship the evil spirit and deities as well as perform sacrifices for good crops and family prosperity.

2 Objectives of the Study

Shifting cultivation has cultural meaning among the tribes of Hill areas. The important factors for shifting cultivation are land and labour. The process requires intensive labour force and extensive land. Yet due to the growing population, nowadays the area of available land is decreasing and that made adverse impact on the Jhum cultivation. Shifting cultivation is different from slash-and-burn. Slash and burn is a land clearing method to open up forest land. On the other hand, shifting cultivation is an integrated farming method that combines agriculture, forestry and robust community social organization (Talukdar 2009). The Indian Forest Act, 1927 excluded the forest of NE Region from protection which is controlled by the communities. In recent years, the Act has been modified for NE India and since then the large forest areas are converted as reserved forest areas. In reserved forest areas there are different restrictions that are imposed in doing shifting cultivation. Different studies have classified shifting cultivation into four types: traditional and distorted, innovated and modified. Traditional agriculture is prevalent in the remote areas, which are sustainable in nature. However, traditional agriculture satisfies the rural agriculture. Distorted cultivation means when the ratio of man and land is declining, thus affects the entire shifting cultivation. Innovative and modified types of cultivation identify the new innovations now as are applied in the shifting cultivation. Innovative methods have been widely adopted to the current availability of resources, which are best suited to present day societal requirements. In some cases, pulse crops are used to increase the fertility of the land. However, modern methods of shifting cultivation are also used in different states of North-Eastern States under the direct supervision of respective state governments (Talukdar 2009). Jhum cultivation is an indigenous and primitive practice of cultivation in the states of North-East India, which occupy more than 84% of their total amount of lands. In these lands, both the Jhum (53%)

and abounded jhum (47%) are included (NRSC 2011). Meanwhile scholars have pointed out that about 0.44 million tribal families are dependent on Jhum cultivation for their livelihood (Yadav 2013). There are continuous debates between anthropological or sociological perspectives as well as the ecological perspective of Jhum Cultivation. Both anthropologists and sociologists have been supportive towards the Jhum cultivation. Jhum cultivation has been part of tribal social life where lands are owned by community wise. Community ownership of land is the opposite discourse of private ownership of land. The peasants who have been involving in the Jhum cultivation have equal rights over their land and as a result the entire process is based on equity and sustainable environment. But with the rising population issues, the situation of accessing equal land has been changing. Issues like land scarcity has been gradually part of tribal people's lived experiences and that creates hindrance in shifting cultivation. Again the method of Jhum cultivation demands the clearing the forest areas by burning of the phytomass and tillage. Ecologists pointed out emission of harmful gases/chemicals in such burning activities and that made adverse effect on the environment. Therefore, there are differences between sociological and ecological perspectives of Jhum Cultivation (Ninan 1992). Jhum cultivation stands as contrary to the neo liberal economy and therefore both centre and state government make law to control the Jhum cultivation. Scientists, environmentalists and other managers of natural resources have a misunderstanding about the term "Jhum." Shifting cultivation is the general term for any agricultural technique in which fields are cleared with the help of fire and cultivate for a shorter amount of time (Conklin 1957). Jhum is not identical with 'slash-burn agriculture' or it is not just a land-use system where forest is reduced and burned to make the land for cultivation for a particular period (Kherkhoff and Sharma 2006). However, Jhum is constituent of traditional agro-ecosystem encompassing practices that has been derived from nature and where tribal's culture and society shows close proximity with the Jhum cultivation (Bhagawati et al. 2015). Different studies have focussed that there are differences between other shifting cultivation and Jhum cultivation. In this paper an attempt has been made to understand the Jhum cultivator's understanding about their everyday life. Understanding the social aspects related to Jhum is another objective of the study. Meanwhile the paper also tries to understand state's aspect towards the Jhum cultivation.

3 Methods and Materials

To understand the lived experiences of the farmers who have been involved in the Jhum cultivation, phenomenological understanding is used. Qualitative research is conducted through intense or prolonged contact with the "field" or "life situation" (Miles and Huberman 1994). In this type of qualitative study, the researcher attempts to capture the data with the help of the "verstehen" method where the individual's first-hand experiences are considered important sources of data. In the phenomenological research approach, in-depth understanding is required to know the everyday experiences of women in the above-mentioned situations. Dilthey mentioned "For social interpretation deep understanding over the subject is required" (Dilthey 1977). Phenomenologists often use interview schedules and transcription to collect the data. In this study also, methods like interview schedules and observation methods have been used for data collection. This research design focuses on thick descriptions where the daily experiences of women are considered. The researcher tries to understand the meaning-making of the narratives as they are told by the respondents. In qualitative research, words are considered as important sources of data collection. Dealing with the lived experiences of women, the researcher tries to understand their perception of weaving, social structure, social and cultural meaning of weaving (Manen 1977) and how such meanings are being connected through the social world. But while doing the fieldwork, being an outsider to her field settings this researcher faced different challenges. These challenges are inevitable because the researcher is unfamiliar to the field. Different post-colonial scholars have questioned the relationship between the researcher and researched subjects. Rose (1997) argues that the power position of the researcher could affect the rapport-building process of the researched people that is it leads to disrupting the process of data collection. Later Rose (1997) suggested the process of reflexivity that could be helpful in maintaining the relationship between the researcher and the researched. Meanwhile, various feminist scholars also pointed out the importance of self-reflexivity techniques for creating rapport between the researcher and the researched. By keeping these facts in mind, the researcher has tried to build rapport with the respondents to minimise insider-outsider issues.

The study is based on both primary and secondary data. The primary data are collected with the help of field notes/journaling—informal and formal interviews are used to further compile data. For empirical evidence 20 people who remain involved in Jhum cultivation in Karbi Society are interviewed. The secondary data for the study has been selected with the help of journal articles, newspaper columns, books and government reports.

4 About the Community

The Karbis, who are known as Mikir, is one of the important tribes in Assam. The Karbis are considered as the majority community in the Karbi Anglong district of Assam—the district was named in after the Karbi community. The Karbis call themselves as the "Arleng" which means man. The Karbis have an ethnically mongoloid origin while linguistically they belong to "Tibeto-Burman Group" (Engtipi and Dhanaraju 2020). Though Karbi Anglong is majorly reported as the residence of mostly Karbi people, yet some of the Karbi also reside in the district of N.C. Hills and some parts of Golaghat, Nagaon, Kamrup district of Assam. There are also members of the Karbi tribe found in some pockets of Meghalaya and Arunachal Pradesh. The Karbis have been considered as the third largest tribal community of Assam after the Bodos and Missings. In the words of Dr. Phukan Ch. Phangcho, the first native doctorate, "The Karbis are a group of people living in the hills, plateaus and plains of the central part of North East India, especially in Assam." Some of them are also reported to be living in Sylhet district in Bangladesh' (Gogoi 2011). The word "Karbi" is derived from the word "Thekar Kibi" or "Mekar Kibi." "Thekar Kibi" is traditional ceremony of purifying a new born, so that he or she can be socially considered as a member of Karbi society. On the other hand, "Mekar Kibi' means keeping spark of fire in the fireplace of the house for performing the ceremony of "Thekar Kibi". Therefore the tribe is known as "Me akar Kibi" which later becomes as "Karbi" (Engtipi and Dhanaraju 2020). The Karbis are divided into four groups namely: Amri, Ronghang, Chinthong and Dumurali (Bey 2004). Amris are originally found in Amri Block—the western most part of the Hamren sub-division, Ri-Bhoi area of East Khasi Hills, low hills and foothills of Southern Kamrup. Later they scattered to the other parts of Kamrup District. They are considered as the oldest settlers among the Karbi tribe. The "Ronghang" section of the Karbis is found in the eastern part of Hamren sub-Division, i.e. the southern part of Hojai sub-division of Nagaon district. They are also found in the parts of North Cachar Hills. The Chinthong Karbis mostly reside in Chinthong development block of Hamren while some of them are found in Diphu township also. Located in the southern region of Nagaon district, the "Dumurali" Karbis are plains people who primarily live from Jamunamukh in the east to Sonapur (in southern Kamrup) in the west (Barua and Kikhi 2016).

5 Findings

5.1 Lived Experiences of Jhum Cultivators

There are continuous debates between the ecology and cultural aspect of Jhum cultivation. Ecologists have pointed out that Jhum cultivation creates major havoc to the environment. The technique of shifting cultivation results in the loss of valuable tree species, shrubs, medicinal plants and small forest products, among other flora and fauna. Meanwhile, ecologists also call attention to other various ill effects of shifting cultivation, for instance intensification of deforestation, wild animals lose their shelters, the entire ecosystem might be affected and as a result there might be heavy rainfall in the surrounding areas. Due to lack of trees on the ground, it might lead soil erosion. In a nutshell, ecology is distorted and disturbed in a manner that can never restored. Yet, within the ecologists, there are different schools of thought, where one of such school agrees with the adaptation to Jhum cultivation, while the other school mentions about the adverse effect of Jhum cultivation over nature. Both the state and some of the non-state authorities also believe that Jhum cultivation has made a severely negative impact on the environment. The arguments against the Jhum cultivation has come from the forest department, DONER as well as the World Bank for trade promotion. Again, some private entities also want to utilise Jhum land for industrial purposes (Dash 2006). In the neoliberal consumption-oriented society, both the state and non-state are interested in profit making which endangers the traditional ethos and culture. This study is conducted in the few pockets of Karbi dominated areas of Assam where Jhum cultivation prevails. For Jhum cultivation both private and common lands have been used. In Orissa, using both private and common lands system is known as *podu* (Dash 2006). There is a pattern of private land ownership among the Khasi, Karbi and Garo tribes in Meghalaya (Deb et al. 2013).

Extensive use of human labour is one of the important characteristics of Jhum cultivation. In most cases, human labour is being mutually exchanged among the cultivators. Such mutual exchange of labour force helps to create solidarity among the people. One of the respondents, who has been involving in these practices over a long time, said *"The cultivation demands labour work and therefore we use to help each other during the cultivation. We love to help our fellow cultivators; it creates We Feelings."* In most cases, the families who have practiced Jhum cultivation—they are used to the exchange of labour tasks and cultivate together. The practice of community cultivation increases productivity. It is said during the harvest time; all the farmers exchange their innovations. These shared techniques eventually help to increase their productivity (Punitha et al. 2018). The shifting cultivators of Nawalparasi village of Nepal show the exchange way of cultivation and where they use different new innovation to enhance their productivity (Kherkhoff and Sharma 2006). Division of labour is another important component of Jhum cultivation. There are mix responses were found where some of the respondents believe that there is still a strict division of labour in the Jhum cultivation, while others don't agree about this. One 60-year-old respondent who has been involved in the Jhum cultivation stated *"Both male and female equally participate in the cultivation process and there is no strict demarcation of gender-based work. Both of them equally share the cultivation work."* Again, from another section, the narrative sounds somewhat different. *"Women are good in seed selection and planting. Clearing the forest and other activities like cutting the trees in the jungles, clearing, burning are generally done by men,"* a 55-year-old respondent said.

Jhum cultivation is considered as an egalitarian cultivation. But in different areas where such cultivation is being practiced, there are sections that have reported a social hierarchy. The village headmen exercises power and authority over their fellow Jhumians (Gupta 1994). Thus, distribution of Jhum Land has been initiated on the basis of social position of the society. Socio-political condition has been important phenomena in distribution of Jhum lands among the people. During the empirical work, one of the respondents stated *"How much land we will get—that is determined by the Headmen of the village."* Within the egalitarian society also, there is constant power hegemony that has been a hidden force in the distribution of land. In Manipur the Hangvu, a customary village authority is also part of the shifting cultivation community. The Hangvu is comprised of chieftain and council of clan elders (Kherkhoff and Sharma 2006).

The traditional institution has been playing very important role in the Jhum Society. In sociological terms, the institutions comprise shared norms, values, traditions, beliefs, religion, rules, regulation, law, civil society organisations and government agencies (North 1990). The traditional institutions mostly are the village councils which have been responsible for managing land resources at community level. The composition of traditional institutions varies from tribe to tribe. There are three tiers of traditional self-governing institutions for the Karbis of Karbi Anglong. The King (also known as Lingdokpo or Recho in the Rongkhang area) is positioned at the top, accompanied by the Pinpomar—his parliament and council of ministers. Twelve Habis or administrators, one for every twelve Longris, are positioned in the middle and the Mei or village council, is positioned at the bottom (Goswami 2002). Giving Social Security to the villagers is one of the prime responsibilities for the traditional institutions of tribal society. Lands have been an important metaphor for the Jhum cultivators as it is considered as social capital for the Jhum cultivators. Bourdieu has coined the term social capital in the discourse of sociology. The term social capital was first used in 1916 by Lyda Hanifan. In contemporary times the theory has its origin in works of Coleman, Bourdieu and Putnam (Smart 2008). Sociologist Coleman conducted an investigation at the individual level, presuming that rational choice theory adheres to methodological individualism. Bourdieu's meaning of social capital is relatively less clear (Bourdieu 1986). It was Putnam, who convinced the masses that social capital is a concept that will help them to solve many of their problems (Harris 2001). For Putnam, social capital is predominantly characteristics of "societies" or "communities." 57-year-old respondent stated,

> Having land is one of the important factors for Jhum cultivation. Land is our emotion, we depend on Jhum cultivation because of our geographical location, in such situations, allotment of land helps us to continue the Jhum cultivation.

Allocating the land among the Jhum cultivators has been an important, characteristic criteria in the Jhum cultivation process. Meanwhile the entire land distribution process ensures the establishment of equality and equity in the tribal society. The allocation of plots is based on the family size and probable consumers.

Cultural and religious ethos have been closely associated with Jhum cultivation. Different annual festivals are part of Karbi society. Before beginning Jhum cultivation, the people offer prayers to their deities during various festivals. Rongker is one such annual festival. This festival is organised once in a year by each individual to appease the local deities. Such activities are associated with the wellbeing of the village community. To get rid of the evil spirits, people habitually offer their prayers. There are various types of Rongker in Karbi society and within that Havar Rongker is performed during the Jhum cultivation. The Havar Rongker is known as "Than Rongker." In Karbi "Than" denotes the domiciling place of the Godhead. "Havar Rongker" is performed by the people who are staying in the Hills with view to cultivate Jhum cultivation for livelihood. Havar means paddy field where people engage themselves in Jhum cultivation. The Havar Rongker is the worship for the blessing so that people can cultivate without sickness (Engtipi and Dhanaraju 2020). A 45-year-old respondent said "*We always offer our prayers before the cultivation.*"

The study area also includes Karbi tribals, who have converted to Christian religion though the people who still use their traditional Karbi language. Another farmer said,

> Though there are difference among the Karbi people, who have converted and those who follow the traditional religion in terms offering prayers before cultivation, still both the sections of Karbi people offer their prayers to their deities before the cultivation.

The indigenous knowledge occupies an important place in shifting cultivations. The customs, beliefs, values and cultural norms have all been integral components of shifting cultivation (Punitha et al. 2018). Indigenous knowledge encompasses a wide range of topics, such as medicine, health care and water management, biodiversity conservation and land management (Chin 2006).

5.2 Social Aspects of Shifting Cultivation

Jhum cultivation has long been practised by farming communities who live in the hills—their geographical location demands it, as they have no other cultivation options. However, the prevalence of shifting cultivation has been declining due to modern agricultural practices. Young farmers with smaller families are not encouraged to practise shifting cultivation. Half the Jhumia households in Tripura still practise Jhuming as a way of life, according to research by Gupta (1994). This could be due to economic pressures. In this study, researcher conducted an empirical study where out of total respondents i.e. out of (100) 73.77% of the participants stated that they received official support to stop the practice. Furthermore, if the practice of Jhuming was to be abandoned, 60.46% of the Jhumias stated that they would need to find an additional source of income. Additionally, it was noted that there were no core Jhumias—that is, Jhumias whose sole source of income is Jhum cultivation—as early as 1982 (Punitha et al. 2018).

The social construction of shifting cultivation is an important factor in this study. The farmers feel a deep bond with Jhum cultivation. Many farmers reported that shifting cultivation has been part of their collective lived experience since the previous generation, but that the previous generation felt more connected with it. Because the current generation perceives farming as being of low status, it is likely that previous generations performed shifting cultivation more successfully than the current generation (Pandey et al. 2020). The newer generation has been less succcessful in inculcating the various mores and customs associated with shifting cultivation. According to respondents in the empirical study, gradually, certain norms and values have declined. Due to this decline in the cultural ethos related to Jhum cultivation, the collective nature of the community is also affected. Also, because of the dynamism of their society, there have also been changes in shifting cultivation. Decisions regarding shifting cultivation are no longer made collectively, even though labour management is still done using traditional techniques (Pandey et al. 2020). Jhum cultivation aids in the preservation of the rich cultural traditions of indigenous peoples. For example, it includes cultural practices such as worshiping gods and goddesses and gifting them freshly gathered fruit in addition to slashing and burning (Lombi 2016). A 55-year-old respondent shared: '*We always offer prayers before the commencement of cultivation; we strongly believe that if we don't offer worship to our deities, they might get disappointed and, as a result, our cultivation might get disturbed'.*

5.3 State's Approach Towards Jhum Cultivation

In colonial times, the state did not encourage Jhum cultivation, and this attitude carries on in the post-Independence era at both the central and state government levels. Verrier Elwin, a noted scholar, was appointed as advisor for Tribal Affairs in the North-Eastern Frontier Agency (NEFA) by the Nehru government. Elwin was very optimistic about tribal development. According to him, modern science would help the development of the economy (Malik 1991). Elwin later asked how contemporary science could coexist with indigenous knowledge systems, are is firmly ingrained in artefacts, sociological conditions, and tribal way of life. In his work, '*The Baiga*', Elwin discussed Jhum cultivation as part of tribal culture.

Jhum cultivation has received constant criticism on the part of the state machinery, as it is believed to destroy the environment. Jhum cultivation impact on the environment is studied by agronomists, economists, policy-makers, and even geographers. Such criticism is the result of a poor understanding of Jhum. It is an example of a collective practice, where domestic modes of production prevail, and it maintains the egalitarian structure of society. Shifting cultivation is based on the Asiatic mode of production, where the concepts of surplus and trade do not arise. Rather, self-sufficiency is the foundation of the Jhum economy. State interference in Jhum cultivation is a holdover from colonialism. The colonial strategy with regards Jhum cultivation was influenced by this view. There were also certain technical challenges related to the practice of Jhum. For example, land revenue departments in NE India were unable to collect taxes on income generated through Jhum (Das 2006). Meanwhile, Jhum cultivation went through various evolutionary processes, again indicating societal evolution from hunting and gathering to a cultivation-based society (Das 2006).

The government has initiated many steps for Jhum cultivation. Most governmental programmes aim to eradicate it (Mertz et al. 1999). The Assam hill district, the Garo hill district of Meghalaya, and Mizoram implemented alternative land-use systems in the 1950s and 1960s with the introduction of crops like cashewnut, rubber, black pepper, and coffee. In this project, over the course of 3 years, 1700 families engaged in shifting cultivation, with each family receiving assistance in the form of 1 hectare of developed plantation land and 1 hectare of terraced land, either irrigated or otherwise (Maithani 2005). River basin schemes were also initiated by the North Eastern Council during the Fifth Five-Year Plan (Punitha et al. 2018). Using lessons learned from central and state government programmes, the seventh—five-year plan developed specific regulations to manage changing crops (Ray and Kamukenthang 1997). Especially in hilly areas and among tribal communities, Jhum is considered the primary mode of cultivation. Pessimistic attitudes towards Jhum cultivation have been increasing due to the rise of neo-liberal economic policies. As part of these neo-liberal policies, both state and central governments have tried to introduce 'Jhum Control Schemes' to control Jhum.

Agriculture is the main occupation of farming communities in the Karbi Anglong district of Assam. They practice both settled and Jhum cultivation. The state has employed a number of development activities to improve the lifestyles of the hill-based people and to encourage them to practise settled cultivation. For instance, the North-Eastern Regional Community Resources Management Project (NERCOMP) is a joint initiative of the Government of India and the International Fund for Agricultural Development (IFAD). NERCOMP aims to improve the livelihoods of vulnerable groups sustainably. Its main agenda is to protect the environment, and at the same time, assist traditional cultivation methods rooted in tribal cultures. This project is being carried out in the districts of Karbi Anglong and North Cachar Hills in Assam, Senapati and Ukhrul in Manipur, and West Khasi Hills and West Garo Hills in Meghalaya. After its implementation, there have been some noticeable changes in the Jhum cultivation practised by the Karbis of Karbi Anglong. Earlier, people mainly depended on Jhum, which gave lower yields and negligible cash incomes. After the project's intervention, villagers were trained in cultivation techniques and methods because of which they saw a marked improvement in yields and cash incomes. While villagers once used most of their produce for personal consumption, now cash crops, such as ginger, turmeric, pineapple, litchi, and broom grass, are sold in the market (Tiwari 2014).

During my empirical work, respondents shared their anxiety about the future of Jhum cultivation. A 59-year-old cultivator said, 'There is constant pressure on Jhum cultivation from the government, as most of the time such cultivation is considered as harmful towards the environment. Therefore, we feel insecure sometimes while thinking about the future of Jhum cultivation.' However, there is no anthropocentric understanding of Jhum cultivation; this can help us understand not only cultivation methods, but also the people's solidarity and community land participation. Verrier Elwin said, 'Jhum is a way of life.' One of the respondents said, 'We cannot think about our culture without thinking of Jhum.'

Since Independence, the state has actively intervened in the tribal way of life. There have been constant attempts to find alternatives to Jhum cultivation—the Indian Council of Agriculture Research (ICAR) and state agencies, such as the horticulture, agriculture, soil conservation, and irrigation departments, have been looking for alternatives. Despite various government policy efforts to stop Jhum cultivation, people do not forget their culture, which is deeply rooted in Jhum cultivation. In India, the Green Revolution began in 1960s. While North-East India was not affected by it, the state encouraged Jhum cultivators to adopt settled cultivation by popularizing the use of fertilizers, canal irrigation, and integrated watershed management (Das 2006). In the post-Independence era, certain policies were initiated for the development of hilly regions. As a result, Jhum cultivators had a chance to improve their livelihoods. During the Second Five-Year Plan (1956–61), the government initiated various Tribal Development Programmes, where special emphasis was given to Jhum cultivators (Sarma 1958). The Integrated Tribal Development Programme was introduced in the Sixth Five-Year Plan with the aim of introducing gradual development in the hills (Chak and Pattnaik 2016). With central support for the state plan, the Watershed Development Project for Shifting Cultivation Areas (WDPSCA) programme was introduced in various North-Eastern states during the Eighth Five-Year Plan. This programme seeks to develop Jhum regions overall (Das 2006). A new discourse was raised during that period, as scientific studies were carried out by agriculturalists, economists, demographers, and hydrologists. On the basis of their data, which they collected through empirical analysis, a new political perspective emerged; it attempted to negate previous views that held Jhum cultivation responsible for environmental destruction. Though there has been a gradual acceptance of shifting cultivation, one cannot deny the constant tension between modern agroforestry and Jhum cultivation.

6 Discussion

The study reflects about the lived experiences of the Jhum cultivators in their everyday life. Jhum cultivation has been considered as one of the primitive modes of production where community participation has been considered prime and imperative factors. A UN study in 1971–1975 observed there were slightly less than three million people primarily or partially dependent on shifting cultivation in India (UN Study 1985). There is a constant debate between sociologists or anthropologists and economists about the approach of the Jhum cultivation. The political economy of Jhum has been one of the important factors in determining the future of this cultivation. Jhum cultivation has been representative of tribal culture as in itself it portrays the egalitarian society where equality and equity persist. The social aspect of Jhum cultivation shows the cohesiveness, cooperation of the people. Meanwhile though there are different criticisms on the shifting cultivation, but the social aspect of such cultivation has been relevant and perseveres in post-modern society. Shifting cultivation demands collective approach, which strengthens the societal bonds of the people who have been involved in shifting cultivation. A functionalist approach is relevant to understand the cooperation among the Jhum cultivators. The study identifies that both the men and women have participated equally in the Jhum cultivated lands. The land of Jhum is allotted by the village head and therefore political economy also is an important phenomenon in the Jhum cultivation. Again, the land has been allotted on the basis of numeric strength of the family. Because Jhum cultivation requires an extensive form of human labour, the numeric consideration is vital.

During the empirical work, gender-based work division has been observed in Jhum cultivation. There is a strict demarcation of work among the males and females in the Jhum cultivation field. As lands are discussed by the headmen of the society, therefore the traditional institution has been very significant in the Jhum cultivation. The study identifies the role of headmen in the studied areas. Intervention of village headmen is recognised by the state's intrusion on Jhum cultivation. On the other hand, the beginning of Jhum cultivation signifies the start of a civilised approach to food consumption. Society has gone through various phases of evolution, for instance from a hunting and gathering society to a modern civilised society. The process of shifting cultivation has been analysed as a part of reflexive sociology. In the past, shifting cultivation extends back to the Neolithic era. It has gone through phases of being subsistence farming and tiny surplus farming. In northeastern India, shifting agriculture is still a vital source of income and food security for rural tribal communities, even in the face of global changes that have had an impact on tribal societies. Once more, changing farming processes have a significant impact on people's cultural identities. Different norms and customs are associated with the shifting cultivation. Tribal people perceive Jhum cultivation as a part of their everyday life. This perception can be understood through Pierre Bourdieu's concept of practice and habitus. For Bourdieu, practice is located in space and more significantly in time. According to Bourdieu practice is not consciously organised and orchestrated. Similarly the tribal people embedded the different socio-religious practices of Jhum cultivation. For them Jhum is not just a shifting cultivation, it's a way of life. This cultivation process can be considered as a form of habitus as it is a mode of life for the cultivators who have been involved in the shifting cultivation. People have embodied the various rules and regulations of Jhum cultivation. Social reality in Bourdieu's view is a process of the dialectic of internalisation of externality and externalisation of internality (Bourdieu 1977). Bourdieu's understanding of practice has been relevant in understanding Jhum cultivation. For social scientists, analysing social activities can also involve integrating agency-structure. Using the concepts of habitus, field and capital, Pierre Bourdieu (1977) suggested dialectic links between objective structure and subjective perception. Social practices are seen by Bourdieu as the result of the dialectic connection between agencies and structures. The idea of habitus and social fields forms the basis of Bourdieu's research and his attempts to unite subjectivism with objectivism. The dialectical interaction between different capitals, including social and economic, cultural and symbolic capitals, was another topic Bourdieu covered. The way Bourdieu's theory of practices unifies these three fundamental ideas is by showing how an actor's habitus—or disposition—depends on shifting cultural capitals in addition to reflecting their actual experiences. The customs and practices have been internalised by Jhum cultivators. With the passage of time, the process of cultivation has become a practice, which is now embedded in society. Again, Bourdieu's concept of cultural capital is relevant as far as traditional knowledge of Jhum cultivation is concerned. During the empirical work, most of the respondents stated how they acquire the traditional knowledge from their ancestors hence it has been considered as cultural capital.

Political economy of state has been one of the determining factors in terms of sustenance of Jhum cultivation. There is debate between social scientists and ecologists about the impact and durability of Jhum cultivation. Political ecology plays an important role in understanding the traditional pattern of shifting cultivation. Political ecology contains both political and apolitical phenomena of ecology where interaction of human with the environment has been studied. Watts (2000) stated political ecology

> … to comprehend the intricate relationships that exist between nature and civilisation by closely examining the many ways that people get and manage resources, as well as the effects that these relationships have on sustainable livelihoods and environmental health.

In shifting cultivation there is a constant human and environment interaction. In western Europe, since the late 1700s the human-nature interaction has become a dominant topic in both academic and political discourse. Meanwhile society-environment relationship is considered as dialectic. Allan Schnaiberg in the work "Social Syntheses of the Societal-Environmental Dialectic: The Role of Distributional Impact" addressed the issue of dialectic relationship between the society and environment. From state's point of view through shifting cultivation there is a constant pressure also on the environment, because such cultivation increases the deforestation process. There is a reciprocal relationship between environment and culture. Jhum cultivation has been part of the culture of tribal people who have been living in hilly areas. The study shows the conflict in the approach of state mechanism towards the shifting cultivation like Jhum cultivation. Thus there is constant interaction going on between societal pressures and human needs as seen through the study of lived experiences. Changing the climate issue has been one of the important issues that affects the agriculture pattern. Nusser et al. (2019) discussed the climate change issues in the different crop pattern in Ladakh area of India.

7 Concluding Remarks

Jhum cultivation is one of the oldest farming methods practiced in the northeastern regions of India and hilly areas of Bangladesh. Over the period of time such cultivation has gone through various changes due to globalisation and neo-liberalism. The Jhum cycle has declined from a healthy 20-year cycle to just a 5-year cycle and even less than that (Das 2006). Again another concern is the decreasing availability of land due to the increasing population. Meanwhile, political economy of the state has been another important factor as far as sustainability of Jhum cultivation is concerned. In most cases the state regime preferred agroforestry instead of Jhum cultivation. Again, the concurrent neoliberal market demands the change of the mode of production—this is itself a challenge for Jhum cultivation. Jhum cultivation reflects the culture of Tribal society, which can be understood with the help of Bourdieu's cultural capital. However the traditional knowledge of Jhum cultivation is also at stake due to lesser involvement of community members compared to the previous generation. The entire question of "sustainable development" of the Jhum-centered land use system needs to be seen from a variety of angles, including issues with land degradation, social organisation and function within a specific cultural landscape, including the rising population pressure on the land along with conservation and management of biodiversity-related natural resources. Das (2006) Jhum cultivation is the representation of little tradition and that might be under the dominion of great tradition. Robert Redfield, noted anthropologist used the concept of little tradition and great tradition. His book, *Peasant Society and Culture* (1956) based on Mexican communities' studies both the types of traditions. Both Robert Redfield and Milton Singer applied these concepts while studying Madras city to understand Indian culture. Little traditions comprise belief patterns, the institutions and anecdotes, folk tales and legends, myth and different cultural traditions that are absorbed from great traditions. Great traditions are called elite traditions, while little traditions are known as folk traditions. However there is constant interaction between little tradition and great tradition. In the Indian context, anthropologists like Yogendra Singh, Indra Dev and S.L. Srivastava have used the concepts little traditions and great traditions to understand Indian societies.

Shifting cultivation or Jhum cultivation in North-East India can't be seen merely within the framework of economics and ecological aspects. Such cultivation in North-East India is beyond the concept of only the Market, it is a part of Tribal culture. Similarly within the Karbi shifting cultivation has been rooted in the society, but nuanced understanding of Jhum cultivation of Karbi Society is required to know the various problems that Jhum cultivators face. In post structuralist society, practice of Jhum cultivation carries its own cultural and social identity, which reflects the traditional knowledge system.

References

Barua P, Kikhi K (2016) Culinary traditions, aesthetics and practices: constructing the cultural identity of Amri Karbi of Northeast India. J Northeast Indian Cult 3(1)
Bey MS (2004) The socio-economic life of the Karbis at a glance. Karbi Lammet Amei, Diphu
Bhagawati K, Bhagawati G, Das R, Bhagawati R, Ngachan SV (2015) The structure of Jhum (traditional shifting cultivation system): prospect or threat to climate. Int Lett Nat Sci 46:16–30
Bourdieu P (1977) Outline of a theory of practice. Cambridge University Press
Bourdieu P (1986) The forms of capital. In: Richardson J (ed) Handbook of theory and research for the sociology of education. Greenwood Press, New York, pp 241–258
Chak H, Pattnaik T (2016) Five-year plans and tribal development in India. Int J Res 3(3):30–45
Chin A (2006) Urban transformation of river landscapes in a global context. Geomorphology 79(3-4):460–487
Conklin HC (1957) Hanunoo agriculture: a report on an integral system of shifting cultivation in the Philippines, vol 2
Das D (2006) Demystifying the myth of shifting cultivation: agronomy in the North-East. Econ Polit Wkly 41(47):4912–4917
Dash B (2006) Shifting cultivation among the tribes of Orissa. Orissa Rev 5(6):76–84
Deb S, Lynrah M, Tiwari BK (2013) Technological innovations in shifting agricultural practices by three tribal farming communities of Meghalaya, North East India. Trop Ecol 54(2):133–148
Dilthey W (1977) Descriptive psychology and historical understanding. Translated by Makkreel Rudolf. F.Netherlands. Martinus nijhoff the hague.Martinus Nijhoff
Engtipi R, Dhanaraju V (2020) Exploring folklore of Karbi tribe in Northeast India. Our Heritage 68(1). ISSN 0474-9030, pp 300–314
Gogoi M (2011) A glance: at Karbi Anglong. Directorate of Assam Institute of Research for Tribals and Scheduled Castes. Jawaharnagar, Guwahati

Goswami A (2002) Traditional self-governing institutions among the hill tribes of North-East India. Akansha Publication

Gupta MD (1994) Continuity and change in tribal societies: a study of the economic aspects of tribal Jhumia society in Tripura. In: Sangma MB (ed) Essays on North-East India. Indus Publishing Company, New Delhi, pp 147–164

Haokip MM (2023) Land use and livelihood: the case of shifting cultivation among the Kukis. J North-East India Stud 13(2):8–30

Harri J (2001) Depoliticising development: the World Bank and social capital. Anthem Press, London

Kherkhoff E, Sharma E (2006) Debating shifting cultivation in the Eastern Himalayas farmers' innovations as lessons for policy. International Centre for Integrated Mountain Development, lib.icimod.org/record/7409

Lombi P (2016) People's perception about shifting cultivation – with special reference to the Galo tribe of West Siang District, Arunachal Pradesh (India). International Justice of Scientific Research Publication 6(7):454–459

Maithani BP (2005) Shifting cultivation in North-East India – policy issues and options. Mittal Publications, New Delhi

Malik B (1991) Hill and forest economies in North-East India – mid-19th century to early 20th century. Unpublished MPhil thesis, Centre for Historical Studies, Jawaharlal Nehru University, New Delhi

Manen MV (1977) Linking ways of knowing with ways of being practical. Curric Inq 6(3):205–228. https://doi.org/10.1080/03626784.1977.11075533

Mertz O, Christensen EA, Hojoskov P, Thomsen BT (1999) Subsistence or cash: strategies for change in shifting cultivation system. https://tidsskrift.dk/index.php/geografisktidsskrift/article/view/2566/4569

Miles MB, Huberman AM (1994) Qualitative data analysis: an expanded sourcebook. Sage

National Commission on Agriculture (1976) Report of the National Commission on agriculture, planning. Statistics and Administration, India

Ninan KN (1992) Economics of shifting cultivation in India. Econ Polit Wkly 27(13):A2–A6

NRSC (National Remote Sensing Centre) (2011) Wasteland atlas of India, Department of Land Resources, dolr.nic..in/dolr/ wasteland_atlas.asp

Nusser M, Dame J, Kraus B, Baghel R, Schimdt S (2019) Socio-hydrology of 'artificial glaciers' in Ladakh India: accessing adaptive strategies in a changing cryosphere. Reg Environ Chang 19:1327–1337

Pandey DK, Kumar DH, Dubey SK, Kumar B, Shivani D, Adhiguru P (2020) Indigenous people's attachment to shifting cultivation in the eastern Himalayas, India: a cross-sectional evidence. Forest Policy Econ 111:1–9. https://doi.org/10.1016/j.forpol.2019.102046

Punitha P, Ansari MA, Pandey DP, Daya R, Datt S, Sharma PK, Aheibam M, Jyothi SSP, Prakash N (2018) Shifting cultivation in North-East India: social dimension, cross cultural reflection and strategies for improvement. Indian J Agric Sci 88(6):811–819. https://doi.org/10.56093/ijas.v88i6.80497

Ray AK, Kamukenthang H (1997) District councils in Manipur Vis-a-Vis sixth schedule. In: Gassah LS (ed) Autonomous District councils. Omsons Publications, New Delhi, pp 243–255

Rose G (1997) Situating knowledges: positionality, reflexivities and other tactics. Prog Hum Geogr 21(3):305–320. https://doi.org/10.1191/030913297673302122

Sarma NA (1958) Economic development in India: the first and second five-year plans. Staff Pap Int Monetary Fund 6(2):180–238

Smart A (2008) Social capital. Anthropological. Can Anthropol Soc:409–416

Talukdar KC (2009) Current situation of Jhum cultivation and analysis of Jhum cycle. Agro-Economic Research Centre for North-East India, Assam

Tiwari BK (2014) Policies and alternative options to shifting cultivation in North-East India – a review 2004–2014. North-East Hill University, Shillong

UN Study (1985) The state of India's environment 1984–85. The second citizens' report, Centre for Science and Environment. New Delhi, pp 168
Watts MJ (2000) Political ecology in a companion to economic geography. In: Sheppard E, Barnes T (eds) A companion to economic geography. Blackwell, Oxford, pp 257–274
Yadav PK (2013) Slash-and-burn agriculture in North-East India. Expert Opin Environ Biol 2:1. https://doi.org/10.4172/2325-9655.1000102

Relevance of Traditional Agriculture Knowledge for Sustainable Agriculture in the Indian Himalayan Region

Gopal Sonkar

1 Introduction

The term 'traditional agriculture knowledge' creates a distinctive identity of what existed originally as native to a region. Traditional agricultural knowledge systems that evolve from Indigenous communities in particular a region is rather one-dimensional. Traditional agriculture knowledge is various types and situated in various systematic frameworks. Traditional agricultural knowledge includes farming practices, cropping patterns, pest control, land management, irrigation patterns and other agriculture-related concerns. Traditional agricultural knowledge (TAK) has been inherited over many generations and is deeply rooted in natives of the Himalayan region. Himalayan indigenous people profoundly understand their region's nature, climate, and soil, which they have integrated into their agricultural methods (Ajeet 2003). There is increasing interest in traditional agricultural knowledge to preserve agricultural biodiversity against climate change (Clarke 1990).

Traditional agricultural methods use natural substances (see Table 2) which do not damage the environment. TAK is also less cost-effective. Traditional knowledge increasingly involves "sustainable management of ecosystems and environmental protection against chemical substances," which is increasingly related as an effective tool to preserve the environment (Krystyna et al. 2011). The Institute of Advanced Studies in the United Nations University identified more than 400 examples of "traditional knowledge" and its role in biodiversity management, including climate change mitigation and adaptation (Parry et al. 2007). The Intergovernmental Panel on Climate Change (IPCC) emphasises traditional knowledge's role in adapting crop varieties (Parry et al. 2007). TAK systems can provide unique insights into how the local environment responds to climate change, such as identifying which

G. Sonkar (✉)
Indian Institute Public Administration, Delhi, India

A. Borthakur, P. Singh (eds.), *The Resilience of Traditional Knowledge Systems for a Sustainable Future*, Advances in Asian Human-Environmental Research,
https://doi.org/10.1007/978-3-031-56858-9_5

crops are most resilient to drought or which areas will be most affected. TAK can help local communities develop adaptation strategies to protect their livelihoods, such as planting more resilient crops or relocating to safer areas, choosing suitable crop varieties, using appropriate irrigation and pest management strategies, and using natural means to protect crops from extreme weather conditions.

In 2010 parties to the United Nations Framework Convention on Climate Change (UNFCC) decided on "enhanced adaptation action," which acknowledged the need to incorporate traditional knowledge alongside the best available science and technology (Krystyna et al. 2011). Despite this increased awareness, decision-makers continue valuing traditional knowledge's role in sustainable agriculture and climate change impacts (ICAR 2002). The negotiating parties in international climate discussions represent nations rather than communities, often viewing traditional intensification of agricultural practices as the route to food security and adaptation. The lack of recognition of traditional knowledge has resulted in traditional communities being excluded from decision-making. This leads to a lack of representation and their voices not being heard in these conversations. As a result, traditional communities cannot effectively advocate for their rights and interests, making it difficult to access resources and support that can help them adapt to climate change impacts. However, Indian Himalayan agriculture practices have helped many local farmers in the matters of inappropriate weather and environmental changes. TAK is important due to modern agriculture and exceptionally high-yielding verities (HYV) seeds, have increased their dependency on external resources (Krystyna et al. 2011).

Twelve states makes the Indian Himalayan Region (IHR), including Jammu and Kashmir, Himachal Pradesh, Uttarakhand (in north central Himalayan) and Arunachal Pradesh, Sikkim, Nagaland, Meghalaya, Manipur, Mizoram, Tripura, and hill districts in West Bengal and Assam (northeast Himalayan). Natural, rich agro-biodiversity resources and various flora and fauna are found here. Kashmir tribal villages grow mixed rainfed crops and livestock, and apple growers rely on the Kashmir valley to sustain their livelihoods. Himachal Pradesh is known for cultivating food grains, fruits, and vegetables. Agricultural production in Uttarakhand is primarily based on rainfed crops and livestock. The Nagaland, in the northeastern Himalayas, slash and burn is the predominant farming technique.

2 Discourses and Studies

When we talk about "traditional agricultural knowledge" in a restricted sense, it refers to eco-friendly modes of relationship between agriculture and eco-biodiversity management. Policymakers, environmental scholars, and social scientists have only recently realised the value of traditional knowledge in areas such as biodiversity, agriculture and medicine. The World Intellectual Property Organization (WIPO) and Convention on Biological Diversity (CBD), both part of the United Nations Environment Programme, have been working on international frameworks to protect such information in recent years (Christoph 2013). The adaptive capability of traditional knowledge can be increased through local communities. Several traditional knowledge systems have maintained help in the sustainable use of natural resources and ecosystem management that are becoming increasingly recognised for their potential as powerful tools in combating climate change (Krystyna et al. 2011). The 'Brundtland Report' defines sustainability as "fulfilling demands of the present without compromising the potential of future generations". However, defining sustainability is difficult due to its broad scope and different stakeholders' goals and objectives. Moreover, agricultural sustainability is challenging to anticipate due to the many drivers of change and ever-evolving agricultural environments. Mountain agricultural systems (MASs) have diverse natures and grow marginally. The sustainable growth of MASs should consider the current balance of environment, economic, and sociocultural factors, as well as stakeholders' opportunities for broader transformative change as predicted by the United Nations Sustainable Development Goals (Bandana et al. 2019).

The Himalayan region contributes significantly to India's overall biodiversity because, its geo-biophysical characteristics. The Himalayan region has a variety of land altitudes, terrains, and climates, which create a wide range of habitats for different agricultural zones. As a result, the region is home to a wide variety of crop and plant species, many of them are no found anywhere else in India. Himalayan Traditional Agricultural knowledge has been passed down through generations but is often overlooked in conservation planning and biodiversity management. The traditional knowledge is often valuable for conserving the region's biodiversity. It can help decide which areas should be set aside for conservation, which crop and plant species should be protected, and how other human activities should be managed. Also, it can make decisions about managing natural resources sustainably so natives can flourish.

Biodiversity-rich Himalayan Mountain landscapes were dominated by traditional agriculture earlier. Land use changes in many parts of the Himalayan region have been propelled in recent decades by "small agricultural landholders" and "forest fuel wood and fodder harvesting." As a result, there are (1) lowland evergreen forests, (2) species-rich moist maple-oak-rhododendron forests, (3) high-altitude rhododendron and (4) alpine conifer associations in the area, all of which are being used to combat deforestation, poverty, and agriculture (Kamaljit and Reinmar 2015). According to "Agriculture Diversification and Development of Mountain Regions with Special Reference to Himachal Pradesh", fruit and off-season vegetable cultivation have helped Himachal Pradesh achieve more excellent socioeconomic progress than other hill states and regions (Chand 1997). International Development Research Centre (2009) argued that "agroecosystem management practices are in the process of transformation due to population pressure, decreasing availability of agricultural lands, climate changes, market demands and globalisation." These can negatively impact the relationship between people and their ecosystems, affecting human health, disease and nutritional status (International Development Research Centre 2009). Lal and Sharma (2005) examined the impact of agricultural modernisation and commercialisation on the socioeconomic status of Himalayan tribal society in the Kinnaur, Lahul and Spiti districts. People have progressed so much that they are uncomfortable identifying as tribal (Lal and Sharma 2005).

The literature does not provide information on traditional agricultural practices in the Himalayan region nor identify the prevalent types of traditional knowledge practices. Moreover, what is the nature of these practices? Also, the literature is unclear whether traditional agriculture still exists in the Indian Himalayas. There are gaps in the literature that relate to all these objectives, and the current effort aims to fill those gaps.

3 Himalayan Traditional Agriculture Knowledge

Several agroecological consequences (including climate change) have emphasised traditional agriculture knowledge in recent years since the solution can be found in traditional knowledge. Therefore, imparting skills and identifying and applying tradition knowledge as soon as possible in agriculture. The traditional agricultural activities of the earlier included selecting the right agricultural inputs, predicting the weather, preparing the land, monitoring soil health and water conditions, identifying pests and diseases and harvest activities. The traditional agricultural knowledge acquired during periods when demographic pressures were less severe, climate change did not have a significant impact, farmland was more extensive, modern technology did not occupy much of the agricultural field, and lastly, commercial crops did not take on the shape it is today.

Mission Mode project from ICAR[1] launched on gathering, documenting, and authenticating traditional agriculture knowledge as part of NATP[2] in 2000 (ICAR 2002, 2003a, 2003b, 2004). The NATP has also developed a list of Indian Himalayan traditional agriculture knowledge according to classification and State. Himalayan has more than 117 types of traditional agriculture practice knowledge such as "'Rain Water Management', 'Methods for preventing water and soil erosion', 'Tillage and Intercultural Management', 'Crop and Cropping Systems', 'Pest and Disease Management', 'Soil Fertility Management', 'Farm Implements', and 'Horticultural Crops'". Name the traditional knowledge and practices places listed in Table 1.

However, this list does not refer many of the traditional agricultural practices used in the region at present, and recent studies do not support these practices (Dhyani 1994; Integrated Mountain Initiative 2019; Chand 1997; Kamaljit and Reinmar 2015; Lal and Sharma 2005). Their significance is not associated with current agricultural practices because land use patterns, cropping patterns, farmland sizes, and climate have significantly changed (Integrated Mountain Initiative 2019). The NATP Traditional Agriculture Practices case studies show that these traditional knowledge practices were prevalent in the past. Some of the findings collected from 53 NATP case studies are summarised in Table 2 in the following.

[1] Indian Council of Agricultural Research.

[2] National Agricultural Technology Project.

Table 1 Names of list Indian Himalayan Traditional Agriculture Knowledge with their Practices, Places or Region

Classification of traditional agriculture knowledge basis their components	**Western & Central Himalayan Traditional Farming Technique Names** (Jammu and Kashmir, Himachal Pradesh, Uttarakhand and Ladakh)	**Eastern Himalayan Farming Technique Names** (Sikkim, Arunachal, Pradesh, Meghalaya, Nagaland, Manipur, Mizoram, Tripura and hill regions of 2 states viz. Assam (Karbi Anglong and Dima Hasao) and West Bengal (Darjeeling and Kalimpong))
Rain Water Management	1. *Khatri* (Kangra et al.) 2. Use tree needles and grasses for mulching to conserve soil moisture (Kinnaur et al.) 3. Curved land ploughing for soil conservation and water retention (Cold deserts of Western Himalayas) 4. Own plots in different pockets for risk distribution (Eastern, Central, and Southern states) 5. *Kuhl* water for running water mills (cold desert of Himachal Pradesh) 6. Irrigation method for the use of *Khul* water (Upper Kinnaur, North West Himalayas) 7. Construction of *Kuhls* (water channels) for irrigation (Himachal Pradesh) 8. Dividing the fields into sub-plots for irrigation water management (Spiti and Kinnaur of Himachal Pradesh) 9. Use Pang (Spang) grass to control seepage and side losses in water tanks and irrigation *Kuhls* (Ladakh, Jammu and Kashmir) 10. Indicators of irrigation requirement (Himachal Pradesh) 11. Moisture measurement with Belcha (spade) (Ladakh, Jammu and Kashmir)	1. Zabo system of rice cultivation (Nagaland) 2. Ownership of plots in different pockets for risk distribution (Eastern states, Central states and Southern states) 3. Kuhl water for running water mills (North Eastern Region)
Methods to Check Soil and Water Erosion	1. Bench terracing for rainwater management (Hilly areas of Western Himalayas) 2. Grassed waterways to avoid water logging in terrace farming (Upper North West Himalayas) 3. Interplot stone bunding for soil conservation (cold deserts of Western Himalayas) 4. Vegetative barriers for the control of sheet erosion (Forest areas of the West Himalaya region) 5. Ash for softening hard soils (Nubra Valley, Himachal Pradesh) 6. Methods for control of gully erosion (Lower areas of Western Himalayas) 7. Loose boulders for soil and water conservation (Lower areas of Himachal Pradesh) 8. Conserving topsoil by autumn irrigation (Cold deserts of Western Himalayas)	
Tillage and Intercultural Management	1. Ploughing in standing maize for better stand establishment (Mandi, Himachal Pradesh) 2. Minimum tillage for rabi crops after the rainy season (Kharif) following Lahaul valley, Himachal Pradesh) 3. Soil management through cultural practices (West Himalayan region) 4. Management of direct seeded rice (Himachal Pradesh) 5. Halod or bushing in standing paddy (Hilly areas of Himachal Pradesh)	

(continued)

Table 1 (continued)

Classification of traditional agriculture knowledge basis their components	**Western & Central Himalayan Traditional Farming Technique Names** (Jammu and Kashmir, Himachal Pradesh, Uttarakhand and Ladakh)	**Eastern Himalayan Farming Technique Names** (Sikkim, Arunachal, Pradesh, Meghalaya, Nagaland, Manipur, Mizoram, Tripura and hill regions of 2 states viz. Assam (Karbi Anglong and Dima Hasao) and West Bengal (Darjeeling and Kalimpong))
Crops and Cropping Systems	1. Determining quantity and uniformity broadcasting of seed (Himachal Pradesh) 2. Crushing of coriander seeds for better germination (Kinnaur, Himachal Pradesh) 3. Reducing inbreeding by selecting seeds from different villages (West Himalayan cold deserts, Himachal Pradesh) 4. Cultural practices for minimising the risk (Palampur region of Himachal Pradesh) 5. Optimise the maise population by appropriate spacing (Hamirpur, Himachal Pradesh) 6. Deep sowing of maise and planking (Himachal Pradesh) 7. Osoconditioning of pea seeds by seeding before the snow (Kinnaur, Himachal Pradesh) 8. Traditional rainfed farming for soil conservation by cultivating close-growing crops (Valley regions, western Himalayas) 9. Establishment of localised greenhouse conditions in grape cultivation (Nubra Valley, Himachal Pradesh) 10. Crop rotation with barley and buckwheat (Lahaul Valley, Himachal Pradesh) 11. Crop rotations for maximising resource utilisation (Lahaul Valley, Himachal Pradesh) 12. Snow management for raising double crops (Lahaul Valley, Himachal Pradesh) 13. Rotational farming with pulse crops for soil fertility and disease management (West Himalayan cold deserts, Himachal Pradesh) 14. Mixed cropping for crop diversification and risk distribution (West Himalayan cold deserts, Himachal Pradesh) 15. Private and community fodder wood blocks (Mid and high hills, foothills and sub-mountainous regions of the Himalayas) 16. Hay/dry grass storage techniques in fields (Lahaul Valley, Himachal Pradesh) 17. Increasing cropping intensity through relay planting cucurbits with wheat (Himachal Pradesh) 18. Increasing cropping intensity through maise + tomato intercropping (Hamirpur, Himachal Pradesh) 19. Control of wild rice by growing a coloured variety of paddy (Himachal Pradesh) 20. Soil management by crop residue harvesting (Zanskar and Ladakh) 21. Nitrogen fixing trees in tea gardens for shade and soil enrichment (Kangra Valley of Himachal Pradesh) 22. Multipurpose trees on the cropland for enhancing land use (Himachal Pradesh) 23. Agri-silviculture with bamboo (Palampur of Himachal Pradesh)	1. Wet rice cultivation on terraces (Subansiri, Arunachal Pradesh) 2. Jhuming or shifting cultivation (Hills of North East India) 3. Nitrogen fixing trees in tea gardens for shade and soil enrichment (Assam and West Bengal) 4. Agri-silviculture with bamboo (Sikkim and other Northeastern hill states)

Pest and Disease Management	1. Bajagun weed reduces the rat menace (Plains of Himachal Pradesh) 2. Combating the menace of birds and monkeys (Himachal Pradesh) 3. Sling and Karban for scaring birds and stray animals (wet Himalayan regions, Himachal Pradesh) 4. Use Bana plant twigs in paddy fields to control pests and diseases (Himachal Pradesh) 5. Protection of rice from epidemic disease and insect-pests by Bara twigs (Himachal Pradesh) 6. Kerosene oil is used to kill borers (Himachal Pradesh) 7. Ploughing, hoeing and basin preparation for insect control (Himachal Pradesh) 8. Hand-picking of pests (Himachal Pradesh) 9. Control of insects in mustard (Tehari Garhwal, Uttaranchal) 10. Use ash as an insecticide (Kumaun Hills, Uttaranchal) 11.'Shangma stone' for weed control (Changthan area of Ladakh in Himachal Pradesh) 12. Control of stemborer in fruit trees by kerosene oil swap (Kumaun Hills)	
Soil Fertility Management	1. Enhancement of soil fertility through animal bones (Himachal Pradesh) 2. Use of river bed soils (Western Himalayas) 3. Use Sachik soil for higher crop yield (Tagloom area, Himachal Pradesh) 4. Use of goat manure for improving millet production (Ladakh, Jammu and Kashmir) 5. Collection and usage of farm yard manure (Kinnaur, Himachal Pradesh) 6. Organic manuring, collection and management (Himachal Pradesh) 7. Technique of uniform distribution of organic manure (Ladakh, Jammu and Kashmir) 8. Nutrient recycling with ash, household waste and human excreta (Ladakh, Jammu and Kashmir) 9. Soil mixing with night soils improves soil fertility (Ladakh, Jammu and Kashmir) 10. Green manuring in paddy with biomass of wild bush (Bharmour, Himachal Pradesh) 11. Nutrient management practices of Himachal Pradesh (Himachal Pradesh) 12. Technology for nutrient management in Jammu and Kashmir (Jammu and Kashmir) 13. Nutrient management technology of Central Himalaya (Central Himalaya)	1. Nutrient management practices of NorthEastern Hill (Kohima district of Nagaland, Arunachal Pradesh), (Northeastern Hill Region)

(continued)

Table 1 (continued)

Classification of traditional agriculture knowledge basis their components	**Western & Central Himalayan Traditional Farming Technique Names** (Jammu and Kashmir, Himachal Pradesh, Uttarakhand and Ladakh)	**Eastern Himalayan Farming Technique Names** (Sikkim, Arunachal, Pradesh, Meghalaya, Nagaland, Manipur, Mizoram, Tripura and hill regions of 2 states viz. Assam (Karbi Anglong and Dima Hasao) and West Bengal (Darjeeling and Kalimpong))
Farm Implements	1. Kudali (Spade) for digging and weeding (Himachal Pradesh) 2. Plough for different farming situations (Himachal Pradesh) 3. Plough with separate shoe and body (Mandi, Chamba, Una and Kangra, Himachal Pradesh) 4. Use broader plough in upper valleys (Ladakh, Jammu and Kashmir) 5. Ploughing with desi plough (Spiti Valley, Himachal Pradesh) 6. Plough with one piece (Kullu, Solan, Shimla, Sirmour, Lahaul, Spiti and Kinnaur of Himachal Pradesh) 7. Use of Maddim for field levelling (Western Himalayas) 8. Roller for breaking clods (Karnal (Haryana), Sirmur (Himachal Pradesh) 9. Mallot for breaking clods (Himachal Pradesh) 10. Leveller (suhaga) for land levelling (Himachal Pradesh) 12. Cultivation of Seabuckthorn (Hippophae) (High hills, river beds, valleys, dry and cold mountain areas of Himachal Pradesh)	1. Nangal plough for land preparation (Nowgong, Cachar and Sibsagar, Assam) 2. Clod crusher -cum-plank (Mai) (Darang, Assam)

Horticultural Crops	1. Inducing walnut fruiting by oozing water from the tree trunk (Himachal Pradesh) 2. Root spreading for surface feeding in vegetable plantation (Ladakh, Jammu and Kashmir) 3. Management of frost damage in mangoes by smoking (Lower areas of Himachal Pradesh) 4. Large cardamom-based agroforestry system (Joginder Nagar of Mandi, Himachal Pradesh) 5. Enhancing germination of coconut by keeping it in water (Una, Himachal Pradesh) 6. Grafting in apricot (Nubra Valley, Himachal Pradesh) 7. Indigenous practices for vegetable crops (Uttaranchal) 8. Control of root disease in fruit trees by applying ash or lime (Kumaun Hills, Uttar Pradesh) 9. Pruning of fruit trees for better growth (Himachal Pradesh) 10. Cow dung and clay fruit trees mixture for controlling pests and diseases in fruit trees (Himachal Pradesh) 11. Control root rot disease in fruit trees and vegetables by ash (Kumaun Hills) 12. Burring the dead animals around the roots of fruit trees to increase growth (Kumaun Hills) 13. Use of wood ash to vegetable crops to supply nutrients and protect the crops (Himachal Pradesh) 15. Enhancing vegetable production using kitchen ash and poultry manure (Nubra Valley, Himachal Pradesh) 16. Quality improvement in potatoes through the application of kitchen ash and goat manure (Nubra Valley, Himachal Pradesh) 17. Spraying ash dust in cucurbits and other vegetable crops for potassium and pest control (Cold deserts of the western Himalayas) 18. Traditional apple plucking (Himachal Pradesh) 20. Apricot drying on the trees (Nalling, Nako, Pooh of Himachal Pradesh) 21. Storage of apples in underground pits (Lahaul and Spiti, Himachal Pradesh) 22. Storage of fruits and vegetables by sun-drying (Shimla, Sirmour, Kullu and Chamba of Himachal Pradesh)	1. Large cardamom-based agroforestry system (Sikkim)

Source: ICAR 2002, "Inventory of Indigenous Technical Knowledge in Agriculture" Mission Mode Project on Collection, Documentation and Validation of Indigenous Technical Knowledge Document 1 Project National Agricultural Technology Project, Indian Council of Agricultural Research New Delhi

Table 2 Summarised 53 Case Studies of the Indian Himalayan Traditional Agriculture Knowledge Practices with their Methods Explanation

Classification of Traditional Agriculture Knowledge basis their Mechanisms	Traditional Agriculture Practices in their Corresponding Region Based on NATP Collected Case Studies	References and Information Collector
Rain water management (mechanisms)		
1. "Use of surface-bed irrigation system to supplement rainfall."	The entire Himachal Pradesh practices surface-bed irrigation. In which excessive rainfall water diverts into beds of manageable dimensions via small canals	"Dr T. R. Nandal Vegetable Scientist RARC Dhaulakuan Sirmour (Himachal Pradesh)"
Soil and water conservation (mechanisms)		
1. "Checking of soil erosion by growing trees along the fields"	Baldhar village in Himachal Pradesh has a long tradition of planting trees along its fields. This practice reduces wind velocity and sunlight intensity and helps reduce moisture evaporation. It has been used for a long time	"Shri Rajeev Kumar Room no 32 D NBH II Hostel Himachal Pradesh University Summer Hill Shimla (Himachal Pradesh)"
Tillage and intercultural management (mechanisms)		
1. "Crushing of clods of soil with a wooden hammer (bharrotta)."	The structure of fine-grained soil is disturbed when cultivated in wet conditions, leading to hard clods. Wooden hammers are used to break up the clods, which can be done prior to sowing. Then, under a levelling board, the tougher clods are crushed, sohaga, which has been practised in Himachal Pradesh since immemorial	"Shri Bivinder Chand Katoch, Research Fellow, Department of Biosciences, Himachal Pradesh University, Shimla (Himachal Pradesh)"
Crops and cropping systems (mechanisms)		
1. "Use of bhang leaves for germination of paddy seed."	Bhang (Cannabis sativa) is a local growing herb that can increase paddy seed germination in temperate areas of Jammu and Kashmir	"Survey Report from Jammu Province J&K State) Documentation and Validation of ITK, 2002. Sher-E Kashmir University of Agricultural Sciences & Technology, Jammu."
2. "Mixing of paddy seeds with cow dung before sowing."	Mixing paddy seeds with cow dung before sowing protects the seed from birds and acts as manure in the Jammu region of Jammu and Kashmir	Same (previous)

(continued)

Table 2 (continued)

Classification of Traditional Agriculture Knowledge basis their Mechanisms	Traditional Agriculture Practices in their Corresponding Region Based on NATP Collected Case Studies	References and Information Collector
3. "Use of brankar leaves in raising paddy nursery."	Adatoda leaves (brankar) improve soil fertility, serve as a pesticide, and facilitate the uprooting of seedlings. This tradition is prevalent in the Mogla region of the Rajouri district in Jammu and Kashmir	Same as previous
4. "Use of deodar branches to control paddy weeds."	Weeds in paddy fields can be controlled using "deodar branches and leaves" as a weedicide. Jammu and Kashmir's hilly regions practice this practice	Same as previous
5. "Growing of maise as grain and fodder crop."	This is a prevalent method in Jammu and Kashmir, where maise is cultivated, "seeds" are sown at a higher rate and thinning is done to pick up undesired plants	Same as previous
6. "Tin is beating for protecting maise crop from wild animals."	Monkeys and Neel cows attack maise cobs in the Kandi areas of the Jammu region, so farmers beat tin peepa at night to protect the crop	Same as previous
7. "Mulching of mustard field with cow dung."	Mulching mustard fields with cow dung helps moderate temperature, protect plants, reduce evaporation, conserve moisture, and improve physical conditions. This practice is most common in Jammu and Kashmir's hilly regions	Same as previous
8. "Mixing of mustard seed with friable soil for the uniform broadcast of seed in the field."	Mixing mustard seed with soil helps maintain uniformity in distribution, resulting in a uniform plant population and no cost. Several hilly areas of Jammu and Kashmir practice this practice	Same as previous

(continued)

Table 2 (continued)

Classification of Traditional Agriculture Knowledge basis their Mechanisms	Traditional Agriculture Practices in their Corresponding Region Based on NATP Collected Case Studies	References and Information Collector
9. "Mixed cropping of maise with soybean."	Mixed cropping of maise and soybean is a valuable cropping system in Himachal Pradesh, Jammu, and Kashmir. It reduces insect pest attacks and allows multiple crops to be grown in a single field. Wheat, mustard, and peas are also popular in Jammu and Kashmir due to their adaptability and capacity for stress tolerance	"Shri Rajeev Kumar Room no 32 D, NBH II Hostel, Himachal Pradesh University, Summer Hill, Shimla (Himachal Pradesh)"
10. "Mixed cropping in Kharif."	The sowing of crop mixes in Himachal Pradesh to diversify food, maximise soil nutrients, increase productivity, and prevent insect attacks	"People of Himachal Pradesh at large Communicated by: Prof L. R. Verma Department of Bio-Sciences Himachal Pradesh University, Shimla (Himachal Pradesh)"
11. "Mixed cropping in rabi."	Rabi is a practice in Himachal Pradesh in which leguminous and non-leguminous crops are planted to increase food diversity and maintain soil fertility	Same as previous
12. "Intercropping of wheat and maise along with mustard and sesame."	Intercropping is a primary agricultural technique farmers in Baldhar, Kangra District, Himachal Pradesh, use to increase crop yield and combat insect-pest attacks. For example, mustard and sesame are intercropped to protect wheat from insect pest attacks and maise from insect pest attacks	"Shri Rajeev Kumar Room no 32 D, NBH II Hostel Himachal Pradesh University, Summer Hill, Shimla (Himachal Pradesh)."
13. "Intercropping of corn and chillies for increased production."	Intercropping involves planting corn and chilli at 100 cm × 15 cm row spacing and applying organic manure with phosphorus and nitrogen (1:2/3) for enhanced soil fertility, parallel cultivation of two crops and additional income	"Shri Deep Ram Verma, Ghaghri, Theog Shimla (Himachal Pradesh)"
14. Use of different crops in rotation:	Himachal Pradesh's cultivators grow vegetables, pulses, and cereals to meet their families' needs, utilising soil nutrients and improving soil fertility and income	"People of Himachal Pradesh at large Communicated by: Prof L. R. Verma Department of Bio-Sciences Himachal Pradesh University, Shimla (Himachal Pradesh)"

(continued)

Table 2 (continued)

Classification of Traditional Agriculture Knowledge basis their Mechanisms	Traditional Agriculture Practices in their Corresponding Region Based on NATP Collected Case Studies	References and Information Collector
15. Growing indigenous tree oei (Albizzia chinensis) and tea crops."	"Oei (Albizzia chinensis)" is grown with tea plants in equal spacing to provide shade and optimum growth at 23–26 °C, utilised throughout the entire Kangra valley	"Shri Bovinder Chand Katoch Research Fellow, Department of Biosciences, Himachal Pradesh University Shimla (Himachal Pradesh)"
16. "Use of edible oil for seed treatment."	Applying edible oil to seeds is popular in Jammu and Kashmir. This is because it acts as a defence mechanism that keeps out lower organisms	"Survey Report from Jammu Province (J&K State) on Collection, Documentation and Validation of ITK, 2002. Sher-E Kashmir University of Agricultural Science Technology, Jammu"
17. "Seed treatment with ash and cow dung."	Ash and cow dung prevent pathogens and pests by creating a physical barrier and reducing moisture in the seed. They are widely used in Jammu and Kashmir	"Dr M. D. Gupta, Director of Extension Education, Sher-E Kashmir University of Agricultural Sciences and Technology, Jammu (Jammu and Kashmir)"
18. "Control of congress grass weed by application of kerosene oil."	Kerosene oil is sprayed on it by farmers in the Udhampur district of Jammu Kashmir. This strategy should be supplemented or incorporated with other weed management techniques	"Survey Report from Jammu Province (J&K State) on Collection, Documentation and Validation of ITK, 2002. Sher-E Kashmir University of Agricultural Science Technology, Jammu"
19. "Management of bio-physical resources"	The Buddist Manpas in Arunachal Pradesh practise settled subsistence hoe farming without constructing terraces. They use sheep droppings and oak leaves as manure and human waste and litter as sprayed. Maintaining soil fertility requires the cultivation of maise and finger millet	"Agricultural Extension Division, ICAR Research Complex for NEH Barapani (Meghalaya) Indigenous technological knowledge followed by the tribal farmers of Northeastern hill in agriculture. Indian J. Hill Farmg 12 (1&2): 134–135(1999)"

(continued)

Table 2 (continued)

Classification of Traditional Agriculture Knowledge basis their Mechanisms	Traditional Agriculture Practices in their Corresponding Region Based on NATP Collected Case Studies	References and Information Collector
Pest and disease management (mechanisms)		
1. "Use deodar oil to control insect pests in paddy nursery."	Arunachal Pradesh's Buddist Manpas practice Deodar oil is used to kill caterpillars in paddy by soaking a stick in it and applying it to the foliage, raising the water level and allowing it to stand for 1–2 h in residing sustenance in the high highlands, cultivating crops by hand without constructing terraces (above 2500 m)	"Survey Report from Jammu Province (J&K State) on Collection, Documentation and Validation of ITK, 2002. Sher-E Kashmir University of Agricultural Science Technology, Jammu"
2. "Use of ash for controlling insect pests in paddy nursery."	Ash applied to paddy nurseries has insecticidal properties and enriches the soil, improving fertility. A similar practice is practised in the Jammu region, where paddy is grown	Same as previous
3. "Insect pest and disease management in paddy and vegetables."	Farmers in the North Eastern Himalayan region use a traditional practice to decompose cow dung urine; garlic and chilli are mixed with water for around 10 days and sprayed at 15-day intervals over paddy and vegetables	"Agricultural Extension Division, ICAR Research Complex for NEH Barapani (Meghalaya)- Indigenous technological knowledge followed by the tribal farmers of Northeastern hill in agriculture. Indian J. Hill Farmg 12 (1&2): 134–135(1999)"
4. "Use of pine leaves for insect-pest control in paddy."	Pine leaves are utilised to defend paddy from the gundhi bug strike under assured irrigation conditions, assuming that the aroma of pine leaves will preserve the crop. The assumption is that the aroma of pine leaves will preserve the crop. North Eastern Himalayan farmers follow a traditional practice	Same as previous
5. "Reduction in intensity of blast in paddy."	Farmers in Bain, Chenani, reduce blast infestation in paddy by postponing urea application 25–30 days before replanting	"Survey Report from Jammu Province (J&K State) on Collection, Documentation and Validation of ITK, 2002. Sher-E Kashmir University of Agricultural Science Technology, Jammu"

(continued)

Table 2 (continued)

Classification of Traditional Agriculture Knowledge basis their Mechanisms	Traditional Agriculture Practices in their Corresponding Region Based on NATP Collected Case Studies	References and Information Collector
6. "Method of saving maise seed from insects."	In order to achieve this, seeds are soaked in cow urine for 10–12 h, and if it does not rain for 20–25 days, the seeds will remain unaffected. As a result, maise seeds have been protected from insects for centuries in the Kasoli area of the Solan district in Himachal Pradesh	"Shri Pawan Kumar Himachal Pradesh Government Press, Shimla (Himachal Pradesh)"
7. "Use of wood ash for controlling cutworm attack in maise."	Farmers in the Mogla region of Rajouri district use wood ash to protect their maise crop from cutworm attacks. This method has proven to be beneficial	"Survey Report from Jammu Province J&K State) on collection, Documentation and Validation of ITK, 2002. Sher-E-Kashmir University of Agricultural Sciences & Technology, Jammu"
8. "Indigenous methods of preventing blister beetle infestation in maise."	"Maize" growers in Jammu and Kashmir use a technique to destroy blister beetles by setting them on fire and exposing them to smoke. It serves as a deterrent for other nuisance insects	Same as previous
9. "Drum beating for control of grasshopper."	Farmers of Kalakot use drum beating to drive grasshoppers away, effectively controlling the insect's attack	Same as previous
10. "Use of cow dung ash for termite control."	Cow dung ash is traditionally used to reduce termite infestations in the Kalakot area of Rajouri district	Same as previous
11. "Use of Eupatorium rugosum as a pest repellent in the rice field."	Farmers in Medziphema employ the Eupatorium rugorum plant as a pest repellant, positioning it 5–6 feet away from pests in their fields. It has a potent odour and is supposed to repel insects	"Shri Khrolie Angami NU SASRD, Medziphema (Nagaland)."
12. "Use of bhang as a pesticide."	"Bhang (Cannabis sativa)" plants, uprooted from the ground and kept in standing water until seedlings appear, are used in paddy nurseries to control thread worms. The worm can be killed by adding crushed leaves to standing water if the problem is severe	"Survey Report from Jammu Province J&K State) on collection, Documentation and Validation of ITK, 2002. Sher-E-Kashmir University of Agricultural Sciences & Technology, Jammu"

(continued)

Table 2 (continued)

Classification of Traditional Agriculture Knowledge basis their Mechanisms	Traditional Agriculture Practices in their Corresponding Region Based on NATP Collected Case Studies	References and Information Collector
Soil fertility management (mechanisms)		
1. "Green manuring of the soil."	Green-manuring is seeding leguminous crops into light soil to increase fertility and water-holding ability. In addition, it provides nitrates that contain nitrogen, which are returned to the soil for use by the next crop	"Dr Sukhdev Regional Research Centre, H P K V Dhaulakau, Sirmour (Himachal Pradesh)"
2. "Enhancing soil fertility by using cattle bones."	Soil fertility is improved by mixing phosphorus with animal bones and burying them in plant basins	Same as previous
3. "Soil sterilisation in nursery areas for raising onion seedlings."	In the Udhampur district of Jammu and Kashmir, onion-growing farmers burn dry leaves/grasses before sowing onion, killing soil insects and pathogens and improving soil quality and fertility	"Shri A.K. Agnihotri Joint Director (Planning) and Sr. Scientist, Directorate of Research, Dr. YSPUHF Naini, Solan (Himachal Pradesh)"
4. "Use of pond sediments as fertiliser for crops."	Farmers in Shimla district use pond sediment as fertiliser to increase soil fertility and improve production	"Shri Chandu Lal Sharma Khamala, Kotkhai, Shimla (Himachal Pradesh)"
5. "Soil reclamation through frequent irrigation."	Saline soils with a pH greater than eight are often found in hilly regions. A frequent irrigation program can wash down the salts to the permanent water level and alleviate the problem	"Shri Chaman Lal Rao Research Fellow, Himachal Pradesh University Shimla (Himachal Pradesh)"
Horticultural crops (mechanisms)		
1. "Intercropping of marigold in Tomato."	The marigold plant protects tomato crops from insect pests and diseases	"Survey Report from Jammu Province (J&K State) on the collection, Documentation and Validation of ITK, 2002. Sher-E-Kashmir University of Agricultural Sciences & Technology, Jammu."
2. "Spraying of gold-dipped water in cucurbits."	"Gold-dipped" water is sprayed on cucurbit plants to help with fruit set, which some farmers believe is normal	Same as previous

(continued)

Table 2 (continued)

Classification of Traditional Agriculture Knowledge basis their Mechanisms	Traditional Agriculture Practices in their Corresponding Region Based on NATP Collected Case Studies	References and Information Collector
3. "Dusting of ash in cucurbits to check insect pests."	In Jammu and Kashmir, cucurbit plants are dusted with ash to make them unpalatable to insects and to prevent damage caused by insect pest infestation	Same as previous
4. "Control of semi-looper in onion."	Kerosene oil and ash are sticking materials to reduce crop loss caused by semi-looper, while ajwain cultivation attracts the pest. (Jammu)	Same as previous
5. "Protection of vegetable nurseries against various insect-pest."	Hukka is common in remote hilly areas of Jammu Kashmir, where farmers smoke hukka for insect repellent	Same as previous
6. "Cultivation of vegetables in the sandy river bank."	The river bank grows vegetables such as bottle gourd, pumpkin, khira, cucumber, tomato, watermelon, capsicum and brinjal in trenches	"Krishi Vigyan Kendra Choudhary Sarankul, Himachal Pradesh Agricultural University Una (Himachal Pradesh)"
7. "Application of hukka water for control of caterpillar in kitchen/ vegetable garden."	Hukka (hubble bubble) is an indigenous technique to control insect pests and skin infections in kitchen gardens. It contains nicotine sulphate and is effective at 80–90%—an indigenous technique developed in the "Bhaderwah" tehsil and Doda district of Jammu and Kashmir	"Survey Report from Jammu Province (J&K State) on Collection, Documentation and Validation of ITK, 2002. Sher-E-Kashmir University of Agricultural Sciences & Technology, Jammu"
8. "Spray of vinegar to increase the yield of vegetables."	The farmers spray A vinegar solution on the field in "The Kotkhaki area of Shimla" district. Vinegar spray increases vegetable production by 15%, reduces cost, increases soil fertility, and increases fruit production	"Shri Chandu Lal Sharma Khamala, Kotkhai, Shimla (Himachal Pradesh) 171 201"
9. "Preservation of vegetables for off-season use."	Farmers in the Bhaderwah and Doda districts of Jammu and Kashmir practice this indigenous vegetable preservation method. Pits are dug in the ground and covered with local herbs, keeping vegetables fresh and devoid of pest and disease infestation. Earthen pots can also be used to store vegetables	"Survey Report from Jammu Province (J&K State) on collection, Documentation and Validation of ITK, 2002. Sher-E-Kashmir University of Agricultural Sciences & Technology, Jammu"

(continued)

Table 2 (continued)

Classification of Traditional Agriculture Knowledge basis their Mechanisms	Traditional Agriculture Practices in their Corresponding Region Based on NATP Collected Case Studies	References and Information Collector
10. "Augmenting shelf life of an apple."	Bhaderwah tehsil of Doda district in Jammu and Kashmir has adopted a technique to increase apple shelf life: detaching the fruit from the tree and removing the spur	Same as previous
11. "Use of semi-baked bricks for control of flower drop and induction of flowering in mango."	In the Nagri area of the Kathua district of Jammu and Kashmir, farmers dig a hole 5 feet deep and fill half with semi-baked bricks and half with soil. This practice induces flowering and prevents mango flower drops	Same as previous
12. "Use of limewater for controlling aphids in mango."	Jammu and Kashmir's Kathua and Jammu districts use limewater to control aphids on mango stems. It is mixed with water and coated on the stem, acting as a repellent and protecting the plant from the scorching sun	Same as previous
13. "Use of raised platform around mango trees."	People in Himachal Pradesh use raised platforms to protect mango trees from stray animals. These trees have lasted for centuries. They are heritage and serve to bring prosperity in rainfed and dry areas	"Shri A.K. Agnihotri Joint Director (Planning) and Sr. Scientist, Directorate of Research, Dr. YSPUHF Naini, Solan (Himachal Pradesh)"
14. "Use of bamboo drip irrigation for plantation crops."	Khasi hills of Meghalaya use drip irrigation to water their plantation crops. It diverts water from upper to lower reaches using gravity flow. Water is transported to the plot location and given to the branches. At the end of the water application process, reduced channel sections and diversion units are utilised	"Indigenous technological knowledge followed by the tribal farmers of Northeasternhill in agriculture. Indian J. Hill Farmg 12 (1 & 2): 134–135 (1999)"

Sources:
1. ICAR, 2003, "Inventory of Indigenous Technical Knowledge in Agriculture" Mission Mode Project on Collection, Documentation and Validation of Indigenous Technical Knowledge Document 2 (Supplement 1) Project National Agricultural Technology Project, Indian Council of Agricultural Research New Delhi
2. ICAR, 2003, "Inventory of Indigenous Technical Knowledge in Agriculture" Mission Mode Project on Collection, Documentation and Validation of Indigenous Technical Knowledge Document 2.1 (Supplement 1) Project National Agricultural Technology Project, Indian Council of Agricultural Research New Delhi
3. ICAR, 2004, Validation of Indigenous Technical Knowledge in Agriculture: Mission Mode Project on Collection, Documentation and Validation of Indigenous Technical Knowledge Document 3: Project National Agricultural Technology Project, Indian Council of Agricultural Research New Delhi

Table 2 presents case studies findings that establish Himalayan farmers have used traditional agricultural knowledge for many generations. Most traditional farmers employ some knowledge in their farming practices related to water management, yield fencing, animal husbandry, and storage. However, most traditional farming practices do not involve plant breeding, preserving landraces, or selecting seeds for preferred and adaptive characteristics, so these pieces of knowledge are gradually disappearing (Dhyani 1994; ICAR 2002, 2003a). Although new technology, HYV seeds, shrinking agricultural lands, commercial crops, changing climatic conditions, farmer migration, demographic pressure, and socioeconomic conditions hindered its practices. Most agriculture communities face the problem of practicing traditional knowledge. Traditional knowledge derived from farmers' knowledge is shared between villages and within villages. Most traditional farming communities share knowledge, but the mechanisms for doing so are eroding (Dhyani 1994).

Many factors contribute to losing traditional knowledge, such as: (1) Modern varieties and technologies are promoted through agricultural policies, subsidies and research at the expense of local knowledge; (2) In the absence of equal protection for farmers' rights over traditional varieties (now intellectual property rights protecting new varieties prevent farmers from maintaining traditional varieties) (3) Modern varieties and foods are promoted through the media, influencing consumer demand and reducing traditional varieties' markets; (4) A limited amount of arable land and small landholdings can force communities to adopt modern varieties with HYV, (5) Modernisation is the weakening of traditional knowledge, change in occupation nature and outmigration, and eroding cultural values and customary practices (Krystyna et al. 2011).

The benefit associated with traditional knowledge is mitigating "climate change impacts" through crop variety selection, which is most pronounced on marginal land. As a result, they can recognise resilient crop species and variabilities that can be adapted. In traditional agriculture, farmers—particularly women and the elderly—maintained "local landraces" and selected seeds for favourite and adaptive characteristics over generations. Several farmers have crossed lines in their quest to improve their crops. Societies often rely on wild areas around farms' yield to improve and domesticate crops, supplement their diets, and provide food when crops fail. Traditional agricultural methods such as water management, soil management, pest management, erosion control and land restoration are critical resources for resilience and adaptation. Farmers can access local weather forecasts, extreme event predictions, and other local information from traditional knowledge. In addition to monitoring climate change, traditional farmers can fill the resolution gap in scientific models by observing climate change in specific locations (Krystyna et al. 2011).

3.1 Past Scenario of Agriculture

R.P. Dhyani has extensively researched Agricultural Planning and Development in the Central Himalayas. He suggested that "central Himalayan has socioeconomic problems unique to its geo-climatic conditions" since he highlighted fundamental agricultural issues impacting traditional farming. In order to develop these hills and improve their conditions as high-lenders, new agricultural technologies, cropping patterns, land use alternatives, soil testing, and appropriate land and water management must be implemented" (Dhyani 1994). Here, Dhyani indirectly indicates that the traditional agricultural pattern is ineffective in the contemporary period; it means that the agro-ecological and cultivator society has changed and needs external support, which he described above. Agriculture was the backbone of people's lives in the early 1980s. It is estimated that only 14% of the land is arable. 41.9% of the working population is engaged in agriculture, accounting for 75.3% of the workforce. More than 71.5% of the operational holdings are estimated to be less than one hectare (Govt. U. P. 1981–1982)."

Dhyani described Himalayan agriculture conditions before 1990: "It is also necessary to consider that natural calamities, such as hailstorms, landslides, and wild animals, can also be considered misfortunes of hill farming. About 18.5% of snow and 52.5% of forest cover were on the entire land. It is estimated that only 10.5% of the land in the Garhwal Himalayas is cultivated, whereas this percentage is closer to 16% in Kumoan Himalayas. In 1990, agriculture was conducted on slopes of 60%, while slopes of over 30% were discouraged. Soil erosion is also caused by quake water runoff, which prevents water from permeating the soil. Fields are small terraced strips because of steep slopes. The topography in this region has several unfavourable effects on agriculture, including high risk, low productivity, and high maintenance costs. The physical environment hinders agricultural development. Inadequate irrigation facilities and soil erosion are acute problems not solved by traditional practices. Erosion of soil occurs on terraced fields due to slope not being considered. There is a gradual decline in soil fertility that occurs over time. Population pressure on agricultural land is relatively high. The meagre income of cultivators makes it impossible for them to save.

Consequently, farmers could not improve their land or adopt improved agricultural methods. There is a shortage of local production compared to local demand. In the Central Himalayas, cropping patterns are influenced by the varying altitude, climate and soil conditions as well as the various agro-economic and cultural practices that have evolved, such as elaborate terracing, crop rotations and 'gul' irrigations, which also serve small water mills like in the Himachal Himalayas. The inhospitality of the physical environment has infused many adaptive skills and unusual physical vigour among the people (Dhyani 1994)." Household cultivators did not consider crop operational costs when calculating production costs. The economics of chillies, potatoes, and wheat were studied. It was found that chilli production was far more profitable than food grains. The pressure of population on land is responsible for extending crop cultivation to unsuitable lands. The land use pattern can be improved only if alternative employment opportunities are offered to people relying solely on land. Fields belonging to a single individual are scattered over several locations surrounding the village (Govt. U. P. 1981–1982).

3.1.1 Lack of Appropriate Methods

The Himalayan agriculture methods that Dhyani pointed out in 1990 have remained static regarding methods and practices. Farmers still used the same inputs as their ancestors decades ago. A new feasible agricultural technology concerning purchasing power, skill and institutional framework available in the area is urgently needed. A new generation of technology must alleviate the constraints of severely limited arable land and water to be a "powerhouse of development." Technology should be equally accessible to farmers of all income levels, as its income distribution effects are economically and socially acceptable. As Dhyani explains, modern technology and knowledge must update traditional agricultural practices. Additionally, he discusses the importance of modern irrigation facilities, fertilisers, HYV seeds, land holding size, soil erosion prevention, and agriculture credit and market facilities (Dhyani 1994).

3.2 Present Scenario of Agriculture

Several agricultural farming systems are prevalent in the Himalayan region (ICAR 2002; Dhyani 1994; Roy et al. 2015; Reimeingam 2017), but a recent study (Integrated Mountain Initiative 2019; Roy et al. 2015; Prasad and Vangchhia 2017) claims the most dynamic systems are: "(1) Rain-fed crop-livestock mixed farming system in a Kashmiri tribal village, (2) Economies and livelihoods of apple growers in Kashmir valley, (3) The fruit and vegetable farming systems of small hill farmers in Kullu valley of Himachal Pradesh, (4) The subsistence crops and livestock in rainfed mixed farming in Almora, Uttarakhand, (5) Nagaland and other northeast Himalayan regions have swidden slash-and-burn farming culture, (6) Ladakh's highlands-Agropastoral and pastoral systems of the Himalayan Nomads."

Traditional agricultural knowledge is not used in most farming systems, and some places still use this knowledge, but it has been upgraded with modern equipment (Reimeingam 2017; Integrated Mountain Initiative 2019).[3] A few NATP studies are experimental and belong to earlier periods, but some studies on traditional knowledge practices in the region are irrelevant in the present context.[4] The Integrated Mountain Initiative (IMI) conducted a field-based study in the "Indian Himalayan Region", and their observations reveal the current scenario of agriculture changes that include changes in land use patterns, changing land holding relationships, changes in cropping patterns, farmer migration, the type of knowledge attributes in the region today, and the changing relationship amongst agriculture, socio-economics, food consumption patterns and agro-biodiversity. The study also describes how agriculture patterns have been altered due to demographic pressure, cash crop culture (commercial crops), and the use of HYV seeds; these three

[3] Bases of analyses from IMI (2019). State of The Himalayan Farmers and Farming: A study about social, economic and ecological factors changing the mountain agriculture scenario; Dhyani (1994, "An Approach to Economic Planning for the rural poor of Central Himalaya.

[4] Due to changes of Ecological niche and methos have not effective (Rai and Mishra 2023).

reasons are the main ones. There are many reasons (issues) described in the study. The following are some of the findings drawn from the Integrated Mountain Initiative 2019 study:

3.2.1 Changing the Dynamics of Agriculture

The traditional agricultural pattern and its inherent limitations have proven ineffective in alleviating "food insecurity and poverty" among farmers. However, new technology, HYV seeds, shrinking agricultural land, commercial crops, changing climatic conditions, and farmer migration issues may make it possible. All types of farming systems have changed. Himalayan agriculture increasingly relies on external inputs, including seeds, fertilisers, labour, and mechanisation. Despite this, varied levels of agricultural commercialisation are observed in different Himalayan states.

There were mostly no harvests in subsistence mixed farming villages because rainfed farming was prevalent on sloped farmlands and faced wild animal threats. Northwest Himalayan villages have subsistence farming conditions that allow only 15% of farmers to produce sufficient food to feed their household members. Farmers in the northeast Himalayas produce enough food to feed their families in over 50% of cases. Some farmers in Himachal Pradesh have abandoned their traditional food crops farming. Food insecurity is a transient issue throughout the Himalayan states, regardless of the farming conditions. There is a strong tradition of nomadic culture in Himalayan Mountain agriculture, and the nomadic situation is also changing due to farmers' migration. A problem with nomadic highlanders is that their land is vast, but there are fewer people due to local families migrating to towns. A similar story can be found with "Changpas of Ladakh, Bakarwals of Jammu and Kashmir, Gujjars of Jammu, Gaddees from Himachal Pradesh, Van Gujjars from Uttarakhand, and Yak herders from Sikkim and Arunachal Pradesh." Nomadic lives are displaying warning signs that the system is shrinking and is in danger of extinction.

Agricultural scenarios in the northeastern Himalayas fall into two categories: (1) traditional shifting cultivation (Jhum) and (2) settled (plain) land cultivation in valleys. Northeastern Himalayan shifting cultivation is the most traditional type of farming. According to estimates, 19.91 lakh hectares of the region's land are under traditional shifting cultivation (83.73%). Shifting agriculture patterns had declined from 2.69 to 1.00% of the entire geographic area and from 2.99 to 1.35% of hill areas between 1974 and 2012. Undoubtedly, it will drastically decrease in the future. Since traditional shifting cultivation has declined in the northeastern region, rarer farming families are practising it. Assam, Nagaland, Arunachal Pradesh, Tripura and Manipur have fertile alluvial plains suitable for settled agriculture. Approximately 70% of the area is hilly North Eastern Himalayan, with most crops being rice, maize, wheat, and oilseeds grown on fertile alluvial plains.

3.2.2 Changing Traditional Cropping Pattern and Its Significance

Himalayan mountain farming systems have traditionally relied on cereal crops and some fruit (lesser amount) before the 2000s (Dhyani 1994). After that period, Himalayan agriculture cropping pattern has changed. It continues to be characterised by new cropping systems, knowledge of agriculture, and advanced technology (traditional knowledge reduction), in which fruit and vegetables play an increasingly important role (Integrated Mountain Initiative 2019). This "cropping pattern" is attributed to lesser traditional knowledge practices and the ability to grow these crops rapidly (vegetables) and provide a steady income for long periods. Small landholdings are more suitable for growing vegetable crops since they are highly capital and labour-intensive, making their adoption by an individual farmer less viable.

Consequently, "new cropping systems are emerging", and "cash cropping" is leading the way now. This system has resulted in a significant increase in employment and the diversification of farm incomes. These new cash crops require farmers to advance agriculture skills management and development with new technology. As a result, new service industries developed that benefited the whole community. From an ecological point of view, diversity is an essential feature of cash crops. One hundred seventy-five vegetable types are grown in the Himalayan states, creating a perception that native agro-biodiversity needs to be protected. Intensive agriculture makes farming gainful for marginal and small hill farms. Some vegetables grow in hilly terrain where cropping concentration has already reached 200% (Solan district) (Integrated Mountain Initiative 2019).

The IMI study highlighted the "Hills and mountains are particularly suitable for horticulture", which has led to a shift in cropping patterns from cereal yields to cash yields (cash crops). There was a growth in the area under fruit from 792 hectares in 1950–51 to 224,352 hectares in 2014–15 in Himachal Pradesh. The fruit production improved from 7.52 lakh tons in 2014–15 to 9.29 lakh tons in 2015–16. Similar changes have appeared in Jammu Kashmir. The Kashmir valley has experienced tenfold growth in the area under fruit trees (one million hectares) over the past two decades. The northeastern Himalayan states (Assam, Nagaland, Arunachal Pradesh, Tripura and Manipur) abandoned or significantly reduced cultivation of food (cereals) crops (20–30%). Traditional swidden farmers (shifting cultivators or jhum) are willing to switch to persistent commercial crops on their swidden yield land holdings. These shifting cultivation areas have become more dominated by cash crops, and there is already potential for them to move into them on a larger scale, which helps reduce traditional swidden farming. The availability of subsidised food grains through the Public Distribution System (PDS) has increased cash cropping throughout the Himalayan region. Many farmers have stopped consuming local grains and abandoned farming indigenous crops due to shrinking land holdings, demographic pressure and access to new HYVs food and cash crops. Current temperature rises have made it challenging to continue yielding certain traditional temperate fruit varieties. The apple is an example of one such fruit variety whose production is being swapped by others, such as pomegranates, persimmons and plums. Climate conditions have been changing rapidly in the Himalayas over the past decade, and farmers have predicted uncertainty for the future due to changing weather patterns. A significant issue with climate change is the accessibility of irrigation, which is essential for cash crop farming (Integrated Mountain Initiative 2019).

Farmers in central Himalayas (Uttarakhand) and some northeast Himalayan states are still tense about cultivating traditional subsistence crops. As a result, 1700 ghost villages in Uttarakhand were abandoned in 2018, and uninhibited cultivating landscapes were left in numerous hill districts. Traditions of indigenous crops have been completely wiped out from the present cropping patterns as traditional agriculture shifted from subsistence to a cash economy. The trend is evident in several areas of Kashmir Valley and Himachal Pradesh. Additionally, Central Himalayan agro-ecosystems have also experienced steady changes in their agro-biodiversity. There has been a significant decline in cultivated areas under many traditional crops in 150 villages (Central Himalaya) in the past two decades, according to a survey done by IMI study from 150 villages. While Jammu & Kashmir and Himachal Pradesh are ahead of the pack regarding traditional diversification, several states in the North Eastern region are also making concerted efforts in this direction. The agro-climatic and socioeconomic conditions of the locality have enabled farmers to utilise the local niches more effectively. Therefore, food grain farming in the Himalayas appears to have been pushed to a lesser extent by the shift from traditional cropping patterns to cash crops. Moreover, the Himalayas are expected to become essential for cultivating cash crops, vegetables, fruits, flowers, herbs, and medicines soon (Integrated Mountain Initiative 2019).

Present Transformation has led to the development of a new class of farmers, one is "New Age farmers", who are advanced and serve as protagonist models and second is farmers who fall into the category of 'late adopters and laggards. Farmers became the centre of attention in the New Age either by adopting "new niche crops with large market potential or developing highly beneficial crops for the local environment". There are plenty of examples of farmers to be found. One example is the 2018 "Most Innovative Farmer Award" (Navonmeshi Kisan) given by the Ministry of Agriculture, Government of India, to farmers in the Kullu district of Himachal Pradesh for growing pomegranates in the Kullu valley. Pomegranates have taken over the lower valley areas vacated by apples due to climate change. Similarly, many other Himalayan farmers have relentlessly worked to develop agriculture and farmers through their modernisation and extension facilities. Many New Age agriculturalists operate organic farms in Jammu & Kashmir, Ladakh, Uttarakhand, Himachal Pradesh, Sikkim and other northern states (Integrated Mountain Initiative 2019).

4 Conclusion

The present study provides a holistic database related to traditional Himalayan agriculture knowledge with names, places and descriptions of nature. Most knowledge was practised at an earlier time in the region. Three discourses are running to make arguments related to traditional agriculture patterns in the Indian Himalayan region. First discourses, there are sufficient studies available about the traditional pattern of agriculture which claim this knowledge is not applicable at present Himalayan agriculture due to profound changes such as land use patterns, land holding, cropping patterns, demographics, food consumption patterns, socio-economics patterns, agriculture technology patterns and agrobiodiversity. Moreover, some issues related to urbanisation, farmer migration, chemical fertilisers, use of HYV seed, cash crop culture, and climate change are the most impactful factors. The second discourse argues that traditional agriculture knowledge is the best remedy for sustainable agriculture development because of the adaptive nature of knowledge to changing climate change, but they forget it applies only to some cases, not all; in the case of Himalayan, they give some examples of sustainable agriculture but, they forget other factors in which describe in first discourses, how much importance their robust application implication methods, second discourse not able to explain it. Third discourse, policy and decision-makers, which were produced databased about traditional knowledge and hurry to the implication on the ground, ignored physical, ecological and socio-economical barriers.

Sustainable development comes through traditional knowledge or patterns; it is not valid in all cases, but in some, it is accurate. It also depends on how traditional knowledge copes with present development with acceptance of social culture and demographic needs. For example, see what happens with Jhuming cultivation in the northeast Himalayas. The type of knowledge attributes in the region today matters, the traditional segment of knowledge introduced is challenging.

The present study maps out changes that have begun in the Himalayan states that appear irreversible, which motivates farmers to divert cultivated lands to commercial crops. Though the scale of diversification varies widely among states, Jammu & Kashmir and Himachal are the leading states, and Arunachal Pradesh, Sikkim, and Nagaland are making serious efforts. Moreover, in the past few decades, the conversion rate of food grains to cash crops has increased, and the trend is expected to continue. As a result, the Himalayan region has been stripped out of traditional knowledge or patterns.

References

Ajeet M (2003) Who owns traditional knowledge? Econ Polit Wkly:4471–4481

Bandana S, Flurina S, Yongping Y, Eklabya S (2019) A multiscale transdisciplinary framework for advancing the sustainability agenda of mountain agricultural systems. Int Mountain Soc Stable 39(3). https://doi.org/10.2307/26915079

Chand R (1997) Agricultural diversification and development of mountain regions with special reference to Himachal Pradesh. M. D. Publication Private Limited, New Delhi

Christoph A (2013) Asian borderlands and the legal protection of traditional knowledge and traditional cultural expressions. Cambridge University Press, vol 47, pp 1403–1433. Retrieved from https://www.jstor.org/stable/24494202

Clarke WC (1990) Learning from the past: traditional knowledge and sustainable development. University of Hawaii Press (The Contemporary Pacific, FALL1990) 2(2):233–253. Retrieved from https://www.jstor.org/stable/23698358

Dhyani RP (1994) An approach to economic planning for the rural poor of central Himalaya. Classical Publishing Company, New Delhi

Govt. U. P. (1981–1982) Parvatiya Kshetra Ka Vikas. Pragati, Utter Pradesh

ICAR (2002) Inventory of indigenous technical knowledge in agriculture: Mission mode project on collection, documentation and validation of indigenous technical knowledge document 1 project National Agricultural Technology Project. Indian Council of Agricultural

Research, New Delhi. Retrieved from Traditional Knowledge in Agriculture: https://icar.org.in/sites/default/files/Inventory%20of%20Indigenous%20Technical%20Knowledge%20in%20Agriculture%20Document%201.pdf

ICAR (2003a) Inventory of indigenous technical knowledge in agriculture: Mission mode project on collection, documentation and validation of indigenous technical knowledge document 2 (Supplement 1) Project National Agricultural Technology Project. Indian Council of Agricultural Research, New Delhi. Retrieved from https://icar.org.in/sites/default/files/Inventory%20of%20Indigenous%20Technical%20Knowledge%20in%20Agriculture%20Document%202.1.pdf

ICAR (2003b) Inventory of indigenous technical knowledge in agriculture: Mission mode project on collection, documentation and validation of indigenous technical knowledge document 2.1 (Supplement 1) Project National Agricultural Technology Project. Indian Council of Agricultural Research, New Delhi. Retrieved from https://icar.org.in/sites/default/files/Inventory%20of%20Indigenous%20Technical%20Knowledge%20in%20Agriculture%20Document%202.1.pd

ICAR (2004) Validation of indigenous technical knowledge in agriculture: Mission mode project on collection, documentation and validation of indigenous technical knowledge document 3: project National Agricultural Technology Project. Indian Council of Agricultural Research, New Delhi. Retrieved from https://icar.org.in/sites/default/files/Validation%20of%20Indigenous%20Technical%20Knowledge%20in%20Agriculture%20Document%203.pdf

Integrated Mountain Initiative (2019) State of the Himalayan farmers and farming: a study about social, economic and ecological factors changing the mountain agriculture scenario. Integrated Mountain Initiative, New Delhi. Retrieved from https://www.mountaininitiative.in/

International Development Research Centre (2009) Ecosystem approaches to research on agricultural transformation. Nutrition and Human Health

Kamaljit SB, Reinmar S (2015) Deforestation and sustainable mixed-use landscapes: a view from the eastern Himalaya. Ann Missouri Botanical Garden Press:141–149. Retrieved from https://www.jstor.org/stable/43916790

Krystyna S, Song Y, Jingsong L, Reid H, Mutta D (2011) Adapting agriculture with traditional knowledge. International Institute for Environment and Development. Retrieved from http://www.jstor.com/stable/resrep01470

Lal H, Sharma KD (2005) Tribal development and agricultural transformation in North-Western Himalayan region—a case study of Himachal Pradesh. Classical Publishing Company, New Delhi

Parry ML et al (eds) (2007) Cross-chapter case studies. Indigenous knowledge for adaptation to climate change. In: climate change 2007: impacts, adaptation and vulnerability. Contribution of working group II to the fourth assessment report of the intergovernmental panel on climate C. Cambridge University Press, Cambridge

Prasad V, Vangchhia L (2017) Food security status in rural areas of Mizoram, north East India. J Mountain Sci 14(4):795–805

Rai SC, Mishra PK (eds) (2023) Traditional ecological knowledge of resource management in Asia. Springer. Retrieved from https://doi.org/10.1007/978-3-031-16840-6

Reimeingam MR (2017) Use of land and labour under shifting cultivation in Manipur working paper 384. The Institute of Social and Economic change.

Roy A et al (2015) Food security in north east region: a state wise analysis. Agric Econ Res Rev 28:259

Part II

Food for a Sustainable Future: The Resilience of Traditional Himalayan Food Systems

Mountain Cuisine in a Changing Climate: The Resilience of Traditional Himalayan Food Systems

Traditional Food Systems in the Indian Himalayas: Perspectives from Climate Science

Archana Bhatt, Sreeram Vishnu, and Shikha

> Food is not rational. Food is culture, habit, craving and identity—Jonathan Safran Foer.

1 Introduction

The mighty Himalayas are stretched from the extreme north to the far east of the Indian subcontinent. These mountain ranges form the primary water source for the thriving agriculture systems across many Asian countries, especially India (Rasul 2010). Most of the population residing in the Himalayan region depends on subsistence agriculture for their livelihood and nutritional needs (Sati and Kumar 2023). Farming in the mountains is not only vital in the form of food production but also provides other benefits in terms of livelihood and sustainable energy resources. Unlike mainstream agriculture, mountain agriculture is an integrated system of crops, livestock, and other natural resources that link various farm activities (Shukla et al. 2019). Therefore, the links between various agricultural activities and the environment and their subsequent results are more evident in the mountains than in other agroecosystems. It integrates diverse practices such as crop management, animal husbandry, horticulture, forestry, etc., and the conservation practices deeply

A. Bhatt (✉)
MSSRF Community Agro Biodiversity Centre, Wayanad, Kerala, India
e-mail: archanab@mssrf.res.in

S. Vishnu
Kerala Agricultural University, Thrissur, Kerala, India
e-mail: sreeram.vishnu@kau.in

Shikha
KVK Ranichauri, VCSG UUHF, Tehri Garhwal, Uttarakhand, India

A. Borthakur, P. Singh (eds.), *The Resilience of Traditional Knowledge Systems for a Sustainable Future*, Advances in Asian Human-Environmental Research,
https://doi.org/10.1007/978-3-031-56858-9_6

ingrained in the culture. Agriculture in the mountains rather than a form of occupation is a way of life for the tightly knitted hill community that sustainably utilizes natural resources. Agriculture in the Central Himalayas is predominantly traditional and subsistence in nature, with most mountain populations residing close by and sharing a harmonious relationship with nature (Mukerjee et al. 2018). These mountains are also home to a diverse range of nutritionally rich plant species that form a part of the traditional food system (Rana et al. 2021; Palni and Rawal 2010). The conventional, traditional, ethnic or indigenous food system can be referred to as a food system specific to a particular geographic region which primarily relies on the native agro biodiversity, mainly native and underutilized species (Keleman Saxena et al. 2016) with the staple food crops such as rice and wheat in the Himalayan context. With the current generation obsessing over superfoods and nutritional diet options, supermarkets have been loaded with innumerable items in the name of health foods. But, traditional foods worldwide have mostly imbibed with loads of health benefits that have lost their relevance in the current world. These conventional food systems, deeply rooted in cultural heritage and indigenous knowledge, are also particularly vulnerable to the impacts of climate change (Bhatt et al. 2023). "Climate change refers to a change in the state of the climate that can be identified by changes in the mean and/or the variability of its properties, and that persists for an extended period, typically decades or longer" (IPCC 2018). Climate Change has become one of our most significant global challenges, affecting various aspects of human life, including our food systems (Owino et al. 2022). Climate variability over the years has significantly impacted the mountain ecosystem's food production system (Holleman et al. 2020). Indicative changes have been observed in the region, mainly in temperature, erratic rainfall, uneven precipitation and unpredictable weather patterns over the past several decades (Tewari et al. 2017; Sabin et al. 2020). The present study reflects the experiences of an Indian Himalayan state, Uttarakhand (Table 1; Figs. 1 and 2), on the traditional food systems (*Pahari* food) in climate change and the associated cultural and dietary aspects.

Table 1 Basic information about the study area

Country	India
Province	Uttarakhand
Size of geographical area	53,483 km^2
Total population	10.086 million
Rural population	7.037 million
Urban population	3.05 million
Dominant ethnicity(ies)	Hindu
Net sown area	620,629 ha
Geographic coordinates (latitude, longitude)	28°43′N to 31°27′N latitudes and 77°34′E to 81°02′E longitudes

Source: Census of India District Handbook (2011), Government of Uttarakhand (2022)

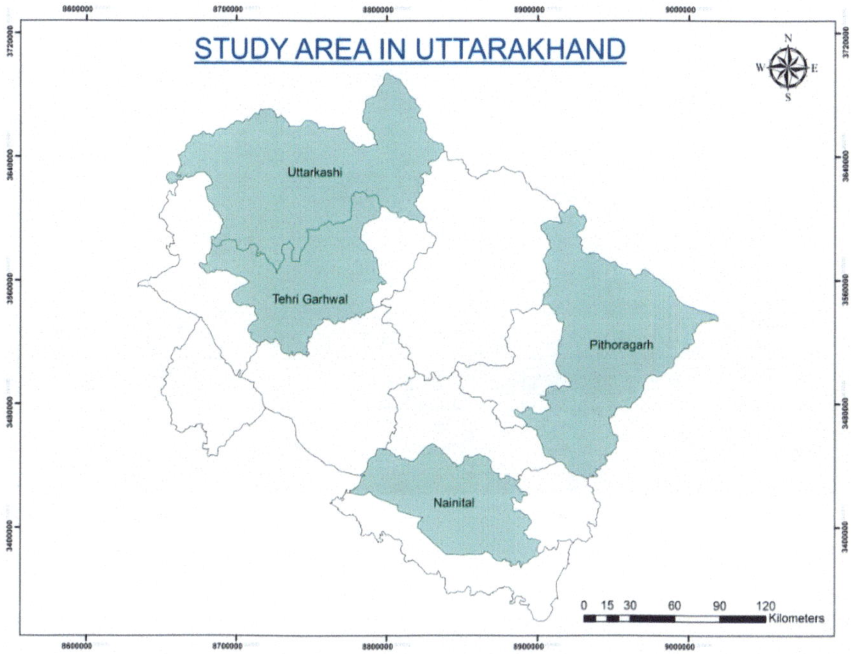

Fig. 1 Map of study area in Uttarakhand, India (Source: Authors)

The majority of the area in the state is hilly, i.e. 46,035 km^2 while the rest is plains, i.e. 7448 km^2 and around 45.44% of the area is covered with forests, i.e. 24,303.04 km^2 (MoEFCC 2019). More than 70% of the population depends on agriculture and allied sectors for livelihood, and the state has a net sown area of 620,629 ha. Subsistent mixed farming is followed in the state, predominantly in the hilly terrains, and a livestock-based crop farming system comprising cereals, pulses, oilseeds, fruits, and vegetables. Major crops grown are wheat, maize, rice, ragi, sawan, barley, urd, lentils, gram, beans, soybean, mustard, sesame, groundnuts, seasonal vegetables and temperate and subtropical fruits (Government of Uttarakhand 2022). The state is divided into two major divisions, i.e. Kumaon and Garhwal, having seven and six districts, respectively. Two districts were selected randomly from each division for the study. Information was gathered through 80 in-depth Key Informant Interviews and focused group discussion on the effect of climate change on agriculture, ethnic foods and their significance in health in 2019. This information is supplemented with a review of relevant secondary sources. The study also looked upon the adaptations in food patterns and diet with variation in climate, i.e., warmer to extreme cold and changing seasons. Respondents from different age groups, genders and socio-economic backgrounds were selected based on their experience. Information on various domains related to ethnic foods of the region, i.e. traditional knowledge of the food, key ingredients, food adaptation according to climate, dietary habits, health benefits and other associated practices, was collected.

Fig. 2 Study area in Uttarakhand (Source: Authors)

2 Ethnic Food Systems in the Context of Climate Change

Agricultural biodiversity has been given to the Himalayas as a form of heritage (Pandey et al. 2022). This diversity has paved the way for the development of healthy and nutritious mountain-based cuisine, i.e. ethnic food or traditional food (*Pahari* Food). Ethnic cuisine includes the conventional food consumed since earlier times by the natives of a region through their indigenous wisdom as part of their cultural heritage (Raúl Matta 2015). Such foods are an inherent part of the culture passed on from generation to generation, along with their recipes and nutritional value (Trichopoulou et al. 2007). The climatic condition of a region hugely influences the cuisine of the community, and such foods may be based on geography, religion, society, culture, etc. The Himalayan region is a temperate region with warm summers in few low-lying areas, and the foods also depict the adaptations based on the same. The ethnic cuisine of the Himalayan belt is remote and unique, and the diet practices are based on the adaptation of the people to the cold climate. Uttarakhand is a Central Himalayan state located in the northern part of the Indian subcontinent with a diverse range of traditional foods, and by sharing borders with Nepal in the east and Tibet in the north, a lot of similarities with the cuisine as well as the crops grown are seen. The state also has diverse geographical features, ranging from snow-clad mountains in the north to subtropical forests in the south, with diverse climates and vegetation. Subsistent mixed farming is followed in the state, predominantly in the hilly terrain, a livestock-based crop farming system comprising cereals, pulses, oilseeds, fruits, and vegetables (Government of Uttarakhand 2022).

However, the region is also facing an ever-growing challenge of climate variability. For instance, a study conducted by Kumar et al. (2023) indicated that over seven decades (1951–2020), the Himalayan region of Uttarakhand has experienced a significant incline in hot incidents, a declining trend of low-temperature days, and an increasing trend in recurrences of temperature and precipitation-based climate indices. It has also been observed that the total precipitation of the state is decreasing. Besides, it is essential to note that the effects of climate change are more prevalent at a higher elevation (Pepin et al. 2015). The days and nights with lower temperatures are getting less and less than in previous years. With a constant rise in temperature in the summer, incidents of forest fires have also been on the rise, as reported by various studies (Singh et al. 2016). There have been occasional incidents of erratic heavy rainfall causing flash floods and landslides in certain pockets of the region. (Khanduri 2020; Dimri et al. 2017; Kumar et al. 2017). Community perception also indicates that the state faces a dichotomy of climate change-induced effects in erratic rainfall patterns, landslides, cloudbursts, flash floods at one end, water scarcity and higher temperatures, and forest fires at the other (Joshi 2018).

2.1 Effect of Climate Change on Agricultural System

Extreme weather events strongly affect agriculture, which forms the livelihood for most people in the mountain ecosystem. As a rainfed agricultural system, farming is intensely linked with weather and climate since the selection and production of crops are determined by the temperature and rainfall (Thadani et al. 2015). Agriculture being highly dependent on climatic conditions is like gambling at the hands of erratic rainfall or droughts, leading to crop loss regularly in terms of declining yield, increased insect pest resurgence and lack of safety net in case of weather-based disasters, all of which prominently affects the small and marginal farm families. There has also been a continuous inclination towards horticultural crops from agronomic crops in the state over time (Dahal et al. 2023). It is also worth noting that with an increase in population, the size of land holdings is declining within families due to division, leading the farmers to cultivate crops that can provide more returns per unit of land. Horticultural crops are gaining prominence in this regard.

Regarding climate variability, the farmers prefer horticultural crops such as off-season vegetables like potatoes, cole crops, and peas, as there is easy market access from the nearby cities. Varietal preferences and cropping patterns are changing rapidly with the changing weather patterns keeping in view the market demand. However, the changes in agricultural systems have an overarching effect on the dietary patterns and the food and nutrition security of the community. In this regard, it is crucial to understand the local climatic risks and their effect on the agricultural system to develop effective adaptation strategies. Below are the effects of climate change on the farming system as observed by the community supported by previous studies conducted in the region (Table 2).

Gaining insights from the points mentioned above, it is evident that climate change's effect on agriculture will significantly affect the traditional food systems. It is also worth noting that there is a bidirectional relationship between food systems and climate change; both are interlinked to each other and affect positively and negatively. While involving the food systems, the effects also have a more profound impact on the environment. The effects on the agricultural system percolate to farmers' financial insecurity, compelling them to follow unsustainable crop management practices, out-migration, declined interest in farming and more ripple effects. There should be suitable policy and socio-economic interventions in this regard.

2.2 Traditional Food System: Key Aspects

Traditional cuisine in the hilly state relies mainly on locally grown crops and livestock. Regions with an extremely cold climate in the upper Himalayan belt that share borders with Tibet and Nepal also have non-vegetarian food and locally made liquors as an essential part of the diet. Food is light on the stomach, with considerably fewer spices and wholesome, balanced meals, including grains, lentils, and green vegetables. Due to the cold climate, food is based solely on cereals and millet, pulses, meat, and dairy, with tea as part of daily life to stay warm at high altitudes.

2.2.1 Dietary Culture

Ethnic food from traditional plant and animal sources has been closely linked with the mountains' climate, culture, heritage, and livelihood. Food is one of the bases of survival in such harsh cold winters by amalgamating nutrient-rich recipes in the diet since earlier times. Traditionally, meals were eaten while sitting cross-legged on the floor, though with time, the culture is being changed to sitting on dining tables, chairs, etc. People in the mountains strongly believe that everything is given to man by the blessings of god, and hence, before taking a meal, some of the food is offered to the almighty for its blessings. Providing food to certain birds (crows) and animals (cows) during religious occasions or special days is also essential to Hindu culture.

Table 2 Effect on agricultural systems due to changes in climate variables

Changes in climate variables	Effect on agricultural and food systems	Remarks
• Rising temperature • Uneven rainfall patterns • Early summers • Reduced winters • Torrential rainfall and cloudburst	Effect on chilling temperature	Apple orchards are being forced to move to higher elevations because snowfall is decreasing, and the correct temperatures for dormancy aren't occurring at the expected times. Besides, the lack of proper chilling temperature causes scab disease, premature leaf fall and infestation of red spider mites in apples (Rai 2015). The area and production of apples have significantly declined, and old varieties have been replaced with low-chilling varieties (Nautiyal et al. 2020). There has also been a shift towards other fruit crops like peach, plum, and kiwi, which give better returns
	Increased incidence of pests and diseases and decline in yield	With a rise in temperature and humidity, the community has observed the emergence of new pests, diseases, and weeds. With continued night rains, cucurbit vines get damaged, leading to no production. More incidences of pest infestation eventually lead to more usage of harmful pesticides, affecting food and nutrition security
	Changes in flowering and fruiting time (phonological changes)	With a shift in temperature, phonologically changes are observed, e.g. early flowering and fruiting in Apple (*Malus domestica*), Buraansh (*Rhododendron arboretum*), Kaphal (*Myrica esculenta*), Hisalu (*Rubus ellipticus*) mainly attributed to reduced winter period and early summers
	Drying up natural water resources	With rising temperatures and declining precipitation, perennial water resources and natural springs are drying up. With uneven rainfall, the flow in the streams has also been affected
	Decline in soil moisture level	Many crops are affected by wilting, especially during critical periods, and yields eventually decline. This is leading to a shift from rainfed cultivation to an irrigated one
	Heat stress in cattle and other livestock	Higher temperatures cause the cattle, especially cross-bred ones, to go under heat stress if the Temperature Humidity Index (THI) goes above the required level. This eventually affects milk productivity as the animal cannot adapt to the changing climate. Milk spoilage has also been observed at higher temperatures
	Increased incidence of forest fire and changes in forest cover	High summer temperatures have increased forest fire incidents, ultimately affecting the community, which is dependent on forest resources for fuel wood, fodder, and other non-timber forest products (NTFPs). Community also reflects a decline in the regeneration capacity of forests due to wildfire. The fast expansion of pine trees (*Pinus roxburghii*) has also made the forest susceptible to forest fire. It also affects the groundwater and survival of other important species like oak (*Quercus leucotrichophora*). Besides, the changes have also affected the availability of fodder trees. Changes in forest cover and climatic conditions have also been reported to have adverse impact on the wild edibles that are consumed by the community e.g. *Rumex nepalensis* (junglipalak or payoom), *Chenopodium foliolosum* (bethusaag/bhethua), *Amaranthus* sp. (janglichaulai), *Diplazium esculentum* (lingra), *Urtica dioca* (Stinging nettle or Bicchoo) (Rautela and Karki 2015)
	Shifts in transplanted paddy	Due to a decline in precipitation in summers and uneven rainfall patterns, transplanted paddy cultivation is reducing. Some farmers are reviving to direct seeded rice and local cultivars like *Laal Dhan* (Red Rice) however, the population is relatively insignificant

(continued)

Table 2 (continued)

Changes in climate variables	Effect on agricultural and food systems	Remarks
	Changes in yield and productivity	Changes in crop phenology due to uneven weather patterns have a direct effect on the yield and productivity of crops. Moreover, torrential rains, hailstorms, etc., during flowering and fruiting can cause significant damage to crops and make them more susceptible to disease and pest outbreaks. Fruit crops like apples, peaches, plums, apricots, etc., get severely affected by heavy rains and hailstorms, which deteriorates the fruit quality, eventually leading to less economic returns
	Shifts in cropping patterns and varieties	Weather patterns highly influence agriculture; hence the cropping pattern is also witnessing changes. Shift from apple to peach, plum, and apricot (mentioned above); shift from cereals to horticultural crops like cool season vegetables (potato, cole crops, pea, etc.). The community has practised early and late sowing or planting certain crops to adapt to climatic changes. Short-duration crops like Urd (*Vigna mungo*), Ugal or Kuttu or Buckwheat (*Fagopyrum esculentum*) are also cultivated if the main crop fails due to less rainfall in the season
	Landslides or flash floods affecting the logistics of food value chain	With intense rainfall/ cloudbursts, transportation gets disrupted, thus affecting market access and crop harvest sales and leading to post-harvest losses. This also leads to higher costs even with lower-quality crop produce. Access to other crop produce can also directly affect the nutritional security of the community, which is dependent on food crops from different regions. With intense rainfall/ cloudbursts, transportation gets disrupted thus affecting market access and sale of crop harvest and leading to post harvest losses. This also leads to higher costs even with lower quality crop produce. Access to other crop produce can also have a direct effect on nutritional security of the community dependant on food crops from other region
	Decline in soil health	Sudden flash floods and landslides erode the fertile topsoil, thus affecting soil health, eventually leading to increased use of fertilizers to enhance crop productivity
	Human-Wildlife Conflict	Wildlife attacks have increased due to various factors, such as the survival rate of certain wild animals with reduced winter, scarcity of natural water sources and loss of forest cover, leading to more intrusion in human habitation and increased crop raiding. The movement of bears, leopards, and wild boar towards human habitation is growing in search of food

Source: Compiled by authors from field survey

Fig. 3 Typical *Pahari* traditional meals (Source: Authors)

In the cold most parts of the state, unlike other regions, non-vegetarian food and liquor are also offered to god for acknowledgement.

The principal meal of the day includes cooked rice (*bhat*) and lentils (*dal*) for lunch and wheat flatbreads (*chapatti*) with some vegetable curry (*sag/sabzi*) for dinner (Fig. 3). Apart from the main course, additional side dishes include chutneys, *raita* (curd and vegetables), buttermilk, salad, etc. The majority of the population is Hindu, and meat consumption is prohibited on religious occasions and also by some people from the Brahmin caste, similar to Buddhists, on account of their respect for the living. It is also imperative to maintain the purity of food while being served and eaten, i.e. it should not be shared on the same plate or in contact with saliva, however, this practice is also slowly being gone. Food is usually prepared in iron utensils through slow cooking to get more health benefits.

2.2.2 Food Security and Health

One major cultural tradition followed in the state is mixed farming, in which the practice of '*barahanaj*' translates to *bara*, i.e. 12 and *anaj*, i.e. grains. However, it

Table 3 Nutritional Profile of main crops under Barahanaj (per 100 g)

Crop	Carbohydrates (g)	Protein (g)	Fat (g)	Energy (KJ)	Ca (mg)	Fe (mg)	P (mg)
Finger millet	66.82	7.16	1.92	1342	162	4.62	210
Barnyard millet	65.5	6.2	2.2	1284	20	5	280
Maize	64.77	8.80	3.77	1398	8.91	2.49	279
Amaranth	61.46	13.27	5.56	1489	162	8.02	412
Soybean	12.79	35.58	19.82	1596	239	8.29	494
Horse gram	57.24	21.73	0.62	1379	269	8.76	298
Kidney bean	48.61	19.91	1.77	1252	126	6.13	409
Sesame	10.29	19.17	43.10	2124	1664	13.9	568
Black gram	43.99	21.97	1.58	1219	86.18	5.97	345
Cowpea	53.77	21.25	1.14	1340	84.10	5.04	378

Source: National Institute of Nutrition (2017)

sometimes went to an even more significant number of grains, as per the availability is worth mentioning. *Barahanaj* is an age-old seed conservation and cultivation practice that ensures the availability of grains and pulses throughout the year (Munir and Adhikari 2003). It is a mixed farming system wherein farmers cultivate 12 different types of grains, pulses or oilseeds crops on the same land. Major crops grown in the system include various types of Soybeans (locally called *bhat*), amaranth (*ram dana* or *chaulai*), kidney bean (*rajma*), Millets, corn, green gram, horse gram, black gram, sesame, buckwheat (*ugal*), black-eyed peas (*lobia*), etc. to name a few. In the system, after the harvesting of 12 grains, fields are kept fallow for some time, and afterwards, rice or some millet is grown, followed by wheat, and the cycle is repeated to ensure crop rotation.

The main principle behind the system is to ensure food security, i.e., even under harsh weather conditions, people can always get something to eat by growing a mix of tolerant and resistant crops. Since the hilly region is mostly rain-fed in nature, farming of less input-intensive crops ensured that people would get food to eat in any weather condition. Apart from food security, nutritional security is also maintained by following the system with the inclusion of cereals, legumes, oilseeds, greens, etc. (Table 3).

Barahanaj is a sustainable farming system where all inputs are available from the household, i.e., seeds from the previous crop, manure from cows, and labour from the family. With a rich diversity of crops in the system, fewer diseases and pests exist. Along with providing food to the family, this system also fulfils the fodder requirement for the cattle by using the hay and straw left over after harvesting the crops. Including legume crops in the system makes the soil fertile for the following year's crop. Overall, this system ensures food security and improves soil fertility and livestock health, which in turn maintains the ecological balance and conserves the environment. The sad reality, though at present, is that the system is slowly vanishing from the hills due to migration to cities, replacement of traditional varieties with hybrids and monocropping being a few of the reasons. Given the growing challenges posed by climate change, such practices need to be sustained as adaptation and resilience mechanisms.

2.2.3 Nutri-Rich Native Agro-Biodiversity with Underutilized Flora

The Himalayas are a treasure trove of unique plants with valuable potential for food and medicine. The temperate, sub-temperate and sub-tropical climate allows a variety of flora to flourish in the mountains. Major plant species (Figs. 4, 5, and 6) with health benefits that are an essential part of ethnic foods in the state are briefly described below.

2.2.3.1 Major Grains, Millets and Pseudo Cereals

Grains like wheat, rice, and maize form the basis of Uttarakhand's staple food—red rice, which is high-altitude rice that used to be traditionally grown in the state. Millets also form an essential part of the cuisine in the state, especially Finger millet (Mandua) and Barnyard millet (Jhangora), along with foxtail millet and proso millet. Prominent pseudo cereals like Amaranth and Buckwheat are also an integral part of the food palate in the region where the seeds and leaves are consumed. Millets and red rice, being climate resilient, especially in low rainfall conditions, can pave their way back into the food system for long-term benefits.

2.2.3.2 Pulse Crops

Pulses form the quintessential part of the daily diet in the region and are savoured mainly with cooked rice. The state has various pulses, including black soybean, horse gram, black gram, kidney bean, black-eyed peas, and standard beans. The community has consumed various delicacies in different forms for ages but is now losing its identity.

2.2.3.3 Vegetables

People in the hills fondly consume a diverse range of leafy vegetables (spinach, radish leaves, mustard leaves, Stinging nettle leaves, colocasia leaves, amaranth, fenugreek, buckwheat, chenopodium, etc.); Tubers {*pinalu/gaderi* (Colocasia spp.), yams like *gethi/chariyal/tarud* (Dioscorea spp.), potato, etc.}; Cucurbits (Cucumber, long melon or *kakri*, Pumpkin, bottle gourd, bitter gourd, *ram karela*, ash gourd, snake gourd, chow chow, etc.) and Wild edible plants (*Linguda, kairua, kwairal, timil*, mushroom, nettle, etc.). It is made sure that at least one meal must have vegetables as a key ingredient daily.

2.2.3.4 Other Herbs and Spices

Bhaang (hemp seeds), *bhangeera* (perilla seeds), *jakhia* (wild mustard), *gandhrein (Angelica glauca)*, ginger, turmeric, onion, garlic, *jamboo (Allium spp)*, shallots, etc. play a crucial role in enriching the flavour of traditional foods besides adding extra nutrition.

Fig. 4 Crop diversity in Traditional Pahari Food (Source: Authors)

Fig. 4 (continued)

Fig. 5 Diversity in Traditional *Pahari* Food (Source: Authors)

Fig. 6 Unique spices in Traditional *Pahari* Food (Source: Authors)

2.3 Traditional Food Systems and Climate Resilience

Traditional food systems have an inherent indigenous knowledge base that makes the cuisine valuable not just from the culinary point of view but also from nutritional and climate resilience. Table 4 explains some of the existing pathways and practices, and modern agriculture and food systems techniques that can help in climate resilience.

2.3.1 Ethnic Foods and Climate Adaptation

Taste with health has always been the main principle behind the foods in the hills. What makes the cuisine unique, however, is the array of unusual flavours from wild *Pahari* (belonging to the mountains) herbs and seeds like the *jakhiya, faran, chippi, bhang* seeds, *gandhreni* (a fragrant bark) and *jamboo*, used for tempering lentils and curries. These traditional foods have been time-tested for generations to make them specific to the region's climate; thus, specific dietary practices and recipes are consumed according to season. Certain whole coarse pulse crops such as Horsegram (Gahat), Kidney beans, black soybean (Bhatt), cowpea, etc., are consumed more in the winter in the form of boiled dahl as they provide warmth to the body. However,

Table 4 Existing and potential climate resilience mechanisms and pathways in Uttarakhand

Crop diversification	Traditional systems of mixed farming or Barahanaj (12 grain) practice can help ensure food security, i.e., even under harsh weather conditions, people can always get something to eat by growing a mix of crops
Animal Energy	Even with access to power tillers, draught animals are still in use for their crucial role in farming operations in the community's uneven, hilly terrain and socio-cultural dimension
Neglected and Underutilized Crops (NUCs)	The Himalayan region is blessed with a rich diversity of edible plant species that are still neglected and underutilized, e.g. Stinging nettle, hemp seeds, millets, tubers, leafy greens, etc.
Traditional processing techniques	Himalayan community has been using traditional processing techniques in the form of sun-drying, pickling, and fermenting for ages to adapt to climate
Adaptive varieties	Development of climate adaptive varieties suitable for low chilling (location-specific spur varieties), tolerant varieties (drought, heavy rains), disease and pest resistance, high-yielding
Carbon sequestration	Horticultural crops, mainly fruit tree crops, can play a crucial role in carbon sequestration and potentially in sustaining livelihood through carbon trading in future for the farming community (Malhotra 2017)
Value addition	Many fruit crops get damaged due to weather vagaries like unseasonal rainfall, hail storms, snow etc. Processing and value addition of such fruits into jams, jellies, spreads, and juices can help in managing post-harvest losses and also add livelihood opportunities
Rejuvenation of natural springs	The Himalayan region is home to many natural springs that can be rejuvenated with community participation in spring shed and watershed project interventions

Source: Compiled by authors based on field survey

similar pulses like horse gram, black gram, and black soybean are consumed in the form of Dubke (a grinded and cooked paste) in the summer as it is considered cooler. Ghee or clarified butter, jaggery, ginger and garlic, and whole spices are more prominent in winter. Mandua or finger millet-based chapatis or porridge is consumed more in the winter, along with leafy greens like stinging nettle. Wheat-based chapatis or porridge and rice are consumed more during the summer season. Tubers like Colocasia (Pinalu and gaderi) and Dioscorea (Chariyal or gethi, Tarun) are consumed mainly in the winters with added hemp seeds or horsegram to provide warmth to the body. There have not been significant changes in the traditional food system concerning climate change. Still, there have been more substantial changes with the advancements in agricultural systems, cropping patterns, and more access and availability of market and processed products. However, in a changing climate, many traditional food practices are essential regarding the resilient nature of conventional crops. Table 5 explains the primary traditional foods and practices for health benefits, uniqueness and weather patterns.

2.3.2 A Brief Account of the Traditional or Ethnic Foods of Uttarakhand (Figs. 7, 8, 9, 10)

a. Common Traditional *Pahari* Foods

Traditional chapatis/roti (flatbreads) and bhat (steamed rice) Unlike *rotis, which are* made of wheat flour, earlier *rotis* were made using different grains, such as millet, rice, pseudo cereals, etc. Common preparations include *mandue ki roti* (ragi flatbread), *rot* (wheat flour, jaggery, and sesame seeds added flatbread usually made in temples), *poori* (small-sized *roti* but deep-fried in oil), *chaulaiki roti* (amaranth flatbread), *chawal ki roti* (rice flatbread), *choyi roti* (wheat flour and jaggery pancakes), *gahat roti* (horse gram stuffed flatbread), etc. to name a few that provided nutritional benefits. Apart from *rotis*, rice is an integral part of the diet of mountain people. Different varieties of rice were used in earlier times, mainly the sticky white rice and red rice, which is now rarely cultivated and has significantly less availability. However, in the context of growing climate change and nutritional insecurity, traditional food systems can play a crucial role in addressing the challenges.
Dubuk/Dubke *Dubke* is a signature dish of the state prepared by slow cooking any soaked grounded pulse such as black soybean, gram, horse gram, etc. Ingredients: Black soybean/Horsegram/Gram, Asafoetida, Cumin seeds, *Jambu*, Coriander leaves, Salt, Ghee (clarified butter), Spices
Chudkani *Chudkani* is simply a pulse curry made with black/brown soybean or horse gram. It's called *chudkani* because of the sound that comes '*chudak*' while roasting the pulses. Ingredients: Soybean or horse gram, asafoetida, wheat flour, mustard oil, cumin, green chillies, crushed ginger and garlic, coriander leaves, salt, and spices Health benefits: It's very rich in protein and iron content. If served with added lemon juice, its health benefits are further increased.

(continued)

Rasbhat

A nutritious clear soup made out of various Himalayan pulses is a very nutritious and light food that everyone cherishes. It is usually served along with steamed rice for lunch.
Ingredients: Horse gram, Black soybean, Black gram, Rajma, Rice, Spices, crushed ginger and garlic, Asafoetida, *Jambu*, Coriander leaves, Salt, Ghee
Health benefits: Very soothing clear soup for winter and a rich iron source. Leftover pulses provide a handful of proteins as well

Khichdi

Khichdi is one of the healthiest food items, is easily prepared, and is light on the stomach. It is usually given when someone is sick or in a hurry to prepare several food items for lunch. It is a one-pot dish that requires the cooking of rice and lentils together along with *ghee* and little spices.
Ingredients: Rice, any lentils of your choice (Green gram, horse gram, pink lentils, etc.), *ghee* (clarified butter), cumin seeds, ginger and garlic (optional), onion (optional), green chillies, asafoetida, salt and spices (minimum).

Jaula/Chasiya and Paliyo

Jaula and *Paliyo* are similar food items prepared with rice, barnyard millet, and buttermilk. *Paliyo* can also have a sweeter version using jaggery in place of salt.
Ingredients: Rice or barnyard millet, buttermilk, Pisi loon (salt grounded with garlic, green chillies, roasted cumin, coriander leaves),
Health benefits: It is considered one of the best remedies for stomach disorders as it is easily digestible and light on the stomach.

Kaapa/Kafuli

Kaapa is a slow-cooked curry made from the paste of boiled leafy greens of spinach, colocasia or mustard leaves. It is very delicious, along with rice.
Ingredients: Leafy greens, cumin seeds, garlic, ginger, oil, salt, spices, rice flour
Health benefits: Curry is highly nutritious, rich in iron and easily digestible. Having this curry is beneficial for the eyes and general health.

Jholi

Jholi is a very popular curd-based curry eaten with steamed rice and is light on the stomach.
Ingredients: *Besan* (gram flour) or wheat flour, whisked curd, onion, green chillies, radish (optional), mint (optional), mustard oil, coriander seeds, fenugreek seeds, salt, and spices.

Shishunak saag/bichoo ki sabzi

Stinging nettle is a wonder plant with innumerable medicinal uses. The leaves are also savoured in the form of tea in many countries, but here, they are mainly consumed as vegetable curry.
Ingredients: Nettle leaves (as per requirement), Onion (1–2), Garlic (6–7 cloves), crushed ginger (1 tablespoon), mustard oil (one tablespoon), salt and spices as per taste.
Health benefits: The plant provides a warm and heating effect on the body, so it is mainly consumed during winter. It is also suitable for curing cough and gastric problems. Stinging Nettle is a wonder plant with immense benefits ranging from curing reproductive disorders in women, rheumatism, joint pain, asthma, and TB.

Bhaang-gaderi ki sabzi

Delicious curry made from large colocasia yam and hemp seeds is one of the savoured traditional recipes taken along with chapatti and rice.
Ingredients: *Gaderi* (Colocasia yam), onion, green chillies, garlic, ginger, salt, hemp seeds, spices, oil, cumin seeds, and coriander leaves.
Health benefits: A good combination of carbs, protein and essential fatty acids provides warmth in cold winters.

Lingude ka saag

Lingude or fiddlehead ferns are a delicacy eaten throughout the Himalayan belt. These are not cultivated but collected from the nearby uncultivated forest area. In some parts of the Himalayan belt, its pickle is also prepared to make it available in the off-season.
Ingredients: *Lingude* (one bunch), Onion, Garlic and ginger (more than usual), mustard oil, green chillies, salt, spices, coriander or cumin seeds.

(continued)

Dried/dehydrated vegetables:
Since Himalayan regions go from a climate that is warm to extremely cold throughout the year, it is essential to avail different vegetables and food options during the harsh climate. Usually, in older times, when people couldn't go too far off places to purchase vegetables, they used to store them by sun-drying. During the seasons when the vegetables were cultivated or in excess, this practice helped utilize them in the off-season or when there was a food shortage. These sun-dried preparations are called *badi,* a mix of any choice of vegetable (usually cucurbits), black gram, and some salt. Sometimes, roughly chopped vegetables were sun-dried alone without any combination, and these are known as *khwair*. These dried preparations were then used later for making curries. It is a very sustainable method to utilize the excess harvest of vegetables and use them in lean periods.

b. Popular Snacks and Other Side Dishes

Patyode
It is one of the popular snacks item eaten in the hilly areas but also requires time and precision to prepare
Ingredients: Tender colocasia leaves, besan, salt, cumin seeds, asafoetida, mustard oil

Aloo/pinalu ke gutke with bhaang ki chutney
Gutke is a common snack item and is just stir-fried boiled potatoes and colocasia tubers with some spices. It is served along with the popular and delicious *bhaang* (hemp seeds) ki chutney.

Kumaoni raita
It is one of the popular side dishes along with chutneys and is made from curd and cucumber or sometimes other vegetables (radish, raw papaya, chenopodium leaves) with mustard paste
Ingredients: Curd, grated cucumber or another vegetable of choice, salt, chopped green chillies, mustard seeds, turmeric and chilli powder (optional)

Bade and puye
Bade and *puye* are salt and sweet fried preparations respectively, made during religious festivals or occasions. *Bade* are ring-shaped fritters made from frying the batter (made from soaked and grounded black gram with asafoetida and added salt and spices).
Puye are also fried snacks made from a sweet batter (semolina soaked in milk along with mashed banana, sugar, dry fruits and cardamom powder) and fried in *ghee* in dumplings.

Jhangore ki kheer
It's a sweet pudding made from *jhangora* (barnyard millet) and milk.
Ingredients: *Jhangora*, milk, sugar, cardamom, *ghee*, dry fruits of choice

Sana neembu
One of the most relished items eaten during winter while sitting in the sunshine. It is very delicious and considered very useful in curing cold.
Ingredients: Big lemon, Radish (optional), Malta or Himalayan sweet orange (optional), Curd, Sugar and Grounded hemp seed salt or *pisilun* (hemp seeds, garlic, cumin seeds, salt, green chillies, coriander leaves). Good for curing running nose and cold.

c. Unique Ethnic Cuisines from the Extreme Cold Regions

The upper Himalayan region of the state goes through extreme winters for 6 months in a year; hence, the people migrate to the lower valleys during those months. In these villages, most cuisine has been adapted to keep the body warm during harsh winters. Also, it has many similarities with north-east Indian, Nepalese, and Tibetan cuisine. A few noteworthy dishes are briefly explained.

Sile kuto with dungcha *Sile kuto* is a pancake made from mandua flour, while *dungcha* is a dry chutney. Ingredients: *mandua* flour, water, *ghee*, salt, garlic, ginger, green chillies, *timur* seeds, and cumin seeds. Health benefits: When travelling in cold winters, mandua flour and ghee provide energy and warmth to the body, and garlic-added salt offers immunity.
***Marja*:** Popular by different names in various Himalayan states in India, *Marja* is a salt-added tea cherished by people in temperate regions. Ingredients: Water, salt, bark of *Taxus baccata*, milk, *ghee*, roasted millet powder, black pepper, walnut kernel powder (optional) Health benefits: Consumption of this tea helps rejuvenate the body after farm work in the fields and also provides warmth and immunity to the body. It also helps cure motion sickness that occurs when travelling in hilly terrain.
Sundarkala/Champion: This is a locally made noodles recipe that is quite popular among the folks and eaten as a snack item. Ingredients: Wheat flour, oil, turmeric, *dungcha, jamboo*, garlic
Bhutwa (intestine sausage): It is one of the rarest non-vegetable food items eaten in temperate regions of the state. It is mainly consumed while travelling to distant areas as cooking vegetables and curry is not feasible. It is considered very good in winter. Ingredients: Whole goat, *mandua* flour, salt, spices, ginger, garlic, *ghee*
***Chartsha*:** It is similar to meat jerky eaten around the globe. Since vegetables and pulses were not readily available in the earlier times in all seasons and people used to travel long distances, thin meat pieces were dried along wires. After drying, the meat pieces become very hard and shrink; one can take them along while travelling.
Chhurpi/Chirbi: It is a traditional form of fermented hardened cheese consumed in the Himalayan regions of India, Nepal, and Tibet. Ingredients: Milk (cow or yak), cloth Health benefits: It is considered a rich protein source and can be stored for a few years.

The Himalayan region has a lot more variety of foods that keep changing as we move from one area to another. The foods mentioned in this paper are just some of the popular ones, but there are still several ethnic foods on the verge of disappearing and need to be documented. However, it's important to note that the community is still not fully prepared to adapt to hotter temperatures regarding the food system. Most traditional foods cater to the colder climate and adapt less to the increasingly warmer temperature. In association with other factors like agricultural and technological advancements, urbanization and modernization, and climate change may affect the relationships of indigenous communities with traditional foods, primarily the access, availability, crop management practices, and ability to store, process, market and use foods in conventional ways (Lynn et al. 2013). Climate change can impact the ethnic food systems in complex ways. Changes in crop phenology can

Table 5 Traditional Pahari food and practices

Traditional foods and practices	Uniqueness	Health benefits	Remarks
More use of coarse and large pulse groups such as Horsegram, Masoor, soybean, Mix pulses, etc.	Mix pulses or 'Khadidhal' is a unique mix of Kidney beans (Rajma), Black-eyed peas (Lobia), Horsegram, Black soybean (Bhatt), Black gram (Urd), etc.	It is highly rich in complex carbohydrates and protein source as well. Horsegram provides warmth to the body and is also a natural cure for the treatment of kidney stone	Garlic and ginger are used extensively in the winter to provide heat to the body, in addition to whole spices. Asafoetida (Hing) or Gandhrein (Angelica glauca) are always added in any preparation with pulses to aid digestion
Finger millet (Mandua) based foods, mainly flatbreads (Chapathi) or porridge (halwa)	Mandua is climate resilient, especially in drought conditions	Mandua is rich in iron and Calcium. Porridge prepared with jaggery can help the body gain warmth in the cold weather	Mandua or Finger millet used to be a staple at one point, but the cultivation has reduced over time
Leafy greens, especially Stinging Nettle (Bicchhoo), Chenopodium (Bathua), Fenugreek (Methi)	The majority of leafy greens are available in the winter season. Stinging nettle is a unique wild leafy green with spines on the leaves and is consumed only in winter	Urticadioca, or stinging nettle, is a rich source of vitamins and minerals and has medicinal benefits. Methi is rich in fibre and anti-oxidants	Nettle and fenugreek leaves help produce heat and stabilize the body temperature in cold winters
Khichadi over plain cooked rice	Khichadi is a unique comfort food wherein rice and lentils are cooked together with spices	It is a good source of complex carbohydrates and proteins and easily digestible	Mostly paired with Ghee (Clarified butter) for warmth and salad and pickle as a side dish
Meat and eggs	Goat and Sheep meat, eggs and chicken are consumed more in winter with whole spices	Rich source of protein	Meat and eggs are preferred to provide heat to the body in the cold winter season
Jaggery over sugar	Jaggery with various combinations like tea, groundnut, ragi or wheat porridge is preferred	Jaggery gives a unique flavour and also extra iron than regular sugar	Consumption of jaggery keeps the body warm
Bhaang or Hemp seeds and Timur or Indian Shichuan pepper	Hemp seeds are an integral part of Uttarakhand cuisine in the form of chutney or added with Colocasia curry	Hemp seeds are rich sources of fibre, proteins, Omega-3 and 6 fatty acids, vitamin E and minerals. Timur adds heat to the body in harsh winters	Adding hemp seeds to the diet, such as roasted powder, chutney, or curry, helps gain nutrition and keep the body warm in cold weather
Tea with spices	Tea, ginger, and other whole spices are preferred for winter. In extremely cold regions, Marja, a salt-added unique tea, is consumed mainly in winters	Ginger and pepper-based tea helps cure cough and cold. Tulsi (Basil) leaves, and cloves are also added sometimes	In winter, many people prefer to consume tea with a jaggery piece
Dehydrated vegetables (Badi and Khwair)	*Badi* is a mix of any choice of vegetable (usually cucurbits) and soaked and grounded black gram along with some salt, which is sun-dried. Sometimes, roughly chopped vegetables were sun-dried alone without any combination known as *khwair*	A rich source of vegetables and legumes in lean season in different flavours	It is a sustainable method to utilize the excess harvest of vegetables and use them in lean periods

(continued)

Table 5 (continued)

Traditional foods and practices	Uniqueness	Health benefits	Remarks
Laddoos	Laddoos are a typical dessert in India, and in the hills, it's mainly made from sesame seeds, amaranth seeds and flour with jaggery and dried coconut	Many of the seeds are a rich source of good fatty acids and minerals	Adding jaggery with sesame seeds helps keep the body warmer in winter
Spiced grounded salt (Pisilyun)	A unique salt mix usually prepared with salt, cumin seeds, garlic, coriander leaves, and green chillies with added hemp seeds (in winter)	Grounded salt helps enhance the flavour of curd, fruits and raw vegetables and increases their palatability	In the cold season, consumption of such salts helps as a reminder to drink enough water as it increases thirst
Curd or butter milk-based preparations	Jholi is a savoury dish made with buttermilk and chickpea flour. Chasiya is another simple rice and butter milk-based dish consumed in summer. Buttermilk is also consumed mainly in the summers	A simple recipe is also a good source of protein through chickpea flour and is easily digestible	Warmer season requires light foods that can be easily digested
Cooler items like cucumber, juices, buttermilk	*Pahari* Kakri is a large cucumber cultivated in the region that the community relishes. Buransh flower juice malta (Sweet orange) juices are also relished in a warmer climate	Juices are rich sources of Vitamins and anti-oxidants	Cool juices help relieve the heat stress in higher temperature
Tubers like Colocasia (Gaderi and Pinalu), Dioscorea (Chariyal or gethi, Tarun)	Tubers in the state are delicious and softer in texture and are cooked solely with horse gram or hemp seeds	Tubers provide high calorific value, while the addition of horse gram and hemp seeds enrich the food with nutrients	Special care is given to cooking tubers by boiling water to reduce the effect of oxalates

Source: Compiled by authors based on field survey

impact various socio-cultural and religious practices of the community as well. Many traditional festivals are associated mainly with sowing or harvesting crops in the region; climate changes can have an overarching effect on the cultural values associated with the community due to the unavailability of traditional crops. Besides, with changes like these, all the actors in the food chain, from conventional varieties, pollinators and other insects, and human beings, all get affected, leading to more complexity in the food web and the agroecosystem.

3 Conclusion

Foods in the Indian Himalayan belt have a uniqueness in terms of using lesser-known nutrient-rich flora and exclusive taste. With rapid urbanization and migration to cities, ethnic foods, although loaded with nutrition, are slowly disappearing due to a lack of awareness among the current generation and documentation. Climate

Fig. 7 Traditional *Pahari* Foods (Source: Authors)

Fig. 8 Traditional *Pahari* Foods (Source: Authors)

Fig. 9 Traditional *Pahari* Foods (Source: Authors)

Fig. 10 Traditional *Pahari* foods (Source: Authors)

change is a troubling element that significantly influences the agriculture and food systems—identifying the traditional food systems, wild edible plants, food diversities, and value additions in conventional foods to make pathways is necessary for food and nutritional security in the context of growing climate variability. It is crucial to shed light on these foods prepared using the abundant nutrition-rich Himalayan flora. Traditional wisdom on local crops, wild edible plants and ethnic foods needs to be conserved to maintain the mountains' sustainability. Consumption of these ethnic foods that use whole grains, millet, and pulses can help improve the present generation's health, coping with numerous lifestyle diseases. It becomes imperative to record, conserve and revive the local wisdom of the mountain community and amalgamate it with improved scientific methods. Research needs to be done on studying the nutritional status of this flora and its promotion for conservation in augmenting its role in climate resilience.

References

Bhatt S, Purohit MV, Balodi KN, Kumar U, Sharma A, Shridhar V, Arunachalam A, Arunachalam K (2023) The ethnic food system of Jaad Bhotiya tribal community in the Gangotri landscape of western Himalaya: a transition in traditional practices. Altern An Int J Indig Peoples 19(2):250–260. https://doi.org/10.1177/11771801231173396

Census of India (2011) Uttarakhand-Series 06-Part XII A. District Census Handbook, Dehradun. dh_2011_0501_part_a_dchb_uttarkashi.pdf

Dahal KR, Dahal P, Adhikari RK, Naukkarinen V, Panday D, Bista N, Helenius J, Marambe B (2023) Climate change impacts and adaptation in a hill farming system of the Himalayan region: climatic trends, farmers' perceptions and practices. Climate 11(1). https://doi.org/10.3390/cli11010011

Dimri AP, Chevuturi A, Niyogi D, Thayyen RJ, Ray K, Tripathi SN, Pandey AK, Mohanty UC (2017) Cloudbursts in Indian Himalayas: a review. Earth-Sci Rev 168:1–23. https://doi.org/10.1016/j.earscirev.2017.03.006

Government of Uttarakhand (2022) Uttarakhand at a glance. Directorate of Economics and Statistics, Department of Planning, Dehradun, Uttarakhand. https://des.uk.gov.in/

Holleman C, Rembold F, Crespo O, Conti V (2020) The impact of climate variability and extremes on agriculture and food security – an analysis of the evidence and case studies. Background paper for The State of Food Security and Nutrition in the World 2018. FAO Agricultural Development Economics Technical Study No. 4. Rome, FAO. https://doi.org/10.4060/cb2415en

IPCC (2018) Annex I: Glossary [Matthews, J.B.R. (ed.)]. In: Global Warming of 1.5°C. An IPCC Special Report on the impacts of global warming of 1.5°C above pre-industrial levels and related global greenhouse gas emission pathways, in the context of strengthening the international response to the threat of climate change, sustainable development, and efforts to eradicate poverty [Masson-Delmotte, V., P. Zhai, H.-O. Pörtner, D. Roberts, J. Skea, P.R. Shukla, A. Pirani, W. Moufouma-Okia, C. Péan, R. Pidcock, S. Connors, J.B.R. Matthews, Y. Chen, X. Zhou, M.I. Gomis, E. Lonnoy, T. Maycock, M. Tignor, and T. Waterfield (eds.)]. Cambridge University Press, Cambridge, UK and New York, NY, USA, pp 541–562. https://doi.org/10.1017/9781009157940.008

Joshi B (2018) Recent trends of rural out-migration and its socio-economic and environmental impacts in Uttarakhand Himalaya. J Urban Reg Stud Contemp India 4(2):1–14

Keleman Saxena A, Cadima Fuentes X, Gonzales Herbas R, Humphries DL (2016) Indigenous food systems and climate change: impacts of climatic shifts on the production and processing of native and traditional crops in the Bolivian Andes. Front Public Heal 4:20. https://doi.org/10.3389/fpubh.2016.00020

Khanduri S (2020) Cloudbursts over Indian sub-continent of Uttarakhand Himalaya: a traditional habitation input from Bansoli, District-Chamoli, India. Int J Earth Sci Knowl Appl 2(2):48–63

Kumar A, Asthana AKL, Priyanka RS, Jayangondaperumal R, Gupta AK, Bhakuni SS (2017) Assessment of landslide hazards induced by extreme rainfall event in Jammu and Kashmir Himalaya, Northwest India. Geomorphology 284:72–87. https://doi.org/10.1016/j.geomorph.2017.01.003

Kumar P, Patel A, Rai J, Kumar P (2023) Environmental challenges and concurrent trend of weather extremes over Uttarakhand Himalaya. https://doi.org/10.21203/rs.3.rs-2509943/v1

Lynn K, Daigle J, Hoffman J, Lake F, Michelle N, Ranco D, Viles C, Voggesser G, Williams P (2013) The impacts of climate change on tribal traditional foods. Clim Chang 120(3):545–556. https://doi.org/10.1007/s10584-013-0736-1

Malhotra S (2017) Horticultural crops and climate change: a review. Indian J Agric Sci 87:12–22. https://doi.org/10.56093/ijas.v87i1.67138

Ministry of Environment, Forest and Climate Change (MoEFCC) (2019) Government of India. India State of Forest Report 2019, Forest Survey of India, Edition 16

Mukerjee P, Sogani R, Gurung N, Rastogi A, Swiderska K (2018) Smallholder farming systems in the Indian Himalayas: key trends and innovations for resilience. IIED Country Report. IIED, London. https://www.iied.org/sites/default/files/pdfs/migrate/17618IIED.pdf

Munir S, Adhikari K (eds) (2003) Globalisation and mountain farmers: tapping opportunities and mitigating threats. South Asia Watch on Trade, Economics & Environment. https://sawtee.org/publications/Book-3.pdf

National Institute of Nutrition (2017) Indian food composition tables. ICMR-Department of Health Research. Ministry of Health & Family Welfare, Government of India

Nautiyal P, Bhaskar R, Papnai G, Joshi N, Supyal V (2020) Impact of climate change on apple phenology and adaptability of Anna variety (low chilling cultivar) in lower hills of Uttarakhand. Int J Curr Microbiol Appl Sci 9:453–460. https://doi.org/10.20546/ijcmas.2020.909.057

Owino V, Kumwenda C, Ekesa B, Parker M, Ewoldt L, Roos N, Lee W, Tome D (2022) The impact of climate change on food systems, diet quality, nutrition, and health outcomes: a narrative review. Front Clim 4:941842. https://doi.org/10.3389/fclim.2022.941842

Palni LMS, Rawal RS (2010) Conservation of Himalayan bioresources: an ecological, economic and evolutionary perspective. In: Sharma VP (ed) Nature at work: ongoing Saga of evolution. Springer, New Delhi. https://doi.org/10.1007/978-81-8489-992-4_23

Pandey DK, Adhiguru P, Momin KC, Kumar P (2022) Agrobiodiversity and agroecological practices in 'jhumscape' of the Eastern Himalayas: don't throw the baby out with the bathwater. Biodivers Conserv 31(10):2349–2372. https://doi.org/10.1007/s10531-022-02440-7

Pepin N, Bradley RS, Diaz HF, Baraer M, Caceres EB, Forsythe N, Fowler H, Greenwood G, Hashmi MZ, Liu XD, Miller JR, Ning L, Ohmura A, Palazzi E, Rangwala I, Schöner W, Severskiy I, Shahgedanova M, Wang MB, Williamson SN, Yang DQ, Group MRIEDWW (2015) Elevation-dependent warming in mountain regions of the world. Nat Clim Chang 5(5):424–430. https://doi.org/10.1038/nclimate2563

Rai R (2015) Implications of changing climate on productivity of temperate fruit crops with special reference to apple. J Hortic 02. https://doi.org/10.4172/2376-0354.1000135

Rana S, Rawal R, Dangwal B, Bhatt I, Price T (2021) 200 years of research on Himalayan biodiversity: trends, gaps, and policy implications. Front Ecol Evol 8:603422. https://doi.org/10.3389/fevo.2020.603422

Rasul G (2010) The role of the Himalayan mountain systems in food security and agricultural sustainability in South Asia. Int J Rural Manag 6:95–116. https://doi.org/10.1177/097300521100600105

Raúl Matta (2015) Edible identities: food as cultural heritage by Ronda L. Brulotte and Michael A. Di Giovine (eds) Anthropology of food [Online], Book reviews, connection on 09 September 2023. http://journals.openedition.org/aof/7824. https://doi.org/10.4000/aof.7824

Rautela P, Karki B (2015) Impact of climate change on life and livelihood of indigenous people of higher Himalaya in Uttarakhand, India. Am J Environ Prot 3:112–124. https://doi.org/10.12691/env-3-4-2

Sabin TP, Krishnan R, Vellore R, Priya P, Borgaonkar HP, Singh BB, Sagar A (2020) Climate change over the Himalayas. In: Krishnan R, Sanjay J, Gnanaseelan C, Mujumdar M, Kulkarni A, Chakraborty S (eds) Assessment of climate change over the Indian region. Springer, Singapore. https://doi.org/10.1007/978-981-15-4327-2_11

Sati VP, Kumar S (2023) Declining agriculture in Garhwal Himalaya: major drivers and implications. Cogent Soc Sci 9(1):2167571. https://doi.org/10.1080/23311886.2023.2167571

Shukla R, Agarwal A, Gornott C, Sachdeva K, Joshi PK (2019) Farmer typology to understand differentiated climate change adaptation in Himalaya. Sci Rep 9(1):20375. https://doi.org/10.1038/s41598-019-56931-9

Singh RD, Gumber S, Tewari P, Singh SP (2016) Nature of forest fires in Uttarakhand: frequency, size and seasonal patterns in relation to pre-monsoonal environment. Curr Sci 111(2):398–403

Tewari VP, Verma RK, von Gadow K (2017) Climate change effects in the Western Himalayan ecosystems of India: evidence and strategies. For Ecosyst 4(1):13. https://doi.org/10.1186/s40663-017-0100-4

Thadani R, Singh V, Chauhan D, Dwivedi V, Pandey A (2015) Climate hange in Uttarakhand: current state of knowledge and way forward. https://www.cedarhimalaya.org/pdf/Climate-Change-in-Uttarakhand.pdf

Trichopoulou A, Soukara S, Vasilopoulou E (2007) Traditional foods: a science and society perspective. Trends Food Sci Technol 18:420–427. https://doi.org/10.1016/j.tifs.2007.03.007

Bearing the Brunt of Climate Change in Eastern Himalayas: Conserving the Traditional Food Systems of Nagaland, India

Noklenyangla

1 Introduction

Before we look into the core argument of this paper, let us briefly discuss about the physical features of Nagaland. Nagaland lies in the northeast region of India and is one of the Himalayan states. The extension of Himalayas in Northeast India is known as Purvanchal. Furthermore, the mountain system of Nagaland can be divided into four distinct ranges namely, the low mountain system, the middle mountain system, the Patkai mountain system and Barail mountain system. The low mountain system runs along the Assam-Nagaland border and it comprises of rather low hills of altitude less than 1000 m. The middle mountain range forms the most dominant mountain system and runs north to south though the middle of the state. It consists of many ranges and some lateral offshoots which run in an east-west direction. Likewise, the Patkai mountain system is the longest mountain range in the state and forms the international boundary between India and Myanmar (erstwhile Burma). Lastly, Barail mountain range runs from east to west direction and becomes lower from east and west. When it comes to climate of Nagaland, it is blessed with subtropical type of climate. The climate of Nagaland can be divided into four distinct seasons, winter season, pre monsoon, monsoon season, and retreating monsoon. Annually, it receives a rainfall of 200–250 cm. The average temperature is 22–27°C and the region's climate is typical monsoon with deviations ranging from tropical to temperate conditions Sebu (2013). Thus, we can say, Nagaland is blessed with distinct physical features.

Coming to its flora and fauna variation, although, Nagaland is a small state; it is rich in flora and fauna. The fauna diversity is prosperous with rare birds and animals and has common animals like Asian elephant, Indian bison, jackal, tiger, sambar,

Noklenyangla (✉)
Independent Researcher, Nagaland, India

A. Borthakur, P. Singh (eds.), *The Resilience of Traditional Knowledge Systems for a Sustainable Future*, Advances in Asian Human-Environmental Research,
https://doi.org/10.1007/978-3-031-56858-9_7

leopard, barking deer, wild boar, sloth bear, serow, hoolock, macaque, common langur, leopard cat, pangolin, wolf, fruit bat, porcupine, cicada, mongoose, jungle cat, squirrel, parrot, hornbill to name a few. Also, the state has abundance of wild plants and fruits. Wild plants amount to a bulk of the diet for the Naga people, there are numerous edible plants which grow in wild and it outnumbers the home grown vegetables, for example, plants like, *Colocasia esculenta and Houttuynia cordata.* Wild fruits are not only found in abundance but are varied too in the forests of Nagaland; this includes wild grapes, apple, cherries, lemon, bananas, walnut, mangoes, mulberry, cherry and a host of indigenous fruits. These fruits come in all sizes, shapes, colours and tastes (Angami n.d.). Some fruits which have a high demand in the market for its health benefits are now domesticated in the state for commercial purposes, pineapple, kiwi, and dragon fruit are some of the examples boosting the economy of the state. Apart from vegetables and fruits, there are numerous flower and plants which are grown for the sole aesthetic purpose and which adds up to the growth of horticulture in the state.[1]

Nagaland is also known as the land of festivals, the land is blessed with a rich diversity of cultures as there are numerous Naga tribes and sub-tribes residing in the state. Like many indigenous cultures of the world, food forms an intrinsic part of Naga culture. Each tribe has their own festival commemorating sowing and harvest seasons, during those festival as part of celebration, they prepare dishes which holds significance to their tribe (Mao and Odyuo 2007). Hence, food forms an integral part of Naga cultural identity. In the succeeding paragraphs, we will see the traditional food system of Nagaland in detail.

1.1 Traditional Food System of the Nagas in Nagaland

Food system generally means web of activities from production to post-harvest storage, transportation to processing, distribution to trade and marketing, regulation to consumption of food and the outcomes of nutrition and health, environment and the socio-economy (Victor et al. 2022). Hence, we can see a complex set up under which the food system operates. On the other hand, traditional food system has been defined as, "*all food from a particular culture available from local resources and culturally accepted. It includes socio-cultural meaning, acquisition, processing techniques, use, composition and nutritional consequences for people using the food.*"[2] Going by definition we can assume that traditional food system means the method of food acquisition and consumption which has social and cultural significance and the usage of which is passed from generation to generation among a given community.

[1] Department of Horticulture. https://hortidept.nagaland.gov.in

[2] See Kuhnlein. Micronutrient nutrition and traditional food systems of indigenous peoples. https://www.fao.org/3/y8346/y8346m04.pdf

It has been noted, traditional food has vital nutritional, economic, spiritual and cultural value for First Nation People,[3] and for thousands of years these first nation peoples have relied on food harvesting strategies like fishing, hunting and gathering. However, it has been argued, due to the dietary transition, there has been decline in the consumption of traditional food which has led to the prevalence of nutrition-based diseases like obesity, Type 2 diabetes and cardiovascular diseases among the indigenous people (Batal et al. 2021). The importance of traditional food system has not only been looked from the importance it bestows to cultural heritage and health benefits but also its contribution in preserving the environmental health through traditional agricultural practices which regenerates biodiversity and thereby supports healthy ecosystem (Deaconu et al. 2021).

Food, cooking, and eating habits play a central role in every culture and Northeast India is not an exception. Northeast India is a home to numerous tribes; the tribal population dominates the region and each tribe has its own culture having their own specific eating habits. The traditional foods of this region are more or less defined. It is worth noting that the food habits of this region differ from the mainland India adding an exotic sense to the food menu of this region. The food menu comprises of rice which is mainstay food item, starchy vegetables, meats and poultry products. True, we live in a society as dominated by preferences including food preferences. Nevertheless, for various reasons there has been instances wherein the food habits of this region has been looked upon with a sceptical appraisal by the people outside the region, thus, creating a cultural hegemonic kind of scene (Zahan 2022).

The tribal women with their traditional knowledge and their practices of ethnic culinarians can be credited for adding to the sustainability of food and nutritional security of tribal societies and ecosystem (Domadia 2017). Every tribal community of Northeast India have their specific food habits based on the location and availability of the forest, crops resources, ecological variability etc. Each North-eastern states has a specific dish which sometimes is marked as the state's popular dish, for instance, *fish tanga* of Assam, *smoked pork* of Nagaland, *iromba* of Manipur, *khazi* of Arunachal Pradesh (Borthakur 2020).

Segmenting further, each Naga tribe in Nagaland has been credited for their special food ingredient and often has been identified as their area of speciality with regard to it, like, Ao Naga is known for their dried yum (anishi), Sumi Naga is known for fermented soya bean (axone), Kyong Naga is known for bamboo shoots.[4] Rice which is a rich source of carbohydrate is the staple food of Naga cuisine which is served with vegetable dishes, fish and meat. When we talk about beverages, tea and coffee were not known then, hence, drinks made from rice or millet, job tears, sorghum and barely were used as beverages in olden times (Neihu 2018). Even to this day, rice beer is widely popular in the state and is used during festivals or for sales. One of the food items, which is widely produced from the Eastern Nagaland

[3] First Nations refers to indigenous peoples, for ethnic groups who are the earliest known inhabitants of an area, as defined by *International Labour Organization*. https://www.ilo.org/global/topics/indegeous-tribal/WCMS_503321/lang-index.html.

[4] See Mao, A.A and Odyuo.

is kholar (beans) and is in high demand for its taste and nutrient value.[5] This diversity of food cuisines even among the Nagas enriches the traditional food system adding up to the diversity of the Naga culture.

Thus, like in every culture, food, cooking and eating habits plays a central role in defining the indigenous culture of Nagaland. We can see that Nagas have a unique traditional food system which centres around home grown food products and it has been for years. When the knowledge of modern day preservation was not known, Nagas relied on their indigenous method of preserving the food; the process of fermentation was one such method and to this date, it fermented food items forms an integral part of food system in Nagaland.[6]

Although traditional food system is still intact among the Nagas, but one cannot ignore the wave of fast food franchise in Nagaland. Globalization which became the buzz word in the 90 s' brought about various transmissions even in the cultural sphere and food was not an exception. Over the last decades, it has been observed that commercialized packaged fast food items which are termed as 'junk foods' has acquired a place in the diet regime of the hill people of Nagaland. Market for such food items has seemingly grown in demand, so much so that apart from brand foods, there seems to be a local way of bringing about generic fast-food meals so much so that food chains like KFC, Domino's Pizza and Pizza Hut has found its own niche in the commercial towns of Nagaland, Dimapur and Kohima[7,8] The interesting part is the region had easily and amicably adopted the wave. The reasons leading to such can be mooted; however, it can be related to the numerous effects of rapid globalization which is felt in the region too. The rise in nature of consumerism has been spontaneous rather than gradual. In due time, the region has explored ways of canning and packaging its own ethnic dishes not only for the domestic market but also for the world market. The demand in the market for such may be small but, in a way, it signifies commercialization, hence, promotion of culture through culinary art can be looked as a positive impact as it leads to the employment generation of the state.[9] However, with growing awareness about the negative impact of fast foods, the wave of getting back to the traditional foods has been growing among the masses with some amount of caution as even some of the Naga traditional foods have been held partly responsible to be the leading cause of cancer.[10]

---

[5] Kholar-the economic pulse of the east. *The Morung Express*. http://morungexpress.com/kholar-economicpulse-east.

[6] See Mao, A.A and Odyuo.

[7] KFC opens its first store in Kohima. *The Morung Express*. http://morungexpress.com/kfc-open-its-first-store-kohima.

[8] Domino's Pizza opened in Kma. *Nagaland Post*. https://nagalandpost.com/index.php/domino-s-pizza-opened-in-kma/

[9] Nagaland's Rich Agri-Horti Sector: Transforming Northeast India's Food Processing Landscape. *Nagaland Tribune*. https://nagalandtribune.in/nagalands-rich-agri-horti-sector-transforming-northeast-indias-food-processing-landscape-2/

[10] Smoked meat: Oncology cancer gravest danger. *Nagaland Post*. https://nagalandpost.com/index.php/smoked-meat-oncology-cancer-gravest-danger/

1.2 Impact of Climate Change in the Traditional Food System of Nagaland

A report by Food and Agriculture Organisation highlights the impacts of climate change in the crop yields. It has been observed that there have been negative effects on wheat and maize yields in many regions and also globally. Leaving aside the major staple crops, there have been fewer studies on the effect of climate change in some food products like tubers, roots and pulses which provides not only nutritional value to our diet but also opens livelihood opportunities to many.[11]

Many fruits and nut trees that require winter chill may continue to decline. The rise in temperature which causes hot nights in many parts of the world damages crops, such phenomenon has been noticed in rice yields and its quality. Although, with the climate change there has been an increase in the CO2 level but crops like cassava seems to be benefitting out of it as it demands high temperature for growth and photosynthesis. Assessments on major crops through various methods and models has been made to determine the effect of climate change on the crop production and it shows a negative impact on the food production on the global front. In spite of the same levels of technology deployed for managing the crop productivity for maize, wheat and rice, there has been negative crop productivity in the low latitude areas. However, it is expected to be beneficial in the high latitude areas due to the warmer temperatures and prolonged growing seasons, which is generally not present in these areas but several other environmental conditions such as quality of the soil in the areas of far north may witnesses only a limited expansion. Keeping in view the difference between low and high latitude regions, there are chances of witnessing widening imbalances in the developing and developed countries, as it has been observed that there are latitudinal differences mainly in the elevation in regional and sub-regional scales.[12]

It is quite clear; the impact of climate change will depend on temperature, precipitation patterns and the increase in atmospheric CO2. It has been studied that climate change will increase variability in crop yields in many regions; this is mainly attributed to the significant increase in the level of CO2 in the atmosphere. There has also been growing worries about the interactions between CO2, nitrogen and high temperature; as such interaction might aggravate the situation concerning the crop yield and its variability. Research study have shown negative sensitivity of extreme temperatures on the crops and to understand the effects of temperature in a crop, one need to have the knowledge of optimum level of temperature required by a specific crop.[13]

---

[11] Climate change and food security: risks and responses. Food and Agriculture Organization of the United Nations (2015). www.fao.org/publications

[12] Ibid.

[13] Climate Impacts on Agriculture and Food Supply. *Climate Change Impacts*. United States Environmental Protection Agency. https://climatechange.chicago.gov/cilmate-impacts/climate-impacts-agriculture-and-food-supply

Apart from the crops, climate change affects the livestock production in numerous ways. This affect is seen through direct and indirect ways. The most important impacts can be witnessed in animal health and its productivity, feed crops and in biodiversity. It has been observed, the rise in temperature and decline in precipitation have direct negative outcomes on yield and also droughts caused by climate change have further led to the decline of the yields. Although, in the past years, the affect on the labour force due to climate change is yet to be quantified but the effects on the feed crops and grasslands for the animals has been examined to some extent. Nonetheless, climate constraints on livestock require assessments as it still lacks the required examination. This is because of various challenges in relation to modelling and quantifying the cumulative impacts due to uncertainties when it comes to modelling livestock productivity as there are variations when it comes to estimating the gazing and pastoral systems and also the grass species and carrying capacity are largely unknown though the studies on regional climate scenarios are becoming more available. Also, though it is clear that climate change affects animal diseases but the effects of it should be thoroughly studied to analyse the pattern of distribution. The availability of groundwater and the impacts of its unavailability should be examined especially in the grazing areas as it indicates the level of productivity in livestock regions. Last but not the least, the management should show accountability and it should be taken into consideration while assessing the impact of climate change on livestock production to have a comprehensive picture. The vulnerability of livestock relies on various factors like location of stock, its exposure to climate change, and requirements like housing system, water, and feedstock, type of breed and health of the animal among many other things. Moreover, there is an increase vulnerability in arid and semi-arid regions that consists of destruction of grazing areas, degradation of rangeland, alteration in land tenure which includes poor access to land and problems attached to it and also the market which covers animal products, by-products and crop residues. Thus, we see apart from the climate change which impacts livestock production there are various other factors which determines its vulnerability.[14]

The impact of climate change can also be witnessed in fisheries and aquaculture development in relation to freshwater and marine environment undermining its sustainability. These are results of changes in physical temperature, storm systems, atmospheric warming, ocean circulation and waves. Additionally, chemical changes which include acidification, thermal stratification, oxygen concentration, salinity content of the aquatic environment was responsible too. Cyclones, deep sea swells and high temperatures which are considered as events extreme in nature affect the ability of ecosystem like coral reef and mangroves to provide important functions for livelihood sustainability and food security. Studies have shown that the rapid 'tropicalization' of mid and high latitude systems is the caused by migration of various species of fish towards pole-ward. Based on researches, there is a forecast of large scale redistribution of global marine fish with an expected average increase of marine fish in the high latitude regions and a decline in the tropics. This, however,

[14] See Food and Agriculture Organization of the United Nations (2015).

is to be seen from the lens of predicted changes in the environmental conditions and also from the angle of habitat types and phytoplankton primary production.[15]

If we look into the impacts of climate change in the food production system in India then we see a picture which is not so positive. The problems of water scarcity have often been an issue of concern, the decline in the rainfall followed by droughts has resulted in drying up of wetlands causing degradation of ecosystem in many regions of the country. There is enormous pressure on the groundwater for irrigational purposes as crops like rice and wheat depend heavily on water resulting in decrease of groundwater level, adversely it will affect the production of crops which will add up to the burden of already existing food scarcity. The variation in the length of the crop growing season and undue occurrences of extreme climate events like droughts and floods no doubt effects the food production but it also affects the net income of the farmers who rely on agricultural productivity for their livelihood (Datta et al. 2022).

As discussed in the above paragraph, climate change will also have adverse effect on the fisheries and forest-dependent livelihood. The negative effect of climate change in India's food production system can lead to rise in hunger. Therefore, policy makers are putting forward ideas and measures to tackle the issue of food production and food security in the face of climate change. One of the ways, suggested is, 'transformative adaption', which is defined as '*promoting long term resilience by changing the fundamental attributes of a system in response to actual or expected climate change and its effects* (Preethan and Ginoya 2021).' However, for the transformative adaption to be adopted efficiently the support of the policy institutions and vulnerable community is important as for a policy to bear results an inclusive approach is necessary and in this case, the role of all the stakeholders is essential.

In recent times some of the scholars and climate change activists have been vocal about the impact of climate change in the traditional food system of the world. They have highlighted in the ways through which climate change hampers the traditional food system, mainly through changes in the fertility of the soil and the nutrient bioavailability which in turn affects the crops yield (Victor et al. 2022). There is a direct co-relation between the biodiversity of plants and animal species consumed and food security. Sadly, there is a loss of biodiversity due to habitat destruction caused by settlement patterns, deforestation, industrialization and also climate change factors like global warming. Indeed, climate change impacts food system and in doing so, it effects the global food production.[16]

However, there has been a going body of research which argues that the indigenous community might provide an alternative model to tackle the effects of climate change on the food system. Indigenous people's food system is considered to have higher resilience to ecological challenges due to the diversity in the system itself. The indigenous population in India makes less than 9% of the country's population, they are found to be residing in the areas which are the biodiversity hotspots of the

[15] Ibid.

[16] Ibid.

Indian subcontinent as well as rich in natural resources. These biodiversity hotspots are rich in agro-biodiversity, which includes the diversity and abundance of animals, plants and micro-organisms that are used either directly or indirectly for the purpose of food and agriculture and which are reared and nurtured by indigenous communities (Mawroh 2023).

It has also been noted that the territories inhabited by the indigenous communities are found to be in the regions which are marked as the place of origin of crops and crop diversity tracing back to 12,000 years ago where it was domesticated and, in the process, the seeds and breeds went through the process of selection, cross sharing and were passed on to the generations providing the genetic materials for the livestock and crops to continue with the adaption of various climatic changes including droughts, temperatures and other conditions. The food systems practised by the indigenous peoples have a higher resilience against diseases and pests, the possible reason being the diversity of flora and fauna from across various landscape areas adding to the rich breeds, species and varieties in this food system which is a contrast to the monoculture agricultural practices. It has also been noted that to a large extend the agricultural practices of indigenous communities have shown resilience and adaptiveness in the past and are continues to do in the conservation and adaptation of agricultural diversities and genetic resources. The contribution of varied food base results in flexibility of food systems especially in the face of impacts caused by various environmental issues.[17]

When we look into the indigenous food system of the Nagas, one should also keep in mind that Nagas are basically cultivators; they grow crops, which is their staple food. The crops commonly grown by them are rice, maize, millet and fruits such as pears, plums, oranges, lemon, pineapple and passion fruit etc. Fermented foods are part of their cuisine be it be yum, soya beans (axone) or fish or bamboo shoots, fermented drink like rice beer is also consumed by some of the population. The type of farming systems practiced in Nagaland is traditional jhum, terrace rice cultivation and wet rice cultivation.[18]

A survey conducted by the Nagaland Economic Survey (2016–2017) showed some interesting findings, rice which is the staple food which is cultivated in about 52.34% of the net area sown with rice production of 478,210 MT during 2015–16. The survey further highlighted, although there has been an increase in food gain production over the years but with the increase in population, per capita food grain availability has been declining. Hence, the survey suggested, state should on increasing productivity though sustainable technologies in all the existing farming system which may enable the state to attain a sustainable agricultural development. Additionally, the survey stated some key areas like research and development, crop protection measures, soil mapping and crop zonation, diversification, rural non-farm activities, reduced dependence on monsoon and crop insurance which needs to be focused on so as to achieve self-sufficient and sustainability in agriculture. The survey also pointed out that although agriculture still remains as one of the largest

[17] Ibid.

[18] See Mao, A.A and Odyuo.

contributors of the state's economy but the share of agriculture and allied sector in the Gross State Value Added which was 30.94% in 2022–12, has declined to 29.44% as per Quick Estimate of 2015–16. Even though, the survey did not focus on the impact of climate change in the agricultural sector of the state but the concerning issues reflect a role of climate change impacting the agricultural productivity of the state.[19]

Sharma (2017) has stated, "*rising temperature, erratic rainfall and occurrence of moderate drought are beginning to impact the food production in Nagaland* (Sharma 2017)." The state is aware of the challenges which would be brought about by climate change; hence, it is now opting towards the forgotten resource that is traditional food crops which has the ability to endure high temperatures and water-stressed conditions. It is embracing measures to revive traditional rice and millet varieties, these two crops are known for its high yields in this hilly state. Although all rice production systems are marked by exposure to heat stress but rain-fed uplands are vulnerable to temperature rise in particular. Warmer climate leads to introduction of new diseases and insects as well as it leads to stunting which affects yields. Hence, farmers are in need of new varieties which are heat and drought resistant. For this reason, traditional rice and millet varieties are the best available options, as they can grow even in marginalised lands are to their merits, are highly diverse in terms of pest resistant, yield, water requirements, time of cultivation and maturity. There are farmers who act as the custodian of knowledge with regard to these traditional varieties for sustenance but over the years they have not been practising it for sustenance. However, in the face of climate change, these varieties will be promoted among the farmers, as these crops are best options available for adapting to changing climate. It is worth noting that State Agriculture Research Station (SARS) at Mokokchung, Nagaland has identified 867 traditional 'land races' of rice which signifies remarkable biological diversity of rice in the state. The rice varieties has been widely categorised as glutinous, brown and aromatic and most of them are grown under the jhum or shifting cultivation system practised by different Naga tribes in the state. In order to promote livelihood and food security as a strategy towards climate change adaption, the state government has started a project titled, '*Gene Pool Conservation of Indigenous Rice Varieties*' under traditional integrated rotational farming system. It is a 24 Cr INR project and will be supported by National Adaption Fund for Climate Change. Under the projects, field gene banks will be developed for conservation and research of promising varieties, and exchange programmes will be carried out with the involvement of communities engaged in both jhum and terrace cultivation. Selected promising seeds varieties will then be disseminated to farmers across the state. It will be implemented in one village in each selected block in five districts of Nagaland namely in Mokokchung, Kohima, Tuensang, Wokha and Zunheboto in a time frame of 3 years. Rice which is the staple food of Nagaland with about 86% of the cultivable area in the state under jhum and terrace rice cultivation systems. Traditional rice varieties are grown in altitude

[19] Food resource a major challenge for Nagaland. *The Morung Express*. https://morungexpress.com/food-resource-major-challenge-nagaland

ranging from 300 to 2500 metres. But lately, farmers are shifting to cash crops making cereal production under stress. High intensity farming of rice and pulses, promoted under several schemes, has not been sustainable. Therefore, cultivation of traditional rice and millet is significant for achieving the goal of food security in the state in the midst of climate change which is impacting the agricultural productivity across the globe.[20]

There was a prediction of Nagaland likely facing drought after 10 years due to delayed rainfall and if the dry spell continued then the rice production in the state was estimated to be reduced by 70%. India shows contradiction in rainfall pattern, for example, in 2019, while a large apart of the country faced drought like conditions, 256 districts across 13 states faced floods due to excessive rainfall in the year 2019, as observed by the researchers. Irregular monsoon rains and water shortage has become a growing concern as it leads to the country running short of the commodity which is most vital for human life and sustenance. As of now, the government is focusing on measures to minimize crop loss and compensate farmers for the crop loss by distributing seeds that can be sown in early rabi season which begins from September. Likewise, other efforts are also taken into consideration like inducting technology in farming activities, focusing on water conservation methods like building check dams, promoting and practising rainwater harvesting. On the other hand, the Nagaland state government has sought assistance from the central government in tackling water shortage for agricultural activities (Kalia 2021). The infestation of fall armyworm which not only destroys maize crops but also affects vegetables and paddy crops has further aggravated the situation. As on March of 2021, it was recorded that fall armyworm has affected 3048. 45 hectares of maize crop in 344 villages. It was observed, the infestation period happened early, usually the infestation begins in the month of May. The change in the infestation period was due to the change in the climatic conditions of the state leading to a period of dry spell creating a drought like situation (Ambrocia 2021).

As mentioned in the preceding paragraphs, the government of Nagaland has identified climate resilient traditional food crops that farmers can take up. Each village have more than 50 varieties of rice which they have sown. Emphasis has also been given to traditional agricultural practices like jhum or shifting cultivation, this cultivation system has been encouraged to adapt to climate change by reinforcing the crop diversity which is an integral part of the state's agricultural practice. The farmers of the state are also practising mixed cropping and it has proved to be highly effective when rains are erratic.

1.3 Policy Measures and Conclusion

The need of hour is to have a critical assessment on the impact of climate change in the traditional food system of Nagaland. Although it can be safely said that there has not been a drastic affect on the former by the latter, however, the institutions involved

[20] Ibid.

in the policy making concerning the food security measures should not wait for the time when things will go out of hand. There has been numerous studies conducted across continents assessing the ramifications of the climate change on the food system of various countries and communities making it easy for others to have information about it and likewise, adapting policy measures to control or fix the pertaining issues. Some of the measures have been highlighted by Food Agricultural Organisation (FAO) in its policy document,

> Build resilience of agricultural systems through increasing the efficiency of scarce resource use in productive system. Adapting measures for crops, fishing and fish-farming practices and management; increasing the diversity within production systems. Investing in resilient agricultural development, resilient agricultural development, and related investment, can support adaptation. Investments in agriculture, and especially in smallholder agriculture, are key to eradicating poverty. Rural and R&D investments, investments of farmers, fishers and forest dwellers need to be supported. Managing genetic resources is another key means of adaptation. Enable adaptation through policies and institution, Institutions that generate and manage public goods are key, Securing land tenure is paramount, Collective management of natural resources, Collective management of natural resources. Market development and better linkages of smallholder and family farmers to domestic, national and regional markets, policies will be needed to reduce financial risks. (FAO 2015).

In the context of Nagaland, local institutions in the state can help in their own capacities to prepare and create measures to adapt the impacts of climate change which in turn can build sustainability in preserving the traditional food system in Nagaland. As DK Chetri, Joint Directors, State Agricultural Research Station (SARS) had pointed out, "*Nagaland is not adequately equipped to current climate change impacts. The state has a large existing adaptation deficit which is a priority of early action. Therefore, there is a need for developing processes for community led adaption that are rooted in local institutions.*"[21]

Workshops for the farmers like Climate Change, Millets and Women-Concerns and Opportunities, Kohima organised by the North East Network (NEN) Nagaland on 2019. Aimed to understand the current status of millet cultivation and consumption in North East region and to identify the key challenges and opportunities in millet based bio-diverse food and farming systems. The workshop also aimed at facilitating farmer to farmer exchange learning on innovative practices, and discussed the way forward for the recognition of millet farmers' knowledge and to bring millets into the agricultural policies of the State.[22]

Traditional millets varieties known for its high yields as discussed before might withstand the impact of climate change and is a part of traditional food system of Nagaland. Hence, efforts are being made through various initiatives for its promotion. As such, the first Millet Festival in Nagaland was held on 30th September, 2021 at Chingmei Village in Tuensang district under the theme, '*Climate Change: Revitalizing*

---

[21] See D.C.Sharma.

[22] Custodian farmers attend 'Climate Change, Millets & Women' workshop in Nagaland. *nesfas*.https://nesfas.in/custodian-farmers-attend-climate-change-millets-women-in-nagaland/

Millet Based Biodiversity.[23] Subsequently, the second Millet Festival was held at Shamator town in Nagaland in the month of August, 2022. The festival was celebrated under a common theme, '*Celebrating common Heritage for a Shared, Resilient Future*'. The objective of the festival was, "*a preparatory towards the International Year of Millet in 2023, millet festivals series has become an important platform to bond communities, to revive, promote and bring together various stakeholders in addressing the current challenge of our time such as, climate change, food security issues that our communities are facing or being impacted with.*"[24] It was done in collaboration with Self Employed Women Association (SEWA) and the Millet Network of India (MINI).[25] This was followed up by third Millet fest in Shamator, Nagaland in the month of August 2023 under the theme, '*Millets for Community Resilience*'. The aim of the event was to promote millet cultivation and harvest and to recognize the importance of millets as a nutritious and sustainable crop.[26] In a way, through such crop festivals, we get a picture of state's awareness in building a way forward in combating the impacts of climate change on traditional food system in particular and food security in general through community participation.

As seen from the various available policy documents, adaptation and mitigation are two of the approaches which India has adopted as measures for combating climate change. Precisely, adaptation, in the context of climate change means the measures taken to minimize the adverse impacts of climate change which includes relocating the communities living to the sea shore, for example, to cope with the rising sea level or switching to crops that can withstand higher temperatures. Mitigation comprises measures to decrease the emissions of greenhouse gases that cause climate change in the first place like for instance, by switching to renewable sources of energy such a solar energy or wind energy, or nuclear energy instead of burning fossil fuel in thermal power stations (Chandran and Sandhya 2013). However, even while adopting these strategies efforts are being made not to take away resources which are core to the developmental needs and growth of and for the country.

Traditional food system plays a vital role in the cultural lives of the Nagas and researches are being carried out to promote those food items, crops basically, which are part of the traditional food system as well as those which can resist the impact of climate change. These studies have been carried out not only to sustain the traditional food system but also enhance the food security of the state as and when needed. Although, gradual progress can be observed on such fronts but there are still many issues which needs to be addressed. As of now, all we have is, to remain optimistic about a holistic policy approach with regard to tackling the impacts of climate change on the traditional food system of Nagaland.

[23] Millet Festival: 'Revitalizing Millet Based Biodiversity'. *The Morung Express*. https://morungexpress.com/millet-festival-revitalizing-millet-based-biodiversity

[24] 2nd Millet Festivals- 6th & 31st August, 2023, Nagaland. *North East Network: A Women's Rights Organisation*.https://northeastnetwork.org/2nd-millet-festival-6th-31st-august-2023

[25] Ibid.

[26] 3rd Millet fest in Shamator. *Nagaland Post*. https://nagalandpost.com/index.php/3rd-millet-fest-in-shamator/

References

Ambrocia M (2021) Fall armyworm infestation: 334 villages affected in Nagaland. EastMojo. https://www.eastmojo.com/nagaland/2021/06.17/fall-armyworm-infestation-334-villages-affected-in-nagaland/

Angami T (n.d.) Flora and Fauna of Nagaland. https://www.angelfire.com/nm/nagalim/flora.html

Batal M et al (2021) Importance of the traditional food systems for first nations adults living on reserves in Canada. Can J Public Health 112(Suppl 1):20–28. http://www.ncbi.nlm.nih.gov/pmc/articles/PMC8239073/

Borthakur P (2020) 15 delicacies from Northeast India that will make you drool. East Mojo. https://www.eastmojo.com/news/2020/11/17/15-delicacies-from-Northeast-India-that-will-make-you-drool/

Chandran VKP, Sandhya (2013) Adaptation and mitigation strategies of climate change: a serious concern. Yojana. https://yojana.gov.in/adaptation-and-mitigation.asp

Datta P, Behera B, Rahut DB (2022) Climate change and Indian agriculture: a systematic review of farmers' perception, adaption, and transformation. Environ Challenge 8:100543. https://www.sciencedirect.com/science/article/pii/S2667010022001007

Deaconu A et al (2021) Promoting traditional foods for human and environmental health: lessons from agroecology and indigenous communities in Ecuador. BMC Nutrition 7:1. https://bmc-nutr.biomedcentral.com/articles/10.1186/s40795-020-00395-y

Domadia P (2017) Empowering tribal women with a taste of authenticity. CSR J. http://thecsrjournal.in/empowering-tribal-women-with-a-taste-of-authenticity/

FAO (2015). https://openknowledge.fao.org/server/api/core/bitstreams/a4fd8ac5-4582-4a66-91b0-55abf642a400/content

Kalia S (2021) Nagaland likely to face severe drought after 10 years due to delayed rainfall. The Swaddle. https://theswaddle.com/nagaland-likely-to-face-severe-drought-after-10-years-due-to-delayed-rainfall/

Mao AA, Odyuo N (2007) Traditional fermented foods of the Naga tribes of northeastern, India. Indian J Tradit Knowl 6(1):37–41

Mawroh B (2023) Climate change and resilience of indigenous food systems: lessons from Meghalaya's food diversity. https://in.boell.org/en/indigenous-food-systems

Neihu K (2018) Nagaland's dying tradition of rice beer making. The Morung Express. http://morungexpress.com/nagalands-dying-tradition-of-rice-beer-making

Preethan P, Ginoya N (2021) Transformative adaption: ways to improve India's long-term food security (commentary). Mongabay-India. https://www.google.com/amp/s/india.mongabay.com/2021/08/commentary-transformative-adaption-ways-to-improve-indias-long-term-food-security

Sebu S (2013) Geography of Nagaland. Spectrum Publications

Sharma DC (2017) Faced with climate change, Nagaland to revive traditional rice varieties. Down To Earth. https://www.downtoearth.org.in/news/climate-change/amp/faced-with-climate-change-nagaland-to-revive-traditional-rice-varieties-59331

Victor O et al (2022) The impact of climate change on food systems, diet quality, nutrition, and health outcomes: a narrative review. Front Clim 4. https://doi.org/10.3389/fclim.2022.941842/full

Zahan SA (2022) Why food has emerged as a divisive factor for people from northeast. Outlook. https://www.outlookindia.com/national/why-food-has-emerged-as-a-divisive-factor-for-northeast-indians-in-india-news-193094/amp

Part III

Challenges in the Himalayas: Impacts on its People, Ecology and Knowledge Systems

Traditional Knowledge Systems in the Himalayas: Why Moving Away from Biomass Cooking Fuel Is So Difficult?

Bishal Bharadwaj, Rajesh Kumar Rai, and Peta Ashworth

1 Why Nepal Struggles to Achieve the Clean Cooking Target of the Sustainable Development Goals

Firewood has been the dominant cooking fuel in Nepal for many years and this has continued to challenge the country's ability to achieve its clean cooking target by 2030. Currently, the government aims to reduce the use of solid cooking fuels to 30% of the total population by 2030. However, in 2021, the majority of households continue to cook their food using firewood. This is despite the total number of households using firewood as a cooking fuel reducing from 64% of total households in 2011 to 51% in 2021. This reduction is higher than the previous decade, which was at an annual rate of 0.2% (Paudel et al. 2021). However, despite the increased annual rate of 1.3% uptake of clean cooking fuel, it may take several decades to achieve the goal of universal access. For example, an empirical study has shown that one quarter of households will still be using firewood for cooking in 2030 under the existing policy scenario (ESCAP 2021).

Location of households, whether rural or urban is one of the key determinants of switching to clean cooking fuel (Rahut et al. 2014). For example, clean cooking uptake is concentrated around urban areas. This may be due to transportation issues, as the distance to market has been found to be one of the main determining factors of clean cooking fuel use in the Himalayas (Dendup and Arimura 2019). In addition, the

B. Bharadwaj (✉) · P. Ashworth
Curtin Institute for Energy Transition, Curtin University, Bentley, WA, Australia
e-mail: bishal.bharadwaj@curtin.edu.au

R. K. Rai
Institute of Forestry, Pokhara Campus, Pokhara, Nepal

School of Forestry and Natural Resource Management, Tribhuvan University, Kathmandu, Nepal

A. Borthakur, P. Singh (eds.), *The Resilience of Traditional Knowledge Systems for a Sustainable Future*, Advances in Asian Human-Environmental Research,
https://doi.org/10.1007/978-3-031-56858-9_8

structure of rural houses is not always seen to be fit for clean cooking fuel use. For example, the proportion of clean cooking households in urban areas increased from one quarter (25%) in 2011 to 39% in 2021. While during this same period, the proportion of rural households using firewood increased from 73% in 2011 to 75% in 2021.

Secondly, the uptake of government subsidies promoting clean cooking fuels has not been successful in rural areas. For example, despite the Nepalese government subsidizing biogas as a cooking fuel for more than 30 years, the recent 2021 Census reports total households using biogas as their main cooking fuel is approximately1.2% (NSO 2023),—one percent down from the 2.3% of total households using biogas in Nepal in 2011 (CBS 2012). However, this reduction in biogas use is not only due to LPG use in urban areas, but biogas users have also reduced in rural areas as well, from 2.6% of total rural households in 2011 to 1.1% of total rural households in 2021. It is important to highlight that the cow dung user is 2.5 times the number of biogas users.

The reduced contribution of biogas to household cooking fuel raises the question of what the effect of has been subsidizing the biogas. Biogas promotion was initially aimed at reducing the pressure on forest areas, particularly around where conservation was seen to be important (Rai 2016). Due to a change in economic structures and also disposable incomes, household are switching from subsistence farming to a market-based economy (Sunam and McCarthy 2016). Many households are reducing livestock rearing, and with this, transitioning to other options like LPG (Shahi et al. 2022). Such a shift in economic structures are also influencing households to transit to cleaner fuels (Baland et al. 2018). However, if the government had chosen to include electric cooking in their subsidy offering, rather than just biogas, it is likely that many households would have directly used electric cooking instead of using biogas.

At the same time, many transitioning households do not completely leave the use of firewood, instead they move to a mixed-fuel use. Therefore, the key challenges of enabling a clean cooking transition in Nepal and for many developing countries leads to the single question: "What makes it so hard for households to move away from firewood use in their homes for cooking?"

This book chapter uses survey data collected by Bharadwaj (2022). The survey respondents were experts engaged or working in the energy and development space in Nepal. Detailed information on the survey is available at Bharadwaj et al. (2021b).

1.1 Techno-Economic Push for the Clean Cooking Transition

Today there are a range of clean cooking technologies (stoves and fuels) available. These modern cooking solutions provide a range of benefits. For example, using clean cooking fuels means households do not need to spend time collecting firewood from the forests. Clean cooking stoves also reduce the time spent cooking food. Clean fuel also reduces household air pollution that is associated with respiratory diseases like pneumonia and asthma, and therefore reducing health costs (Jaiswal et al. 2022). As well, household members can use the saved time for income generating work. All this adds to the economic return that can arise from clean cooking fuel use. These health and economic benefits of modern cooking technologies are put forward as a reason why government and development partners in developing countries should accelerate the clean cooking transition (IEA 2023).

To explore why households in rural areas continue to stick with using firewood for cooking, using an online survey platform, we conducted a survey among 40 experts linked to cooking practices. The experts were working in the environment sector including climate change, community development and research and were from non-government organisations (NGOs), local governments and civil services. The survey used an online format and included a range of questions such as ranking of cooking fuel, level of agreement on a range of statements and listing the services linked to traditional cooking practices.

The experts were asked to rank the use of different cooking fuels in rural and urban context using the question "Rank the following cooking option for rural/ urban, Nepal. To rank, please drag and drop to the right box". Based on the mean rank values from the lowest value being the most favoured cooking fuel, firewood ranked as the most favoured cooking fuel in the rural context and the least favoured cooking fuel in urban areas (Fig. 1). This ranking of firewood as the most favoured cooking fuel in the rural context indicates the influence of the external context on cooking fuel choice. As such, the techno-economic reasons such as scalability and efficiency of clean cooking fuels for adoption in the rural context may not reflect the socio-cultural needs of households (Herington et al. 2017; Velvizhi et al. 2023).

There are two broad reasons that could explain why firewood is preferred over the other clean cooking fuels. Firstly, households in the rural context have unique needs that they manage to fulfill from their cooking practices (Bharadwaj et al. 2021b), that the techno-economic benefits do not consider as benefits arising from traditional cooking. While arguing for the clean cooking transition, researchers and energy planners put forward the benefits arising from modern technology and ignore the benefits rooted in the traditional practices of firewood use.

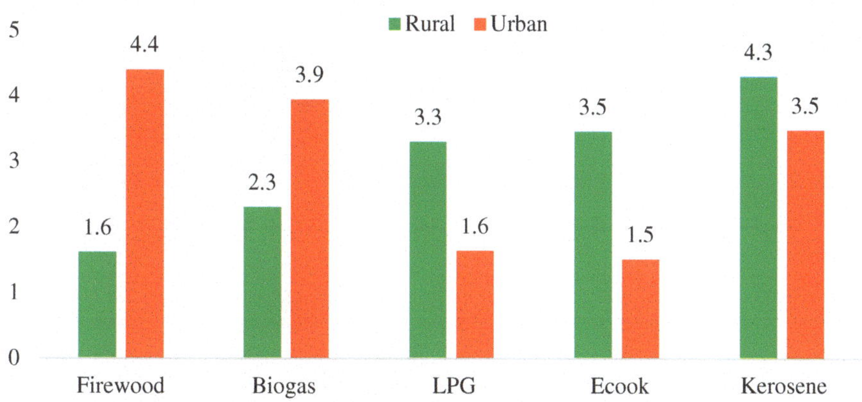

Fig. 1 Mean rank value for different cooking fuels in the rural and urban contexts of Nepal

Rural households' perspective are quite different to many, and they might not emphasize what outsiders think is important. For example, 53% (21 experts) agreed on the statement "Firewood collection reduces pests entering the farm", 22% (9 experts) were neutral and eight experts disagreed with the statement. Also, when considering the saved time and opportunity cost, households use of firewood household heating is not fully considered. When asked about the statement "Households use firewood to heat the house", of 40 experts, 33 (83%) strongly agreed. Confirming that firewood provides a range of co-benefits linked to the cultural aspects of rural living.

Secondly, the perceived value of techno-economic benefits may differ for households in rural contexts, deviating from what the existing literature presents. Economic opportunities in rural areas are often limited, making it challenging for households to find paid work that would directly result from saved time. For instance, the use of solar home systems increases the likelihood of female members working on farms, but it reduces their chances of engaging in paid work (Bharadwaj et al. 2021a; Subedi et al. 2023). In regions with constrained economic opportunities, the time saved through clean cooking fuels might not necessarily translate into the additional income that could offset the cost of adopting fuels like LPG or electricity for electric cooking (Meeks et al. 2018). Therefore, understanding whether the saved time leads to tangible income gains becomes a pivotal question. Undoubtedly, firewood collection time can be linked with the formation of social capital as the women go with neighbours into forests when gathering wood and also get information about their neighbours at the same time.

In rural areas of developing countries, households may opt to utilize firewood collection as an extended economic activity to manage their cooking fuel needs. By doing so, they avoid the expense of purchasing clean cooking fuels and can invest that saved money elsewhere. Figure 2 illustrates the level of agreement among experts regarding the statement "Firewood collection is a working opportunity for

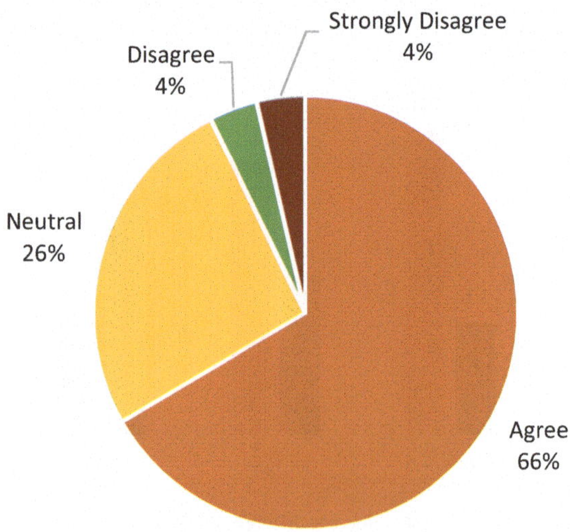

Fig. 2 Firewood collection is an opportunity for households to use their time for saving costs from cooking fuel

households to avoid the cost of cooking fuel." Notably, one third of experts strongly agreed, and 45% agreed with the statement. Only 3% of experts strongly disagreed.

The evidence supporting the health implications of using firewood is compelling, with 36 out of 40 experts acknowledging its link to respiratory health problems. Equally, the fact that 29 of the 40 experts agree, and five remained neutral, on the notion that "People ignore the effect of firewood smoke on their health". It highlights the perceived value of adopting clean cooking solutions varies significantly based on social practices and local context (Malakar and Day 2020). Or, it may be true that people are unaware of the health issues related to the use of firewood for cooking.

It is apparent that the economic considerations of households play a significant role in their choice of cooking fuel. The benefits of saved time from clean cooking must align with available economic opportunities for it to truly make a difference in rural households' livelihoods. The existing studies showed that the income of households is positively associated with using cleaner cooking fuels (Rahut et al. 2014). Understanding these nuances is crucial in designing targeted and effective clean energy initiatives that cater to the unique needs and circumstances of rural communities (Bharadwaj et al. 2022a).

2 Influence of Context on Cooking Fuel Use Behaviour

Himalayan countries are renowned for their rich and diverse cultural heritage, closely intertwined with their unique cuisines. In these regions, firewood cooking has been an integral aspect of culture, cuisine, and livelihoods for centuries. These practices have evolved to cater to contextual needs and provide essential services to households beyond just cooking.

Figure 3 presents the various co-benefits associated with firewood use, spanning before, during, and after its use for cooking purposes. Before its use, firewood offers benefits such as waste management, formation of social capital during firewood collection, supports forest management, provides opportunities to earn a living by selling firewood, and even symbolizes wealth when stacked outside the house. While being used for cooking, firewood serves numerous other purposes within

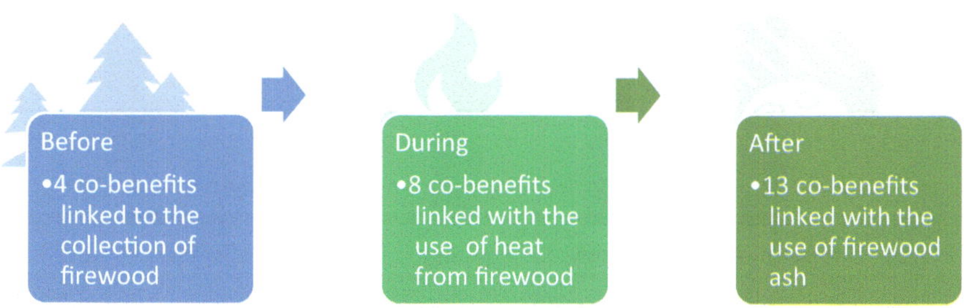

Fig. 3 The range of benefits firewood provides to households in rural Nepal

households. We have recorded eight different services it fulfills. For example, meat is often hung on the ceiling just above the firewood to dry and preserve it. Bamboo baskets are commonly used to store items like chillies and garlic, preserving them as seeds for the next season. Moreover, sitting around the firewood creates a warm and inviting space for family dialogue and social exchange. Additionally, the smoke produced by firewood serves as a natural repellent against termites and bugs, protecting the house, particularly with the use of wooden construction materials.

Furthermore, firewood generates ash and charcoal, which have multiple uses in various livelihood aspects. Fine ash is employed for cleaning purposes, while blacksmiths use unburnt charcoals as a heating source. Charcoal is also utilized as a water filter. Additionally, ashes serve as a pesticide to protect plants from fungal infestations and are mixed with grains and seeds to safeguard them for future planting.

The multi-faceted role of firewood in the Himalayan culture and livelihood is evident through its co-benefits, which extend well beyond its primary use as a cooking fuel. Understanding and appreciating these interconnected benefits can help inform sustainable and context-specific approaches to energy access and environmental conservation in the region.

All these co-benefits may not be true in scientific terms, however, social beliefs and norms makes it a part of ethno-cultural identity (Mazzone et al. 2021). These social beliefs and norms also set the context for clean cooking practice (Shankar et al. 2020). An example is the taste of food. Figure 4 shows how much experts agree that food cooked in firewood is tastier than food cooked in modern cooking fuels like LPG. However, the younger generation, who are not used to eating foods with smoke and traditional cooking, may not find the difference in the taste of the food, instead they might be irritated by the smoke.

Many ongoing conflicts are rooted in ethno-cultural identities, indicating the significant value people place on their cultural and social lives. It is essential to recognize that traditional cooking practices, as well as fuel preparation, storage, and

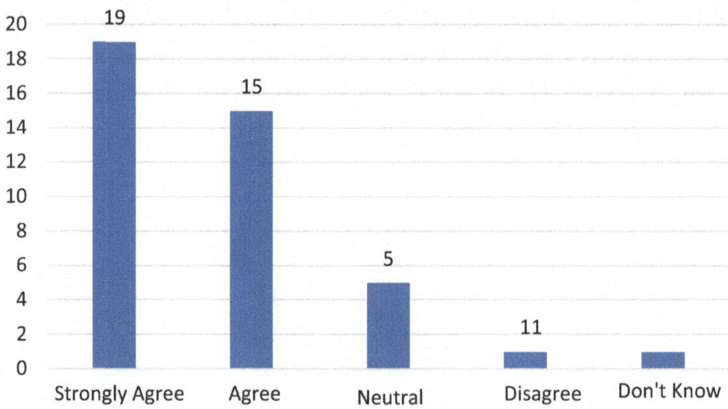

Fig. 4 There is the widespread perception that food cooked on firewood is tastier than cooked in other cook stoves like LPG

usage, are heavily influenced by cultural norms. For instance, different cultural groups consider firewood stacked as symbol of social status. Moreover, some cultural groups cook outside, allowing everyone to participate, while for others, cooking is a sacred activity restricted to key household members. Additionally, there exist cultural practices of worshiping fire as god and offering food to the fire before eating (Shrestha and Wilgenburg 2006). Fire and food are strongly linked in socio-cultural practices in the Indo-Himalayan region (Hegde et al. 2018; Sen 2019). These diverse cooking practices have evolved alongside ethno-cultural identities, which people are often reluctant to relinquish. We asked respondents about their agreement regarding the suitability of firewood for preparing cultural foods. Figure 5 illustrates the traditional cooking practice of firewood is linked with cultural cuisine with 60 precent strongly agreeing and 10% agreeing. Some recipes require fire such as *'Wachipa'—a traditional cuisine* for cleaning and burning chicken feathers. Similarly, woodfired corn/barbecued corn *(Poleko Makai)* and potato needs fire. In addition, in rural settings, the cattle sheds are not fully prepared for using clean cooking fuels, and they require firewood for cattle feed.

However, as the context changes, adhering to cultural practices tied to ethno-cultural identity becomes more challenging. This ultimately compels households to transition to more suitable cooking fuels. For example, household selection of cooking fuel changes when they change their house from a traditional house made of straw and wood to more a modern tin or concrete house dwelling. Despite prevailing beliefs that smoke enhances wood durability and deters termites, the same smoke blemishes paint and makes tin ceilings dark and unclean. Additionally, having concrete-built homes eliminates the need for wood-supported house structures, thus disconnecting the house from the practical application of the belief that smoke strengthens wood. These factors contribute to acceptance of energy technologies through multiple channels like affect, attitude, cost and perception (Lozano et al. 2022). This illustrates how changes in household contexts, such as transitioning from rural to urban settings, weakens the bonds between livelihood and firewood use, while simultaneously strengthening the perceived benefits of adopting clean cooking fuels.

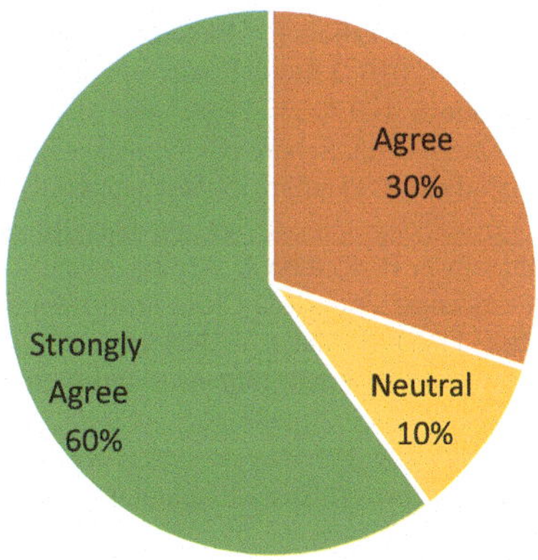

Fig. 5 Firewood is best suitable to prepare cultural foods such as Selroti and roast maize (Poleko makai in Nepali); and liquore such as alcohol

3 Accelerating Clean Cooking Transition in the Heterogenous Context of Nepal

In conclusion, cultural practices and context, deeply influence cooking habits. Households may face pressures to adapt to more appropriate and practical cooking fuel choices with changing contexts. These contexts could be income, education, neighbourhood and information (Rahut et al. 2014; Dendup and Arimura 2019), supply of better energy systems (Bharadwaj et al. 2023b; Subedi et al. 2023) or changes in policy like subsidy (Bharadwaj et al. 2022b). Understanding the intricate connections between ethno-cultural identities, cooking practices, and changing contexts is essential for crafting effective energy access initiatives that respect cultural values while addressing evolving needs and preferences.

While the health benefits of transitioning to clean cooking fuels are well-founded, it is crucial to recognize that traditional cooking practices have their merits too. People's perception of the impact of firewood smoke on their health can influence their willingness to switch to cleaner alternatives. Factors such as information, cultural beliefs, lifestyle choices, and accessibility to clean cooking technologies all play a role in shaping the perceived value of adopting clean fuels (Bharadwaj et al. 2023a, c). If people see value in traditional beliefs, and experiences outweigh the value of health benefits, then switching to a cleaner cooking fuel is almost impossible.

In light of this insight, it becomes essential for clean energy initiatives to be holistic and culturally sensitive. Effective campaigns promoting clean cooking should not only emphasize health benefits but also acknowledge and address the importance of traditional practices within local contexts. But this does not mean to promote stove stacking at household level. By understanding and accounting for the diverse social factors at play, policymakers and stakeholders can design interventions that better resonate with communities and encourage the adoption of cleaner and healthier cooking practices.

References

Baland J-M, Libois F, Mookherjee D (2018) Forest degradation and economic growth in Nepal, 2003–2010. J Assoc Environ Resour Econ 5(2):401–439

Bharadwaj B (2022) Does context matter for energy access in Nepal?, The University of Queensland, School of Chemical Engineering

Bharadwaj B, Dendup N, Pattanayak SK, Ashworth P (2021a) Impacts of solar subsidy: evidence from geographic regression discontinuity design in Nepal. Duke Global Working Paper Series

Bharadwaj B, Kambo A, Arratia Solar A, Ashworth P (2023a) Awareness of biogas increases its social acceptance. J Clean Prod 138432:138432

Bharadwaj B, Malakar Y, Herington M, Ashworth P (2022a) Context matters: unpacking decision-making, external influences and spatial factors on clean cooking transitions in Nepal. Energy Res Soc Sci 85:102408

Bharadwaj B, Pattanayak S, Ashworth P (2022b) Space matters: reducing energy disparity in Nepal through spatially equitable renewable energy subsidies. Environ Res Comm 4

Bharadwaj B, Pullar D, L. Seng To, Leary J (2021b) Why firewood? Exploring the co-benefits, socio-ecological interactions and indigenous knowledge surrounding cooking practice in rural Nepal. Energy Res Soc Sci 75:101932

Bharadwaj B, Subedi MN, Malakar Y, Ashworth P (2023b) Low-capacity decentralized electricity systems limit the adoption of electronic appliances in rural Nepal. Energy Policy 177:113576

Bharadwaj B, Weder F, Ashworth P (2023c) More support for hydrogen export than its domestic application in Australia. Human Soc Sci Comm 10(1):7

CBS (2012) National Population and housing census 2011 (National Report). G. o. N. Central Bureau of Statistics—national planning commission Secratariat. Central Bureau of Statistics, Kathmandu

Dendup N, Arimura TH (2019) Information leverage: the adoption of clean cooking fuel in Bhutan. Energy Policy 125:181–195

ESCAP (2021) Energy transition pathways for the 2030 agenda: SDG7 roadmap for Nepa, economic and social Commission for Asia and the Pacific (ESCAP)

Hegde S, Nair LP, Chandran H, Irshad H (2018) Traditional Indian way of eating—an overview. Journal of Ethnic Foods 5(1):20–23

Herington MJ, Lant PA, Smart S, Greig C, van de Fliert E (2017) Defection, recruitment and social change in cooking practices: energy poverty through a social practice lens. Energy Res Soc Sci 34:272–280

IEA (2023) A vision for clean cooking access for all. International Energy Agency

Jaiswal KK, Chowdhury CR, Yadav D, Verma R, Dutta S, Jaiswal KS, Sangmesh B, Karuppasamy KSK (2022) Renewable and sustainable clean energy development and impact on social, economic, and environmental health. Energy Nexus 7:100118

Lozano LL, Bharadwaj B, de Sales A, Kambo A, Ashworth P (2022) Societal acceptance of hydrogen for domestic and export applications in Australia. Int J Hydrog Energy 47(67):28806–28818

Malakar M, Day R (2020) Differences in firewood users' and LPG users' perceived relationships between cooking fuels and women's multidimensional Well-being in rural India. Nat Energy 5(12):1022–1031

Mazzone A, Cruz T, Bezerra P (2021) Firewood in the forest: social practices, culture, and energy transitions in a remote village of the Brazilian Amazon. Energy Res Soc Sci 74:101980

Meeks R, Sims KRE, Thompson H (2018) Waste not: can household biogas deliver sustainable development? Environ Resour Econ 72(3):763–794

NSO (2023) National population and housing census 2021. (National Report), National Statistics Office

Paudel D, Jeuland M, Lohani SP (2021) Cooking-energy transition in Nepal: trend review. Clean Energy 5(1):1–9

Rahut DB, Das S, De Groote H, Behera B (2014) Determinants of household energy use in Bhutan. Energy 69:661–672

Rai S (2016) In: Gyawali D, Thompson M, Verweij M (eds) Biogas: buoyant or bust? Aid, technology and development the lessons from Nepal. Routledge, London

Sen CT (2019) Hinduism and food. In: Kaplan DM (ed) Encyclopedia of food and agricultural ethics. Springer, Dordrecht, Netherlands, pp 1521–1527

Shahi N, Bhusal P, Paudel G, Kimengsi JN (2022) Forest—people nexus in changing livelihood contexts: evidence from community forests in Nepal. Trees Forest People 8:100223

Shankar AV, Quinn A, Dickinson KL, Williams KN, Masera O, Charron D, Jack D, Hyman J, Pillarisetti A, Bailis R, Kumar P, Ruiz-Mercado I, Rosenthal J (2020) Everybody stacks: lessons from household energy case studies to inform design principles for clean energy transitions. Energy Policy 141:111468

Shrestha BG, Wilgenburg WV (2006) Filming fire rituals in Nepal. IIAS N e w s l e t t e r, International Institute for Asian Studies, p 42

Subedi MN, Bharadwaj B, Rafiq S (2023) Who benefits from the decentralised energy system (DES)? Evidence from Nepal's micro-hydropower (MHP). Energy Econ 106592:106592

Sunam RK, McCarthy JF (2016) Reconsidering the links between poverty, international labour migration, and agrarian change: critical insights from Nepal. J Peasant Stud 43(1):39–63
Velvizhi G, Nair R, Goswami C, Arumugam SK, Shetti NP, Aminabhavi TM (2023) Carbon credit reduction: a techno-economic analysis of "drop-in" fuel production. Environ Pollut 316:120507

"Unorganized Tourism Practices and the Allied Challenges in the Himalayas"

Mohammad Vikas Ashraf, Surya Misri, Shreekar Pant, and Shoeb Ahmad

1 Introduction

The story of tourism in the Himalayas revolves around a struggling economy that ventured into the world's most challenging mountains (Fig. 1). The start of tourism activities has been very well summarized by Singh and Kaur in 1986. They described the evolution of Himalayan tourism in three broad phases: Religious, Middle-class, and Social. The first two types of tourism showed some consideration for the stability of the ecosystem, while the third type, known as "Modern Tourism," had a different approach (Singh and Kaur 1986).

Initially, traders and pilgrims journeyed through the high and trans-Himalayan regions, trading essential goods and traversing difficult mountain paths on ponies or yaks. This landscape, inhabited by cattle breeders and pastoralists, often witnessed pilgrims trekking to sacred destinations. This landscape remained untouched by human intervention for centuries, preserving its natural state. As time passed, the tradition of pilgrimage grew stronger, attracting an increasing number of visitors to Himalayan shrines. This pilgrimage culture contributed to the local economy and fostered the participation of the local communities, benefiting both the hosts and the guests. Environmental conservation was intertwined with the observance of rituals.

M. V. Ashraf · S. Ahmad (✉)
Department of Biotechnology, School of Biosciences and Biotechnology, Baba Ghulam Shah Badshah University, Rajouri, Jammu and Kashmir, India
e-mail: shoebahmad@bgsbu.ac.in

S. Misri
Section of Microbiology, School of Biosciences and Biotechnology, Baba Ghulam Shah Badshah University, Rajouri, Jammu and Kashmir, India

S. Pant
Centre for Biodiversity Studies, School of Biosciences and Biotechnology, Baba Ghulam Shah Badshah University, Rajouri, Jammu and Kashmir, India

A. Borthakur, P. Singh (eds.), *The Resilience of Traditional Knowledge Systems for a Sustainable Future*, Advances in Asian Human-Environmental Research,
https://doi.org/10.1007/978-3-031-56858-9_9

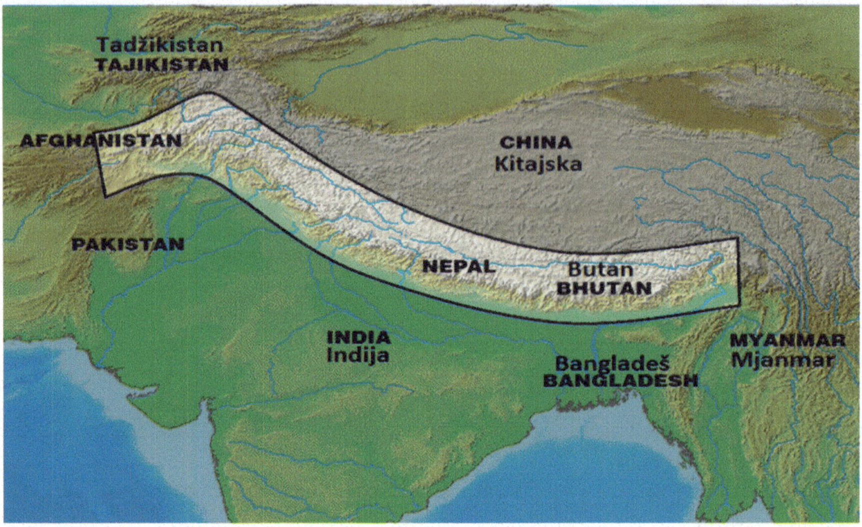

Fig. 1 The Himalayan Mountain range (Photograph Source: *Wikimedia commons* https://w.wiki/7isn; File:Himalayas Map Slovenian.png by Sven Manguard, Wikimedia Commons contributors; available under Creative Commons CC0 1.0 Universal Public Domain Dedication)

Although remnants of this scene can still be observed, the forces of modernization are rapidly eroding it (Sharma et al. 2022; Singh and Kaur 1986).

The second phase of Himalayan tourism, which preceded the emergence of modern tourism, began in the early nineteenth century when the British established climatic resorts and sanatoria in the middle ranges of the Himalayas. This marked the entry of roads into the high mountains, leading to urbanization and other forms of development. Over time, these urban centers expanded amidst the predominantly agricultural and pastoral landscape. Socio-economic environments experienced new waves of growth, and interactions between the highland and lowland regions increased. Despite these changes, the resilient ecosystem of the higher Himalayas remained untouched, except for occasional disruptions caused by local festivities, pilgrims' chants, or visits from royalty for hunting or romantic getaways (Sharma et al. 2022).

The third phase of Himalayan tourism, which emerged after India's independence in 1947, witnessed the promotion of tourism on a large scale by the public sector. The demand for tourism exceeded the available supply, and tourism flourished with all its characteristics. However, the unplanned and unregulated nature of this tourism brought negative consequences. Mass tourism surpassed the elite and middle-class resorts, attracting visitors with different lifestyles and demographics. Consequently, they ventured into the fragile and ecologically sensitive higher Himalayan regions, which became accessible due to the construction of roads after the Sino-Indian conflict in 1962. This event had significant implications for the development of the greater Himalayas in terms of strategic considerations and regional disparities. Tourism was seen as a potential solution to address these disparities, but inadequate resource analysis, planning, and development goals resulted in the exploitation of the region. Negative impacts such as commercial deforestation, overgrazing, mining, steep slope terracing, and dam construction in vulnerable areas became prevalent. Despite these activities being attributed to tourism, they diminished the attractiveness and amenities available to tourists in the area (Sherpa n.d.).

2 Tourist Footfall in Himalayan Republics in the Twenty-First Century

The Himalayan region, which is one of the greatest assets provided by nature to human beings, harbors almost 1.5 billion people from different countries such as India, Pakistan, Nepal, China, Bhutan, Myanmar, etc. which depend upon these mountain ranges for food and shelter. The Himalayan region provides cultural as well as socio-economic support to the population and has been a major tourist attraction since ancient times, either in the form of pilgrimage, holy sites, or tourist destinations such as hill stations (Ashok et al. 2022).

The high tourist inflow in this region can be related to the mantra of "*Atithi Devo Bhava*" followed by the people of this region, which means "Guests are similar to God and should be treated in the same manner as God". Despite holding such importance in the life of a huge population, the Himalayan region was much ignored until the arrival of Britishers in India, who inhabited the region and built some famous tourist destinations such as Darjeeling, Nainital, Mussoorie, etc. The Himalayas were further introduced to the world by mountaineers such as Tenzing Norgay and Edmund Hillary, who were the first to climb Mount Everest in the year 1953. In continuation, people started getting attracted to the Himalayas for spirituality and meditation activities, leading to a high tourist influx in this region in the late twentieth century and twenty-first century (Ashok et al. 2022; Singh and Kaur 1986). Figure 2 represents the prevalence of religious tourism in the Himalayan region.

In the Indian Himalayan region, it is estimated that the native population has grown by almost 30% from the year 2001, reaching nearly 50 million in the year 2019. The tourist influx in this region has doubled, from 60 million in the year 2001 to almost 130 million in the year 2019. The tourist to resident ratio has also increased from 56 tourists per 100 residents in the year 2001 to 159 tourists per 100 residents in the year 2019 (Kapoor and Wangdus 2022).

According to the figures provided by the Nepal tourism department, the tourist inflow in the Himalayan region has increased from almost 3.5 lakhs in the year 2001 to nearly 12 lakhs in the year 2019. The trekking and mountaineering activities have also gone up from 1 lakh tourists in the year 2001 to 2 lakhs in the year 2019 (Nyaupane and Chhetri 2009). Figure 3 portrays the abundance of tourism activities in the Solang valley of Himalayas.

Fig. 2 Tourists visiting Kedarnath Temple in the Himalayan region (Photograph Source: *Wikimedia commons* https://w.wiki/7igm; File:Kedarnath Temple in Uttarakhand, India, photo by Yogabrata Chakraborty.jpg/ Wikimedia Commons contributors; available under Creative Commons Attribution-Share Alike 4.0 International license)

Fig. 3 Tourist footfall in the Himalayas (Photograph Source: *Wikimedia commons* https://w.wiki/7ig$; File:Solang valley.jpg by Arupamdas/Wikimedia Commons contributors; available under Creative Commons Attribution-Share Alike 4.0 International license)

2.1 Surplus Tourist Inflow and Unorganized Activities

The Himalayan region, with its rich cultural as well as ecological diversity, is a major tourist attraction for several activities such as pilgrimage, mountaineering, trekking as well and sightseeing. In recent years, the Himalayan region has experienced a high frequency of tourists due to which the demand for meeting the needs and requirements of the tourists has also skyrocketed. This has led to an increase in the number of construction and transportation activities in the Himalayan region (Chauhan et al. 2022; Pandit and Grumbine 2012). Indiscriminate cutting of trees for development activities such as roads, tunnels, and highways, the use of vehicles such as buses, cars, motorcycles, and airplanes for transportation, and the exploitation of local resources are some of the major challenges faced by the Himalayan region in the present scenario. Lack of planning in chalking out these proceedings has resulted in the Himalayan region facing serious problems such as global warming, pollution, melting of glaciers, and climate change, which ultimately lead to the loss of life and property (Wani et al. 2023).

2.1.1 Unregulated Tourism and Construction

Tourism constitutes an important part of the socio-economic structure of the Himalayan region. Tourist activities in the Himalayan region have brought with them, a plethora of positives, including employment and economy. In a country bounded by the Himalayas such as Nepal, tourism contributes around 10% to its GDP. However, these tourist activities should be responsible as well as sustainable and not unregulated. The unregulated tourist activities in the Himalayan region and the indiscriminate burden on natural resources in this sensitive region are leading to serious environmental problems in this region such as pollution, global warming, and melting of glaciers (Rather et al. 2022; Wani et al. 2023).

Higher tourist footfall demands several construction activities, including blacktop roads, highways, and tunnels, which make travel easier and cut time, raising hotels and restaurants for accommodation. These construction activities put extra pressure on the ecology of the Himalayan region, leading to landslides, flash floods, and other related environmental hazards (Brandt et al. 2019). Figure 4 depicts the negative impacts of infrastructural development, including construction of roads and highways, on the fragile environment of Himalayas. According to a report published by the National Disaster Management Authority (NDMA) of India, the massive inflow of tourists in the Uttarakhand region in an unregulated manner is leading to various disastrous effects in the region, such as flash floods, etc. Unplanned and unmanaged development projects and uncontrolled tourism in the Himalayan region are considered the root causes behind recent natural calamities in the Himalayan regions of Uttarakhand (Dimri et al. 2016; Singh et al. 2023a).

2.1.2 Carbon Emission and Melting of Glaciers

High tourist footfall in the Himalayan region in the twenty-first century has led to an increase in transportation activities in these areas. Transport facilities used by tourists as well as the same required for meeting the growing need for resources in

Fig. 4 Highway construction leading to landslides in the Himalayas (Photograph Source: *Wikimedia commons* https://w.wiki/7ik9; File:Road construction causing landslides in the Himalaya, AJTJ.jpg by A. J. T. Johnsingh, WWF-India and NCF/Wikimedia Commons contributors; available under Creative Commons Attribution-Share Alike 4.0 International license)

these areas have led to an increase in the levels of gases such as carbon dioxide and other greenhouse gases in the Himalayan region. This sudden increase in the levels of carbon and greenhouse gases is a matter of concern as it leads to global warming, resulting in the melting of glaciers in these areas, ultimately leading to a rise in seawater levels (Thind et al. 2022).

The release of black carbon in the atmosphere as a result of the burning of fossil fuels for cooking as well as transport facilities is one of the major causes of the melting of glaciers in the Himalayan region. According to a report published by the Ministry of Science and Technology, an increase in the levels of carbon emission due to activities like the burning of fossil fuels in states such as Uttarakhand is leading to the fast melting of glaciers such as Gangotri glacier (Unregulated Tourism in the Himalayas Comes at the Cost of Ecological Hazards (thewire. in). The Himalayas, with the world's third-largest glacier reserves, are also known as the "Third Pole." Researchers have found that due to unregulated and uncurbed tourism in this region, drastic levels of carbon emission have led to a 10 ten-fold increase in the rate of melting of glaciers in the last four decades (Thompson 2010; Zhang et al. 2020).

2.1.3 Deforestation

Tourist activities in the Himalayan region have brought with them, an era of development, including roads, highways, healthcare facilities, residential hotel facilities, etc. However, development in this region has also resulted in negative environmental impacts, including deforestation. A major result of tourist activities in the Himalayan region is deforestation. An increase in the footfall of tourists in the Himalayan region demands space for different kinds of facilities such as hotels, medical facilities, roads, etc. which leads to the clearing of forests in and around the tourist places. Deforestation, thus, is a matter of grave concern as it may lead to changes in rainfall patterns, increased chances of landslides as well high risk of soil erosion (Sultan et al. 2022). Figure 5 illustrates the scenario of massive deforestation in the Himalayas.

The forest land is being cleared in and around the hill stations of the Himalayan region for construction activities, which is creating an ecological imbalance in the region. The growing economy due to tourism has led to the over-utilization of forest resources such as wood by residents. According to some studies, 75 kg of firewood is used on an average basis by the residents for daily needs. This over-dependence on forest resources has increased the rate of deforestation in the Himalayan region, leading to the loss of habitat for wild animals. The increased income of farmers as a result of tourist activities has encouraged the farmers to increase the area under farming and also resulted in the clearing of forests for grazing activities. It is estimated that due to these activities, the Himalayan region is losing 1.5 million hectares of forest cover on an annual basis (Brandt et al. 2019) (Environmental Impact of Tourism on the Himalayas: Deforestation and Tourism (environmentalissuestt.blogspot.com).

Fig. 5 Deforestation in Himalayas (Photograph Source: *Wikimedia commons* https://w.wiki/7ihw; File:Deforestation-62,486.jpg by Hans/Wikimedia Commons contributors; available under Creative Commons CC0 1.0 Universal Public Domain Dedication)

2.1.4 Pollution

The Himalayan region, with its rich cultural as well as natural diversity, is the center of attraction for tourists from all over the world. Every year, millions of tourists visit the Himalayas for different purposes, including religious pilgrimage, mountaineering, and trekking. The high frequency of tourists has become a major source of pollution in this region. Burning of wood, biomass, and emissions from vehicles are some of the prime reasons behind increased levels of particulate matter in the air, hence, leading to the pollution of air. This particulate matter, when settles on the glaciers, absorbs more amount of sunlight, leading to an increase in the temperature in this region, ultimately disturbing the ecological balance of the Himalayan region (Wani et al. 2021).

Tourism also encourages the use of packaged food materials such as soft drinks, potato chips, and non-biodegradables such as single-use plastics for multiple purposes. The use of polyethylene as well as other non-biodegradable materials by the residents, as well as tourists in these areas, is the main cause of soil as well as water pollution in this region (Azam et al. 2018). Since the beginning of tourism in the Everest region, waste disposal has been a significant problem. The climbers leave behind their high-tech climbing gear, plastics, food, tins, oxygen tanks, aluminum cans, clothes, glass, and papers after either, successfully ascending Himalayan mountains or trying to do so. According to a report published by the Gati Foundation of Uttarakhand, major contributors to soil pollution in the Uttarakhand Himalayas were single-use plastics. The tallest mountain peak on the Earth, Mount Everest, may soon receive a new moniker of "World's Highest Garbage Dump." Mountains of trash have been generated from the base camp up to the high "death zone" due to the massive increase in visitors, which has greatly strained the ecology (Farooq et al. 2023; Sharma and Kumar 2023). Figure 6 depicts the garbage ridden water bodies in Himalayan regions.

The indiscriminate disposal of chemicals in the water bodies by the industries, as well as factories set up in the Himalayan region, has put enormous pressure on the health of water bodies. The unplanned and indiscriminate disposal of single-use plastics around the water bodies and drainage pipes that have been opened in these rivers by the local population, for homes as well as residential facilities has led to the deterioration in the levels of all important parameters such as biochemical oxygen demand of the water bodies in the Himalayan region, which directly or indirectly, affects the biodiversity of this region (Chakraborty et al. 2016).

2.1.5 Overexploitation of Natural Resources

Tourism in the Himalayan high mountains have been a major source of socio-economic development in the region. However, due to unplanned and indiscriminate tourism, over-exploitation of natural resources is a matter of growing concern. The over-dependency of the tourism sector on the natural resources of the Himalayan region negatively impacts the provision of food, water, and healthcare facilities in the long run and also increases the risk of natural disasters. High tourist footfall demands high consumption of food, water, and residential as well as healthcare facilities, thus, putting an additional constraint on available land and water resources,

Fig. 6 Plastic pollution in the Himalayan rivers (Photograph Source: *Wikimedia commons* https://w.wiki/7iiJ; File:Slum and dirty river.jpg by meg and Rahul/ Wikimedia Commons contributors; available under Creative Commons Attribution 2.0 Generic license)

leading to problems such as deforestation, pollution, and depletion of water resources (Lampert 2019).

Dependence on natural resources such as forest material, and forest wood for purposes such as cooking and increased water consumption puts pressure on resources such as forests, groundwater, etc., thus, resulting in the depletion of natural resources in this region. Environmentalists have warned that if the exploitation of resources in the Himalayan region continues at the same pace, there could be a potential threat of more disasters like the Kedarnath disaster of the year 2013 (Champati Ray et al. 2016; Singh et al. 2018).

2.1.6 Disease Outbreaks

The tourism industry in the Himalayan region has developed significantly over the past few years. However, significant tourist inflow in the Himalayan region has also increased the risk of serious, communicable disease outbreaks in this region, which puts the local population at risk of getting affected by these diseases. Human beings act as potential carriers of several diseases such as sexually transmitted diseases, respiratory diseases, and streptococcal diseases. These outbreaks lead to the loss of lives as well as the livelihood of the people who solely depend upon the tourism industry (Haq et al. 2023; Martolia et al. 2009; Sultan Khuroo et al. 1988). Individuals carrying these diseases on coming in contact with healthy individuals spread infection, thus, increasing the risk of epidemics in that region. Tourist activities in the Himalayan region have led to many folds increase in the risk of disease outbreaks in this region. The recent COVID-19 outbreak and the quarantine guidelines following the same have resulted in a decrease in tourist activities in these areas, thus, indicative of the fact that tourists may act as major carries of infectious diseases to the Himalayan region (Haq et al. 2023).

3 Tourism Activities in the Context of Environmental Degradation

The Himalayan Mountain region, due to its abundant ecological diversity and scenic beauty, is of paramount importance for the tourism industry. The main reason behind this strong relationship between the tourism industry and the Himalayas is the refreshing, unpolluted, and unadulterated environment surrounding the Himalayan mountainous region. However, due to human intervention and unplanned tourist activities, this region is subject to irreversible environmental degradation. Unplanned tourist activities are a potential threat to the sensitive environment of the Himalayas, accompanying several environmental complications including soil erosion, deforestation, pollution, and loss of the natural habitat of wild animals (Mallick et al. 2023) (Fig. 7).

Environmental degradation caused by the pollution of water, soil, and especially air as a result of tourist activities is a matter of grave concern as increasing levels of such pollutants in the air are proving to be hazardous. Tourists use different modes

Fig. 7 Solid waste pollution in the Himalayas (Photograph Source: Wikimedia commons https://w.wiki/7ios; File:Plastic Pollution.jpg by Janak Bhatta, Wikimedia Commons contributors; available under Creative Commons Attribution-Share Alike 4.0 International license)

of transport to visit the destination points in these Himalayan mountains. Several industries are also set up in these regions to meet the needs of the local population and in turn, incoming tourists. The smoke coming out from vehicles and industries and the burning of fossil fuels such as firewood in these tourist destinations generate enormous levels of pollutants. These pollutants settle down upon the glaciers, attract more heat, and cause a rise in the temperature, thus, leading to the fast melting of glaciers in the Himalayan region. These glaciers upon melting, increase the level of water in the water bodies, thus, proving hazardous for the communities settled around these water bodies. The pollutants present in the air and water also deteriorate the health of the individuals residing in the surrounding areas and disturb the aquatic flora as well as fauna of the region (Stefanica et al. 2021).

Depletion of water resources, as well as degradation of land resources, are the negative impacts of tourism in the Himalayan zones. Overuse of freshwater from lakes and rivers by hotels, restaurants, and residents results in the shortage of fresh water in these regions and it is also the root cause behind the generation of high amounts of wastewater, thus disturbing the ecological balance of the region and acting as a potential threat to the aquatic life. Infrastructural developments going on in the Himalayan region including the construction of roads, hospitals, residential facilities, and tourist destinations demand more land, which comes at the cost of deforestation. Clearing of forest areas for such activities is also proving disastrous as loss of forest cover leads to soil erosion, landslides, flash floods, and habitat loss, and in turn, results in improper rainfall patterns (Mallick et al. 2023). The ever increasing rate of climate change has resulted in the accelerated rate of melting of glaciers, which are an important source of fresh water reservoirs (Fig. 8).

3.1 Deforestation

Deforestation in the Himalayan region is a growing concern for environmentalists all over the world as it adversely affects climate, rainfall patterns, soil erosion, and directly or indirectly, the livelihood of the human population residing in this area. Several factors related to tourism such as economic growth, changing demographic patterns, and the need for resources aggravate deforestation. Economic growth as a result of the high frequency of tourists in the Himalayan region results in increased demands of the local population such as expanding living areas, and construction of more shops for business, which in turn demand land area. This demand is met by clearing forests, thus, resulting in deforestation. Change in demographic patterns also demands more land area which is fulfilled by cutting trees and clearing the land area under forests (Brandt et al. 2019).

Due to the high tourist influx in the Himalayan region, the land is cleared for different purposes such as building resorts, hotels, and road and tunnel construction. This leads to a decrease in the forest cover of the Himalayan region which directly as well as indirectly affects the human and animal life in this area, ultimately leading to a change in the geographical landscape of this region as well as the destruction of local flora and fauna (Sharma et al. 2022) (Fig. 9).

Fig. 8 Glaciers in the Himalayas melting at a faster rate due to climate change (Photograph Source: *Wikimedia commons* https://w.wiki/7ikr; File:Glacial melt water carving the ice, river source Himalayas India.jpg by Sharada Prasad CS from Berkeley, India, Wikimedia Commons contributors; available under Creative Commons Attribution 2.0 Generic license)

Fig. 9 Deforestation in the Himalayas (Photograph Source: *Wikimedia Commons* https://w.wiki/7inN; File:SaurabhSawant EvergreenForestTreeFelling ArunachalPradesh India IMG 5967.jpg by Saurabhsawantphoto, Wikimedia Commons contributors; available under Creative Commons Attribution-Share Alike 4.0 International license)

3.2 Urbanization

The Himalayan belt is considered one of the most sensitive and ecologically diverse zones of the world with an abundant human population. Due to the dense human population and soaring frequency of tourism in this region, rapid urbanization is the need of the hour. However, this expansion is poorly regulated. The rapid rate of unplanned urbanization is leading to the indiscriminate cutting of trees and mountains and the clearing of forest land for the construction of well-furnished roads, highways, tunnels, hotels, healthcare facilities, and residential accommodations. These urbanization activities are ultimately leading to problems such as deforestation, pollution, and habitat destruction, making the Himalayan belt highly vulnerable to natural disasters and climate change. Due to climatic changes in the Himalayan region, there has been variability in the rainfall patterns in these regions, ultimately resulting in flash floods (Singh et al. 2023a).

Urbanization is one of the main reasons behind the high carbon footprints in the Himalayan region. These carbon compounds get easily deposited on the glaciers and act as heat absorbents, making the region highly prone to temperature changes, leading to the melting of glaciers. The growing demand for urbanization and high tourist influx in the Himalayan region has also increased the demand for freshwater. It has also increased the complications of wastewater and sewage disposal in the water bodies. This increased demand for freshwater and sewage disposal in the water bodies is putting additional stress upon the water bodies, increasing the BOD levels of the water bodies, and rendering them prone to high levels of pollution and destruction of aquatic life (Sharma et al. 2022).

3.3 Habitat Destruction

The Great Himalayas are considered the epitome of diversity, having rich flora as well as fauna. Due to indiscriminate development works going on in this region, forest area is being wiped off for recreational activities, leading to the destruction of the habitat of both flora as well as fauna of the Himalayan region. Loss of habitat is a matter of concern as it will ultimately lead to the extinction of the native flora as well as fauna of this region. According to some reports, around 75% of the original Himalayan habitat has been lost due to recreational activities. This habitat loss in the Himalayas is attributed to gathering fuelwood and fodder, which changes the composition of the habitat, and leads to species extinction. In the Himalayan regions of India and Nepal, where the population is expanding, forests are being converted into agricultural as well as grazing lands, thus, depriving the native species of these regions of their natural habitat (Kapoor and Wangdus 2022; Kattel 2022).

3.4 Human-Animal Conflict

Deforestation activities in the ecologically sensitive zones of the Himalayan region for human settlements as well as agricultural practices have proven to be ignorant towards wildlife. These activities have increased the chances of human-animal conflicts as animals come out of their natural habitat in search of food and other resources which depletes as a result of a decrease in the forest cover. These conflicts are dangerous for the long-term survival of endangered animal species in the Himalayan ranges. Human-animal conflict may lead to injuries as well as the death of both animals as well as human beings (Oommen 2021).

3.5 Species Extinction

Recreational activities leading to deforestation, and urbanization in the Himalayan region are a major threat to the natural habitat of a wide range of native flora and fauna of this region, leading to the extinction of that particular species. Unorganized tourist activities in this region also lead to the introduction of exotic, non-native floral species in this region, which replace the native floral species and lead to the extinction of that particular species. According to a recent study done by the Wildlife Institute of India (WII), headquartered in Dehradun, due to the high frequency of carbon emissions the upper Himalayas, the upper Himalayas are warming, which has put the local flora and fauna in grave danger. It has also depicted that the area's temperature has increased by at least 0.5° C, and the humidity levels have changed by roughly 10%. Flowering patterns and leaf budding patterns have also changed from May to June. Rainfall patterns have also been altered. Insect behaviors also changed. Birds need to adjust their behavioral response. Himalayan birch, White lily, and Tibetan sea buckthorn are among the affected plant species (Iqbal et al. 2020; Kattel 2022).

4 Unorganized Tourism and Climate Change in the Himalayas

Tourism is a fast-growing sector all over the world, especially, in the Himalayan region, acting as a major contributor to the economy of the region. In the Indian subcontinent, tourism contributes around 7% to the total GDP. The Indian Himalayan region, with its rich natural and astonishing landscapes and multiple pilgrim sites, is a major tourist attraction all over the world. Tourism in this region is one of the major sources of employment and economy for the people. The high tourist frequency in the Himalayan region is related to climatic and religious factors, attracting tourists yearly. However, change in the climatic conditions of the Himalayan region due to the negative impacts of tourism has a direct or indirect effect on the survival of life in this region (Khah et al. 2011). Climate change is a broad term, which includes changes in temperature conditions and rainfall patterns, and

Fig. 10 Change in the ice cover of the Himalayas due to climate change (Photograph Source: *Wikimedia commons* https://w.wiki/7iqW; File: Everest nubtse.jpg by Pavel Novak, Wikimedia Commons contributors; available under Creative Commons Attribution-Share Alike 2.5 Generic license)

alterations in seasons. The climatic changes in the Himalayan region are leading to problems such as changes in the habitat of species, extinction of species, and changes in the patterns of vegetation. Tourism is considered to be a highly climate-sensitive economic activity due to its close links to the environment. Tourism in this region has resulted in an over-exhaustion of resources for infrastructure, transportation, and meeting the needs of locals as well as visitors. As with all industries, tourism contributes to greenhouse gas emissions. These greenhouse gases are leading to abrupt warming of the Himalayan region, thus, posing a threat to the Himalayan glaciers and the survival of life in this region (Kattel 2022; Nyaupane and Chhetri 2009) (Fig. 10).

5 Implications of Unregulated Tourism in the Himalayas: Lessons from the Past

With the growing tourist activities in the Himalayan region and the insensitive and irresponsible behavior of both the tourists as well as the administration toward the ecological fragility of the Himalayas, the ecological imbalance created in this region is getting worse. Increase in the overall temperature in the Himalayan mountainous region due to the emission of carbon as well as other greenhouse gases, indiscriminate cutting of trees due to unplanned urbanization, and over-exploitation of the local resources of the region to meet the requirements of the people are some of the major challenges faced by the Himalayas. However, implications faced by this region are severe, including the melting of glaciers, leading to the rise of water levels of the Himalayan water bodies, habitat destruction of the native flora as well as fauna, leading to species extinction, the introduction of exotic species in this region, decrease in the forest cover and high levels of pollutants in the air, thus, decreasing the air quality index of the Himalayan belt (Mallick et al. 2023). Due to such negative impacts, the region is facing major disasters such as flash floods, earthquakes, landslides, and land subsidence. The degree of resource productivity is severely constrained by geo-environmental constraints, making the Himalayas an ecologically vulnerable and economically underdeveloped region. Examining the terrible impacts of unregulated tourism activities is necessary, and there are important lessons to be learned from the Himalayan region's overcrowded tourist hotspots (Wani et al. 2023).

5.1 Floods

The Himalayan River systems provide the Himalayan belt with a significant risk of flooding. These river systems are fed by glaciers. Tourist activities have led to high carbon emissions in the region, have raised the temperature in the hilly areas, stimulated more snow to melt, and increased the river flow, which results in flooding (Ghosh and Ghosal 2021). Uttarakhand is a Himalayan state, well-known for its spiritual and religious tourism, ecological diversity, and cultural ethos. However, it is also well-known for its increasingly frequent and severe natural disasters, as well as for the fragility of its ecological and geological systems. Uttarakhand saw its greatest tragedy in the month of June 2013, which resulted in significant human casualties and widespread destruction. The catastrophe was referred to as the "Himalayan Tsunami" and it resulted in widespread loss of human as well as animal life, destruction of property, infrastructure, and the environment, as well as effects on the local economy and way of life (Champati Ray et al. 2016) (Fig. 11).

One of the many factors contributing to an increase in human vulnerability to natural catastrophes is the growing population combined with yatra tourism in Uttarakhand's mountainous regions (Patel et al. 2022). According to the census, Uttarakhand had an average population density per square kilometer of 159 in 2001 and grew to 189 in 2011. The population rose from 84.89 lakh in 2001 to 100.86 lakh in 2011. From 2000 to 2010, the number of tourists visiting Uttarakhand increased by 300%, from 1.11 crores to 3.11 crores (Kumar et al. 2013). The infrastructure, however, has not grown proportionately to accommodate this tourism rise. Numerous unplanned hotels and other amenities have been built due to the increase in tourist footfall, even in environmentally sensitive areas like some floodplain areas. Construction of roads, bridges, buildings, and other infrastructure, as well as several other developmental initiatives, were encouraged to serve tourists and pilgrims better. The growing ecological imbalances in the area can be attributed to the effects of these development activities on the local environment and land

Fig. 11 Uttarakhand floods of 2012 (Photograph Source: *Wikimedia commons* https://w.wiki/7iqh; File:Flash floods in Uttarakhand 2012.jpg by European Commission DG ECHO, Wikimedia Commons contributors; available under Creative Commons Attribution-Share Alike 2.0 Generic license)

usage, which stimulated such devastating flash floods in this region (Champati Ray et al. 2016).

5.2 Forest Fires

Unwanted wildfires are referred to as "forest fires," and they are one of the world's current environmental problems. The distribution, structure, and composition of vegetation in numerous ecosystems can be influenced by forest fires. A multitude of factors, including well-known culprits like global warming and climate change, uncontrolled development, and greater human pressure, are contributing to a rise in forest fires. Negligence is a major cause of forest fires in India, with examples including throwing lit matches after lighting cigarettes and bidis or using their stubs (straw, husk, etc.) to clean up agricultural fields. Visitors and tourists frequently engage in campfires and other activities for fun which also increases the chance of forest fires (Bhattarai et al. 2023; Kumar and Kumar 2022; Singh et al. 2023a) (Fig. 12).

At least 8 people have died as a result of "the worst in a decade" wildfires in the Indian states of Himachal Pradesh and Uttarakhand, which have blanketed Himalayan Mountain regions in a dense haze (Singh et al. 2023b). In the neighbouring country of Nepal, where wildfires blazed across many districts and forced a four-day school shutdown, seven people lost their lives (Mishra et al. 2023). These fires not only pose serious risks to human health but also to the preservation of the

Fig. 12 Forest Fire in Almora forests of Uttarakhand (Photograph Source: *Wikimedia commons* https://w.wiki/7ir3; File:Forest fire in Almora District 2016.jpg by Ramwik, Wikimedia Commons contributors; available under Creative Commons Attribution-Share Alike 3.0 Unported license)

forest and wildlife. In addition to putting lives in danger, wildfires produce carbon dioxide, the most pervasive greenhouse gas on the planet and a major contributor to climate change. According to an estimate, the wildfires in Uttarakhand alone in 2021 released 0.2 megatons of carbon dioxide, which is the highest since 2016 (Singh et al. 2023a).

5.3 Landslides

The Himalayan region is highly vulnerable to landslides due to its fragile ecosystem. Tourism in the Himalayan region has stimulated landslides and increased their intensity. Several factors related to tourism are responsible for increasing incidents of landslides in the Himalayan region. The first is unauthorized building and unchecked hilltop leveling near the roadways. Rapid and unplanned urbanization, deforestation, encroachment on steep hill slopes, irrational road cutting, and water-intensive agriculture practices are also some of the major contributors. Rapid urbanization leads to the cutting of trees, mountains, and hill slopes for roads, highways, tunnels, and residential facilities. Hillocks are leveled to make room for housing developments, businesses, etc. These activities severely disturb the ecological balance of this region, thus making the region prone to landslides (Fayaz et al. 2022; Haigh et al. 1995) (Fig. 13).

The second is an overall decrease in the green cover due to deforestation, which arises as a result of an increase in the number of people dwelling in the region, sometimes, beyond the carrying capacity of the region. This contributes to soil erosion and puts the region at additional risk of landslides. Water is not absorbed in the soil and gets carried away rapidly as a result of the absence of green cover and the steep gradient of the slope (Haigh et al. 1995).

Fig. 13 Landslides on major highways and link roads is a common sight in the Himalayas (Photograph Source: *Wikimedia commons* https://w.wiki/7irF; File:Friends waiting (47402647571).jpg by solarisgirl, Wikimedia Commons contributors; available under (Creative Commons Attribution-Share Alike 2.0 Generic license)

5.4 Changing Land Use Pattern

Himalayas are one of the most crowded, ecologically vulnerable, tectonically unstable, and commercially underdeveloped mountain ecosystems on the earth. These mountain ranges are extremely sensitive to large-scale tectonic movements, landslides, flash floods, and other natural disasters because of their highly fragile ecosystem. Recent years have seen several changes in the traditional land and resource use patterns, mostly as a result of population pressure and the resulting rise in demand for natural resources. With the extensive change in the demography of the Himalayan region, the demand for food, water, and residential accommodations has also increased. Consequently, this has led to the extensive clearing of forests in the Himalayan region to accommodate the overburdened human population, thus, leading to substantial changes in land-use patterns (Mondal and Zhang 2018). Land under the forest is also being cleared for the extension of agricultural activities to fulfill the food demands of the people. This is creating an unstable ecosystem in the fragile Himalayan mountains, making them highly susceptible to collapse, leading to loss of life and property (Mondal and Zhang 2018; Rasool et al. 2021). Tourism has also increased the economy of the people residing in the Himalayan region. Due to this, there has been an increase in the number of domestic animals such as cows, buffaloes, and goats being reared for dairy and related purposes. Consequently, the demand for grazing land has also increased, leading to the clearing of forest land in the Himalayan region. These rapid and extensive land use changes have harmed the livelihood security of rural communities by drastically reducing forest cover, degrading the land, and decreasing groundwater recharge throughout the entire region, in addition to causing environmental instability and climatic changes in the Himalayas (Tiwari 2008).

5.5 Biodiversity Loss

The Himalayan Mountain region, despite being remote and inaccessible, is suffering from human-induced biodiversity loss. The Himalayan highlands have been inhabited since ancient times, but in recent years, high-frequency tourist footfall and soaring population have increased the demand for the region's natural resources. The natural habitat of the Himalayan region is currently dispersed. The hotspot's steadily expanding population has led to extensive logging and the clearing of forests and meadows for agriculture (Brandt et al. 2019). In Nepal, as well as the Indian states of Sikkim, Darjeeling, and Assam, large-scale deforestation and habitat fragmentation have resulted from the conversion of forests and grasslands for agriculture and urbanization. In many areas, the remaining habitat of the Himalayan hotspots is seriously destroyed. This is accredited to the fact that domestic animals, such as cattle, and yak, typically overgraze in these fragile ecosystems, leading to the destruction of habitat and ultimately, endangering native species or resulting in their extinction. The delicate flora of the Himalayan mountains has been overutilized in traditional medicines and they are also threatened during activities such as deforestation, grazing, and forest fires. These activities ultimately stimulate the loss of biodiversity in the Himalayan region (Kattel 2022).

6 Recommendations for Sustainable Tourism in the Himalayas

The Himalayas, which share their boundaries with five countries, namely India, Pakistan, Nepal, Bhutan, and China, are the greatest gifts of Mother Nature to mankind. They are home to some of the world's most exotic species of flora and fauna. Some of the world's largest rivers such as the Ganga, Brahmaputra, and Indus also find their origins in the Himalayas. They are also a major pilgrimage site for a large population of the world (Ashok et al. 2022). Due to such eco-cultural diversity, it is a major tourism hub, which has shown a significant upward trend in the region over the past few decades. However, due to the negative impacts of tourism such as climate change, land degradation, overexploitation, and natural disasters, the mountains face a surfeit of difficulties that have far-reaching and disastrous effects on both mountain communities and those residing downstream. Tourism, which is one of the fastest-growing industries in this region, has the potential to be the catalyst for future development in the Himalayan region. This will only be possible if it is carried out by the goals of sustainable development proposed by the United Nations (Azam et al. 2018).

It is possible to stop the over-exploitation of the Himalayan ecosystems triggered by a high frequency of tourist influx through the creation of more tourist destinations, as well as through targeted afforestation and solid waste management techniques. The UN Environment Programme and the Responsible Tourism Society of India co-hosted a summit in June 2022 when the Union Ministry of Tourism unveiled the National Strategy for Sustainable Tourism and the Responsible Traveler Campaign. It places a strong emphasis on capacity development, governance, and the encouragement of socio-cultural, economic, and environmental sustainability. There is also recognition of the need for greater investment in green infrastructure, which includes the preservation of open spaces and historic sites, improved transportation systems, and cleaner air (Ashok et al. 2022).

6.1 Regulated Tourism Practice

In response to the dramatic spike in visitors to the Himalayan region over the past decades, which has almost doubled, a report from the Environment Ministry pushes for "controlled tourism". The Himalayan region has seen an enormous boom in the tourism industry, which has over time culminated in a degradation in the quality of the land, water, and air due to increased trash production, a large number of automobiles, and waste discharge into rivers. The rapid growth of infrastructure and tourism in the Himalayan ecosystem is often blamed for a variety of environmental problems, including carbon emissions that contribute to global warming and ultimately climate change (Thompson 2010).

Promoting regulated tourist practices and pushing sustainable agendas are essential for the Himalayan region. This may be done by ensuring that each tourist location in the Himalayan region has an appropriate proportion of visitors, which would largely reduce the generation of solid waste, pollution levels in the water and air, and biodiversity destruction (Thakur et al. 2021).

6.2 Vigilance and Patrolling

The Himalayas have been serving as India's northern defense line since the beginning. Even though India lacks China's Great Wall, the Great Himalayas function precisely in the same manner. However, the ecological diversity of the Himalayas is facing threat from several anthropogenic factors, including tourism and it requires strict vigilance as well as patrolling. Protected areas in the Himalayan region, such as Hemis National Park and the Karakoram Sanctuary in Ladakh, which are rich in biodiversity, need to be carefully observed and regularly patrolled to reduce undesired human-animal interactions as well as habitat destruction due to deforestation and transportation activities (Prashyanusorn et al. 2010).

6.3 Early Warning System

The Early Warning System is a highly essential, unified system of monitoring, forecasting, and estimation of natural hazards, disaster risk assessment, communication, and preparedness activities that enable individuals, communities, governments, businesses, and other entities to take swift action to reduce the risks of disasters before catastrophic events. It minimizes loss of life and property before catastrophic events like storms, tsunamis, and droughts, and uses methods for early warning of multiple hazards (Sahana et al. 2023). Increased accessibility of multifaceted early warning systems and disaster risk information is one of the seven worldwide goals set forward by the Sendai Framework for Disaster Risk Reduction 2015–2030 (Wang et al. 2022).

To establish an early warning system in the Himalayan region against major and sudden floods, rockslides, landslips, glacier lake bursts, and avalanches as well as to prevent Chamoli-type disasters in the future, the National Geophysical Research Institute (NGRI) of the Council of Scientific and Industrial Research has started field studies. The experts at the institute have chosen a few spots in Uttarakhand's higher elevation areas for the "installation" of seismometers and river gauges and in turn, increasing their number from 60 to 100. The goal is to closely monitor river flows in particular catchment areas to look for any unexpected increase in water levels or flooding threshold that could pose a risk. These tools are essential for the security of significant infrastructure projects (Manglik 2023).

6.4 Regional Cooperation

The Himalayan mountainous region, which is spread over an area of 2400 km and shares its boundaries with five countries in South Asia is facing serious ecological disturbances. Indiscriminate emission of black carbon from industries, cooking, and transportation is accelerating the melting of the Himalayan glaciers by amplifying the effects of climate change. The stability of water supplies in South Asia as a whole, as well as the people who live at their feet, are seriously threatened by melting glaciers and the loss of seasonal snow. According to a report published in Glaciers of the Himalayas, black carbon deposits and airborne particles produced by incomplete combustion from brick kilns, diesel exhaust, and the burning of biomass, are speeding up the process of melting of glaciers and snow in these ranges, in addition to changing temperatures and precipitation patterns (Timothy 2003).

It is beyond the responsibility of the policymakers of any one nation to fully address the problems brought on by the melting of Himalayan glaciers. To develop collective adaptation strategies, regional cooperation would be required. Sharing information regarding the concerns connected to glacier change and evolution could be a first step. To enhance regional cooperation, delegates from each of the eight Hindu Kush Himalayan (HKH) states signed a declaration at the Ministerial Mountain Summit on October 15, 2020. In an HKH Call to Action document, all ministers agreed that national cooperation at both the individual and collective levels was necessary to preserve mountain habitats and enhance livelihoods. The HKH region is experiencing the effects of climate change, biodiversity loss, and pollution on a transnational scale that transcends national political boundaries. It is crucial to move away from national viewpoints and towards larger, transboundary thinking that includes eco-regions for the region to reverse or even absorb the effects of these crises. Due to the similarities in the difficulties that the HKH member nations confront, unified strategies are needed to solve them (Mishra et al. 2019).

6.5 Area-Specific Sustainable Plan

The Himalayan region is facing a serious threat of climate change, which is posing a serious threat to its rich biodiversity. Climate change is leading to several challenges in this region, including the fast melting of glaciers, changes in precipitation patterns, etc. To address the challenges of climate change, the Government of India launched the National Action Plan on Climate Change (NAPCC) in 2008. The National Mission on Sustaining the Himalayan Ecosystem (NMSHE) is an area-specific mission among the NAPCC's eight National Missions. The core agenda of this mission is to give a deeper comprehension of the relationship between the Himalayan ecosystem and climate factors and contribute to the sustainable development of the region while simultaneously addressing the need to safeguard a vulnerable environment because the Himalayas are vital, not only to India but to the entire Asian continent (Majra and Gur 2009) (SUSTAINABLE TOURISM IN THE HIMALAYAS I WWF India), (Do you know about National Mission for Sustaining the Himalayan Ecosystem (jagranjosh.com).

6.6 Promote Ecotourism

The breathtaking beauty of the Himalayas is the center of attraction for tourists all over the world. However, the ecosystem of the area is crucial for tourism. If the ecosystem suffers, tourism will suffer as well. However, if improperly managed, tourism can have negative effects on the environment. Due to the high rate of tourist influx in the region, there has been an increase in the economy of the people residing in the Himalayan region. The most obvious route to economic growth in the Great Himalayas is through ecotourism, high-altitude travel, and adventure, biodiversity, and travel. The practice of sustainable tourism ensures the survival of ecotourism in the Himalayas. The term "ecotourism" is the one that is used to describe tourism the most (Fig. 14).

In India, it is crucial to promote ecotourism since people are searching for locations to unwind their brains while they consider making a fresh start. Everything done in the tourism business without causing harm is promoted by ecotourism. By not exploiting them and by exploring them, ecotourism—or sustainable tourism—helps to preserve natural habitats and pristine locations. Benefits of ecotourism include cost effectiveness, pleasurable experiences, and raising cultural

Fig. 14 Wooden huts: Promoting ecotourism in Himalayas (Photograph Source: *Wikimedia commons* https://w.wiki/7ish; File: Ecotourism Doodhpathri—Tourist lodges and other facilities southwest Jammu Kashmir India.jpg by Ankur P from Pune, India, Wikimedia Commons contributors; available under Creative Commons Attribution-Share Alike 2.0 Generic license)

understanding (Ashok et al. 2022; Brandt et al. 2019; Chaudhary et al. 2022) (Ecotourism—Top States Promoting Ecotourism in India—Himalayan Gypsy (Himalayan-gypsy. in).

Giving back to the community without causing any environmental harm is a component of ecotourism in the Himalayas. Treating nature as a vital element of human survival is one of the best strategies for encouraging eco-tourism in the Himalayas. However, the transition to ecotourism needs to be carefully promoted. There is, however, hope for sustainable tourism once both travelers and tour operators understand the severity of the value of upholding the equilibrium of nature for future generations (Brandt et al. 2019).

7 Conclusion

The Himalayan region is a gift of Mother Nature to human beings. It is home to one of the world's most exotic species of flora and fauna and is the source of livelihood for a large human population. With its rich biodiversity and cultural ethos, it is a center of attraction for tourists across the globe. However, the sudden increase in the tourist activities in this fragile ecosystem has led to a plethora of problems, including deforestation, climate change, melting of glaciers, and changes in the paths of rivers, ultimately, resulting in several catastrophic events in the region such as floods, flash floods, increase in the water level of rivers, and earthquakes, causing loss of life and property. Thus, environmentally friendly tourism is the need of the hour. Tourism, which does not exceed the carrying capacity of the region, which supports the sensitivity of the region, and which does not lead to the over-exploitation of resources of the Himalayas should be promoted so that the Himalayas can be sustained and preserved for generations to come.

References

Ashok S, Behera MD, Tewari HR, Jana C (2022) Developing ecotourism sustainability maximization (ESM) model: a safe minimum standard for climate change mitigation in the Indian Himalayas. Environ Monit Assess 194(12):914. https://doi.org/10.1007/s10661-022-10548-0

Azam M, Mahmudul Alam M, Haroon Hafeez M (2018) Effect of tourism on environmental pollution: further evidence from Malaysia, Singapore and Thailand. J Clean Prod 190:330–338. https://doi.org/10.1016/j.jclepro.2018.04.168

Bhattarai H, Wu G, Zheng X, Zhu H, Gao S, Zhang Y-L, Widory D, Ram K, Chen X, Wan X, Pei Q, Pan Y, Kang S, Cong Z (2023) Wildfire-derived nitrogen aerosols threaten the fragile ecosystem in Himalayas and Tibetan plateau. Environ Sci Technol 57(25):9243–9251. https://doi.org/10.1021/acs.est.3c01541

Brandt JS, Radeloff V, Allendorf T, Butsic V, Roopsind A (2019) Effects of ecotourism on forest loss in the Himalayan biodiversity hotspot based on counterfactual analyses. Conserv Biol 33(6):1318–1328. https://doi.org/10.1111/cobi.13341

Chakraborty P, Khuman SN, Selvaraj S, Sampath S, Devi NL, Bang JJ, Katsoyiannis A (2016) Polychlorinated biphenyls and organochlorine pesticides in river Brahmaputra from the outer Himalayan range and river Hooghly emptying into the bay of Bengal: occurrence, sources and ecotoxicological risk assessment. Environ Pollut 219:998–1006. https://doi.org/10.1016/j.envpol.2016.06.067

Champati Ray PK, Chattoraj SL, Bisht MPS, Kannaujiya S, Pandey K, Goswami A (2016) Kedarnath disaster 2013: causes and consequences using remote sensing inputs. Nat Hazards 81(1):227–243. https://doi.org/10.1007/s11069-015-2076-0

Chaudhary S, Kumar A, Pramanik M, Negi MS (2022) Land evaluation and sustainable development of ecotourism in the Garhwal Himalayan region using geospatial technology and analytical hierarchy process. Environ Dev Sustain 24(2):2225–2266. https://doi.org/10.1007/s10668-021-01528-4

Chauhan D, Thiyaharajan M, Pandey A, Singh N, Singh V, Sen S, Pandey R (2022) Climate change water vulnerability and adaptation mechanism in a Himalayan City, Nainital, India. Environ Sci Pollu Res 29(57):85904–85921. https://doi.org/10.1007/s11356-021-15713-5

Dimri AP, Thayyen RJ, Kibler K, Stanton A, Jain SK, Tullos D, Singh VP (2016) A review of atmospheric and land surface processes with emphasis on flood generation in the southern Himalayan rivers. Sci Total Environ 556:98–115. https://doi.org/10.1016/j.scitotenv.2016.02.206

Farooq M, Nisa FU, Manzoor Z, Tripathi S, Thulasiraman AV, Khan MI, Khan MYA, Gani KM (2023) Abundance and characteristics of microplastics in a freshwater river in northwestern Himalayas, India–scenario of riverbank solid waste disposal sites. Sci Total Environ 886:164027. https://doi.org/10.1016/j.scitotenv.2023.164027

Fayaz M, Khader SA, Rafiq M (2022) Landslides in the Himalayas: causes, evolution, and mitigation—a case study of National Highway 44, India (pp. 43–58). https://doi.org/10.1007/978-3-030-89308-8_4

Ghosh M, Ghosal S (2021) Climate change vulnerability of rural households in flood-prone areas of Himalayan foothills, West Bengal, India. Environ Dev Sustain 23(2):2570–2595. https://doi.org/10.1007/s10668-020-00687-0

Haigh MJ, Rawat JS, Rawat MS, Bartarya SK, Rai SP (1995) Interactions between forest and landslide activity along new highways in the Kumaun Himalaya. For Ecol Manag 78(1–3):173–189. https://doi.org/10.1016/0378-1127(95)03584-5

Haq SM, Yaqoob U, Hassan M, da Silva RJ, Calixto ES (2023) Predicted impacts of government policies and actions on the SARS-CoV-2 disease in the northwestern Himalayan region, India. J Public Health 31(4):635–643. https://doi.org/10.1007/s10389-021-01541-x

Iqbal J, Meilan R, Khan B (2020) Assessment of risk, extinction, and threats to Himalayan yew in Pakistan. Saudi J Biol Sci 27(2):762–767. https://doi.org/10.1016/j.sjbs.2019.12.033

Kapoor S, Wangdus J (2022) Impact of overtourism on sustainable development and local community wellbeing in the Himalayan region. Int J Leisure Tour Market 7(3):199. https://doi.org/10.1504/ijltm.2022.122454

Kattel GR (2022) Climate warming in the Himalayas threatens biodiversity, ecosystem functioning and ecosystem services in the 21st century: is there a better solution? Biodivers Conserv 31(8–9):2017–2044. https://doi.org/10.1007/s10531-022-02417-6

Khah SA, Rao RJ, Wani KA (2011) Ecotourism and the impact of the conventional tourism on the fragile. Ecosystems 2(3):432–442

Kumar DS, Rana G, Mairaj H (2013) Status and scenario of tourism industry in India a case study of Uttarakhand. SSRN Electron J. https://doi.org/10.2139/ssrn.2953118

Kumar S, Kumar A (2022) Hotspot and trend analysis of forest fires and its relation to climatic factors in the western Himalayas. Nat Hazards 114(3):3529–3544. https://doi.org/10.1007/s11069-022-05530-5

Lampert A (2019) Over-exploitation of natural resources is followed by inevitable declines in economic growth and discount rate. Nat Comm 10(1). https://doi.org/10.1038/s41467-019-09246-2

Majra J, Gur A (2009) Climate change and health: why should India be concerned? Ind J Occup Environ Med 13(1):11–16. https://doi.org/10.4103/0019-5278.50717

Mallick SK, Rudra S, Maity B (2023) Unplanned urban built-up growth creates problem in human adaptability: evidence from a growing up city in eastern Himalayan foothills. Appl Geogr 150:102842. https://doi.org/10.1016/j.apgeog.2022.102842

Manglik A (2023) INSA national report on seismological research in India: 2019. INSA IUGG-IGU National Committee:2019–2022. https://doi.org/10.22541/au.167907615.52090686/v2

Martolia HCS, Hutin Y, Ramachandran V, Manickam P, Murhekar M, Gupte M (2009) An outbreak of hepatitis E tracked to a spring in the foothills of the Himalayas, India, 2005. Indian J Gastroenterol 28(3):99–101. https://doi.org/10.1007/s12664-009-0036-x

Mishra A, Appadurai AN, Choudhury D, Regmi BR, Kelkar U, Alam M, Chaudhary P, Mu SS, Ahmed AU, Lotia H, Fu C, Namgyel T, Sharma U (2019) Adaptation to climate change in the Hindu Kush Himalaya: stronger action urgently needed. In: The Hindu Kush Himalaya assessment. Springer International Publishing, pp 457–490. https://doi.org/10.1007/978-3-319-92288-1_13

Mishra B, Panthi S, Poudel S, Ghimire BR (2023) Forest fire pattern and vulnerability mapping using deep learning in Nepal. Fire Ecol 19(1):3. https://doi.org/10.1186/s42408-022-00162-3

Mondal PP, Zhang Y (2018) Research Progress on changes in land use and land cover in the Western Himalayas (India) and effects on ecosystem services. Sustain For 10(12):4504. https://doi.org/10.3390/su10124504

Nyaupane GP, Chhetri N (2009) Vulnerability to climate change of nature-based tourism in the Nepalese Himalayas. Tour Geogr 11(1):95–119. https://doi.org/10.1080/14616680802643359

Oommen MA (2021) Beasts in the garden: human-wildlife coexistence in India's past and present. Front Conserv Sci 2. https://doi.org/10.3389/fcosc.2021.703432

Pandit MK, Grumbine RE (2012) Potential effects of ongoing and proposed hydropower development on terrestrial biological diversity in the Indian Himalaya. Conserv Biol 26(6):1061–1071. https://doi.org/10.1111/j.1523-1739.2012.01918.x

Patel P, Thakur PK, Aggarwal SP, Garg V, Dhote PR, Nikam BR, Swain S, Al-Ansari N (2022) Revisiting 2013 Uttarakhand flash floods through hydrological evaluation of precipitation data sources and morphometric prioritization. Geomat Nat Haz Risk 13(1):646–666. https://doi.org/10.1080/19475705.2022.2038696

Prashyanusorn V, Kaviya S, Yupapin PP (2010) Surveillance system for sustainable tourism with safety and privacy protection. Procedia Soc Behav Sci 2(1):74–78. https://doi.org/10.1016/j.sbspro.2010.01.016

Rasool R, Fayaz A, Shafiq M, Singh H, Ahmed P (2021) Land use land cover change in Kashmir Himalaya: linking remote sensing with an indicator based DPSIR approach. Ecol Indic 125:107447. https://doi.org/10.1016/j.ecolind.2021.107447

Rather RA, Bano H, Padder SA, Baba TR, Ara S, Lone FA, Nazir S (2022) Impact of anthropogenic pressure on Physico-chemical characteristics of Forest soils of Kashmir Himalaya. Bull Environ Contam Toxicol 108(6):1088–1097. https://doi.org/10.1007/s00128-022-03458-x

Sahana M, Patel PP, Rehman S, Rahaman MH, Masroor M, Imdad K, Sajjad H (2023) Assessing the effectiveness of existing early warning systems and emergency preparedness towards reducing cyclone-induced losses in the Sundarban biosphere region, India. Int J Disaster Risk Reduct 90:103645. https://doi.org/10.1016/j.ijdrr.2023.103645

Sharma R, Kumar A (2023) Determinants of quantification of tourism waste in the hilly terrain of Himalayas: a spatial approach. Waste Manag 166:46–57. https://doi.org/10.1016/j.wasman.2023.04.039

Sharma S, Joshi PK, Fürst C (2022) Exploring multiscale influence of urban growth on landscape patterns of two emerging urban centers in the Western Himalaya. Land 11(12):2281. https://doi.org/10.3390/land11122281

Sherpa Y (n.d.) Mountain tourism: a boon or a bane? Impacts of Tourism on Himalayan Women

Singh G, Rawat M, Pandey A (2023a) Debris flow simulation and modeling of the 2021 flash flood hazard caused by a rock-ice avalanche in the Rishiganga River valley of Uttarakhand. Environ Monit Assess 195(9):1118. https://doi.org/10.1007/s10661-023-11774-w

Singh RD, Gumber S, Sundriyal RC, Ram J, Singh SP (2023b) Chir pine forest and pre-monsoon drought determine spatial, and temporal patterns of forest fires in Uttarakhand Himalaya. Trop Ecol 65:32–42. https://doi.org/10.1007/s42965-023-00306-9

Singh RK, Hussain SM, Riba T, Singh A, Padung E, Rallen O, Lego YJ, Bhardwaj AK (2018) Classification and management of community forests in Indian eastern Himalayas: implications on ecosystem services, conservation and livelihoods. Ecol Process 7(1):27. https://doi.org/10.1186/s13717-018-0137-5

Singh T, Kaur J (1986) Paradox of mountain tourism: case references from the himalaya. Tour Recreat Res 11(1):19–28. https://doi.org/10.1080/02508281.1986.11014413

Stefanica M, Sandu CB, Butnaru GI, Haller AP (2021) The nexus between tourism activities and environmental degradation: Romanian tourists' opinions. Sustainability (Switzerland) 13(16). https://doi.org/10.3390/su13169210

Sultan Khuroo M, Ali Zargar S, Mahajan R, Javid G, Lal R (1988) An epidemic of hepatitis D in the foothills of the Himalayas in South Kashmir. J Hepatol 7(2):151–156. https://doi.org/10.1016/S0168-8278(88)80476-8

Sultan H, Zhan J, Rashid W, Chu X, Bohnett E (2022) Systematic review of multi-dimensional vulnerabilities in the Himalayas. Int J Environ Res Public Health 19(19):12177. https://doi.org/10.3390/ijerph191912177

Thakur A, Kumari S, Sinai Borker S, Prashant SP, Kumar A, Kumar R (2021) Solid waste Management in Indian Himalayan Region: current scenario, resource recovery, and way forward for sustainable development. Front Energy Res 9. https://doi.org/10.3389/fenrg.2021.609229

Thind PS, Kumar D, Singh S, Chohan JS, Kumar R, Sharma S, Li C, Di Bona G, Forcina A, Silvestri L (2022) Impact of unsustainable environmental conditions due to vehicular emissions on associated lifetime cancer risk in India: a novel approach. Int J Environ Res Public Health 19(11):6459. https://doi.org/10.3390/ijerph19116459

Thompson LG (2010) Climate change: the evidence and our options. Behav Anal 33(2):153–170. https://doi.org/10.1007/BF03392211

Timothy DJ (2003) Supranationalist alliances and tourism: insights from ASEAN and SAARC. Curr Issue Tour 6(3):250–266. https://doi.org/10.1080/13683500308667956

Tiwari P (2008) Land use changes in Himalaya and their impacts on environment, society and economy: a study of the Lake region in Kumaon Himalaya, India. Adv Atmo Sci 25(6):1029–1042. https://doi.org/10.1007/s00376-008-1029-x

Wang X, Wang T, Xu J, Shen Z, Yang Y, Chen A, Wang S, Liang E, Piao S (2022) Enhanced habitat loss of the Himalayan endemic flora driven by warming-forced upslope tree expansion. Nat Ecol Evol 6(7):890–899. https://doi.org/10.1038/s41559-022-01774-3

Wani MA, Mishra AK, Sharma S, Mayer IA, Ahmad M (2021) Source profiling of air pollution and its association with acute respiratory infections in the Himalayan-bound region of India. Environ Sci Pollut Res 28(48):68600–68614. https://doi.org/10.1007/s11356-021-15413-0

Wani MS, Bhat MS, Alam A, Mir SA (2023) Assessing indigenous community's perspectives and attitudes toward tourism development impacts in the northwestern Himalayas, India. Socio-Ecol Pract Res 5(1):63–78. https://doi.org/10.1007/s42532-022-00134-6

Zhang Y, Gao T, Kang S, Sprenger M, Tao S, Du W, Yang J, Wang F, Meng W (2020) Effects of black carbon and mineral dust on glacial melting on the Muz taw glacier, Central Asia. Sci Total Environ 740:140056. https://doi.org/10.1016/j.scitotenv.2020.140056

Unveiling Anthropogenic Environmental Burdens Impacting Ecosystem Services in the Himalayas

Shivangi Singh and Abhinav Yadav

1 Introduction

Unfortunately, ecosystem services are facing a significant decline on a global scale. Currently, around 60% (15 out of 24) of these essential services are being degraded or used in an unsustainable manner (Rietveld et al. 2022). The degradation of ecosystem services poses a threat to human well-being and undermines the natural assets and wealth of countries (Chen et al. 2023). This detrimental combination of increasing demands and deteriorating ecosystems severely hampers the prospects for achieving sustainable development (Murtagh et al. 2020). Additionally, the lack of knowledge regarding the sustainable utilization of ecosystem services further exacerbates their degradation. The impact of human activities on ecosystem services manifests in both negative and positive effects (Fang et al. 2021). In the region under discussion, India, serious challenges such as poverty and a growing population pose significant threats. Nearly half of India's vast population resides within 310 miles of the Himalayan range along the Gangetic plains (Nandy Rekha Pant n.d.). The need for food, shelter, and industries to support modern lifestyles places immense pressure on nature. The negative effects of human activities have weakened ecosystem services by altering habitats, ecosystem structures, and biogeochemical cycles (Sharma and Sharma 2019). Several factors contribute to this degradation, including industrialization, urbanization, climate change, grazing, hunting, international trade, and agriculture, all of which have been reported to influence risks to ecosystem services (Ahmad et al. 2023) (Table 1). Therefore,

S. Singh · A. Yadav (✉)
Institute of Environment and Sustainable Development (IESD), Banaras Hindu University (BHU), Varanasi, Uttar Pradesh, India
e-mail: abhinav.yadav@bhu.ac.in

A. Borthakur, P. Singh (eds.), *The Resilience of Traditional Knowledge Systems for a Sustainable Future*, Advances in Asian Human-Environmental Research,
https://doi.org/10.1007/978-3-031-56858-9_10

Table 1 List of important ecosystem services obtained from Himalayas

Ecosystem Service	Description
Climate Regulation	Regulation of regional and global climate patterns
Water Supply	Provision of freshwater through rivers and streams
Biodiversity Support	Habitat for numerous plant and animal species
Carbon Sequestration	Absorption and storage of carbon dioxide
Soil Formation	Creation and maintenance of fertile soils
Pollination	Vital for reproduction of plants through pollinators
Flood Regulation	Natural flood control through wetlands and forests
Erosion Control	Prevention of soil erosion and landslides
Nutrient Cycling	Recycling and distribution of nutrients in the ecosystem
Medicinal Resources	Source of various medicinal plants and herbs
Wood and Timber	Provision of wood for construction and fuel
Non-timber Forest	Products like fruits, nuts, and resins
Wild Food	Edible plants and animals for local communities
Ecotourism	Attraction of tourists for nature-based experiences
Cultural Heritage	Preservation of traditional cultures and practices
Spiritual Values	Sacred sites and connections with nature
Recreation	Opportunities for outdoor activities and leisure
Research Opportunities	Rich environment for scientific studies and discoveries
Aesthetic Value	Stunning landscapes and scenic beauty
Climate Resilience	Buffers against extreme weather events and climate change
Water Purification	Filtration and purification of water sources
Wildlife Conservation	Protection of endangered and endemic species
Air Quality	Contribution to clean air through vegetation
Renewable Energy	Potential for hydropower and other renewable sources
Education and Awareness	Environmental learning and awareness initiatives

there is an urgent need for the conservation of the Ecological services for the betterment of the Himalayan Ecosystem to function well in the near future.

The Himalayas, known as the highest mountain range in the world and often referred to as the Third Pole, possess nine out of the world's ten highest peaks, including the iconic Mount Everest. These majestic mountains serve as the source of major rivers in Asia and play a crucial role in regulating the planet's climate (Chaudhary and Bawa 2011). However, a significant percentage of the population residing in the Eastern Himalayas lives below the poverty line and relies on crop agriculture, livestock rearing, and the utilization of non-timber forest products for their survival (del Perez-Pena et al. 2021). Cash crops are extensively cultivated, and resources like firewood, fodder, and thatch grass are extracted from the forests, all of which have had a profound impact on the natural habitats of the region (source: wwf.panda.org). The loss of habitat in the region is extensive, with over 75% of the original Himalayan habitat having been destroyed or degraded. The Eastern Himalayas harbor an astonishing diversity of life, including 163 globally threatened species. Among them are Asia's three largest herbivores—the Asian elephant, the greater one-horned rhinoceros, and the wild water buffalo—as well as its largest carnivore, the tiger. The region is also home to an impressive array of flora and

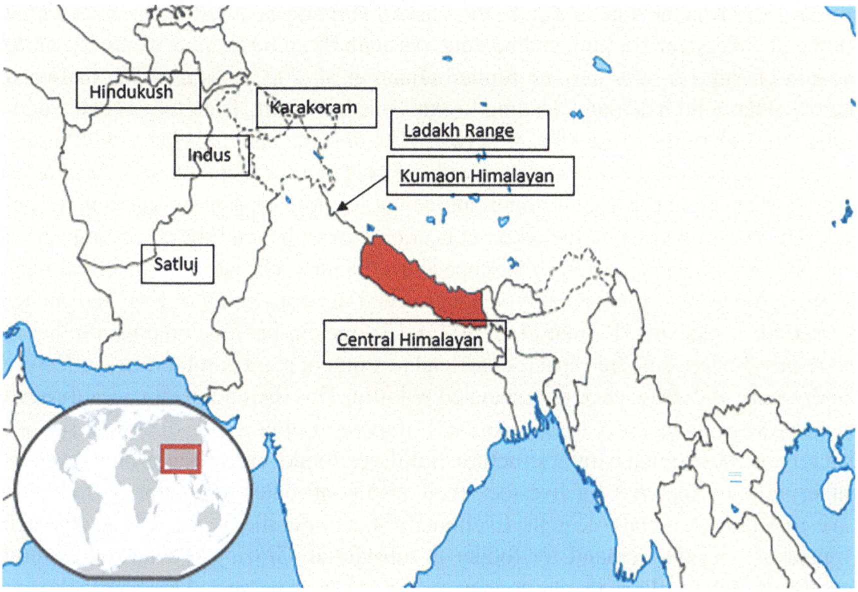

Fig. 1 Representing various different Himalayan Regional Divisions

fauna, such as 10,000 varieties of floral species, 300 placental animals, 977 avian species, 176 reptiles, 105 amphibians, and 269 freshwater fish (source: www.worldwildlife.org) (Fig. 1).

The Himalayan grasslands boast the highest population density of Bengal tigers, which coexist with Asian elephants and one-horned rhinos. These mountains provide a sanctuary for red pandas, golden langurs, and takin. Moreover, it is the only known location in the world where Bengal tigers share a habitat. Many communities' ancient traditions and livelihoods are intricately woven into the sustainable use of natural resources (Ali et al. 2022). These communities rely on these resources for their survival and highly value ecosystem services such as freshwater, erosion control, and agricultural and subsistence harvests. By practicing small-scale agriculture and effective community management, they ensure the sustainability of their traditional activities (Karaeva et al. 2023).

1.1 Several Causes Contribute to the Loss of Habitat in the Himalayas

(a) **Fuelwood and fodder collection**:

Fuelwood and fodder collection are significant drivers of habitat degradation in the Eastern Himalayas, with far-reaching consequences for the region's ecosystems. These activities involve the extraction of wood for fuel and gathering fodder for livestock, both of which profoundly impact the composition of habitats and contribute to species loss (Khanal et al. 2022). In the Eastern Himalayas, fuelwood is an essential source of energy for cooking and heating, particularly in rural areas where access to modern energy services may be limited (Jiang et al. 2022). As a result, fuelwood fuelwood has a high demand, leading to extensive harvesting from forests and woodlands. This relentless extraction of fuelwood disrupts the natural balance within ecosystems. Trees are cut down, often without proper management or reforestation efforts, resulting in the loss of important habitat for numerous plant and animal species. The consequences of fuelwood collection extend beyond the direct impact on trees. Removing trees from the landscape alters the microclimate and soil conditions, affecting the growth of understory vegetation and the availability of food and shelter for various organisms (Parmar et al. 2023). Changes in habitat composition due to the removal of specific tree species can lead to shifts in plant communities, affecting the diversity and abundance of associated wildlife. This disruption cascades through the ecosystem, potentially leading to the decline or disappearance of certain species that rely on the original habitat structure. Similarly, fodder collection, which involves gathering plant material for livestock feed, also contributes to habitat degradation. Livestock play a crucial role in the livelihoods of communities residing in the Eastern Himalayas, and the demand for fodder is substantial. Grazing pressure on natural grasslands and meadows increases as more vegetation is harvested for livestock consumption (Jalil et al. 2021). Overgrazing can lead to the degradation of grassland ecosystems, altering plant species composition, reducing biodiversity, and depleting important forage resources for wildlife. The loss of native grasses and other plant species due to overgrazing affects herbivores and the entire food web. Reductions in plant diversity can impact pollinators, herbivorous insects, and other small animals, which in turn can have cascading effects on higher trophic levels (Estoque 2023). Additionally, overgrazing can result in soil erosion, further degrading habitats and impairing the capacity of ecosystems to support diverse and healthy plant and animal communities. Fuelwood and fodder collection collectively disrupt the delicate balance of the Eastern Himalayan ecosystems. The alteration of habitat composition and the loss of key plant species have implications for a wide range of organisms, including mammals, birds, reptiles, amphibians, and insects. Protecting and sustainably managing these habitats, while meeting local communities' energy and livelihood needs, is crucial for maintaining the ecological integrity and biodiversity of the Eastern Himalayas (Vujanovic et al. 2021).

(b) **Land conversion**:

Land conversion has become a pressing issue in the Eastern Himalayas due to the expanding population and the resulting need for agricultural lands and settlements (Qin et al. 2022). As human settlements grow and urbanization advances, the conversion of forests and grasslands into cultivated areas and infrastructure is escalating rapidly (Foggin et al. 2021). This conversion process has far-reaching and detrimental consequences for the region's ecosystems, destroying vast areas of biodiverse habitats and crucial wildlife corridors. The increasing demand for food, fueled by population growth and changing dietary preferences, drives the expansion of agricultural lands (Zhong et al. 2020). Forests and grasslands, which are rich in biodiversity and provide essential ecosystem services, including habitat for numerous plant and animal species, are cleared to make way for croplands and livestock grazing. The clearance of natural vegetation disrupts the intricate web of life that depends on these habitats, triggering a chain of environmental impacts. The conversion of forests into agricultural lands directly affects the diverse array of species that call these ecosystems home. Many plants and animals are specially adapted to forest environments and rely on the resources and shelter they provide (Xu 2022). The loss of forest cover results in habitat fragmentation, isolating populations and reducing their genetic diversity. Species that require larger territories or have specific habitat requirements are particularly vulnerable, as their habitats are fragmented or completely destroyed (Yadav et al. 2022). Moreover, the destruction of forests and grasslands affects wildlife corridors, which are vital for the movement and migration of species across the landscape (Zheng and Chou 2023). These corridors enable genetic exchange, facilitate species dispersal, and support the survival and resilience of ecosystems (Theodora and Spanogianni 2022). As these corridors are encroached upon or severed by human activities, the ability of species to adapt to environmental changes and disperse to suitable habitats is severely hindered. This fragmentation of habitats can lead to population declines, increased vulnerability to extinction, and disruptions in ecological processes such as pollination and seed dispersal (Kamble and Bahadure 2022). The conversion of natural habitats also disrupts the intricate balance of nutrient cycling, hydrological processes, and carbon storage. Forests, for instance, play a critical role in regulating water flows, reducing soil erosion, and maintaining water quality (Kattel and Wu 2023). Their removal impairs these vital services and exacerbates the risks of floods, landslides, and other natural hazards. Additionally, forests are significant carbon sinks, sequestering and storing vast amounts of carbon dioxide. The conversion of forests into agricultural lands contributes to greenhouse gas emissions and exacerbates climate change (Zahraee et al. 2020). The cumulative impact of land conversion in the Eastern Himalayas is not limited to the loss of biodiversity and ecosystem services. It also affects the cultural heritage and traditional knowledge of local communities who have relied on these natural resources for their livelihoods and sustenance for generations. The destruction of biodiverse habitats and wildlife corridors undermines the region's ecological integrity and threatens human and non-human communities' long-term well-being and resilience. To address the challenges posed by land conversion, it is imperative to adopt sustainable land-use practices and promote conservation strategies that balance the needs of human populations with the preservation of biodiversity and ecosystem functions (Zeigermann 2021). This involves implementing land-use planning that takes into account ecological considerations, supporting initiatives that promote agroforestry and sustainable agriculture, and establishing protected areas and corridors to safeguard critical habitats (Barlagne et al. 2023). Collaboration between local communities, governments, conservation organizations, and other stakeholders is crucial for ensuring the sustainable management of land resources and the protection of the Eastern Himalayas' invaluable natural heritage.

(c) **Overgrazing**:

Overgrazing, resulting from widespread grazing by domesticated livestock, significantly threatens habitat and biodiversity in the Eastern Himalayas (Rana et al. 2021). The region's high-altitude alpine meadows, renowned for their rich species diversity, bear the brunt of this ecological challenge (El M'hadi and Cherkaoui 2023). In particular, overgrazing by domesticated yaks has led to the domination of a few unpalatable shrub species, endangering the species that rely on these meadows and jeopardizing herders' livelihoods. Livestock rearing, particularly the traditional practice of yak herding, plays a vital role in the livelihoods and cultural heritage of communities in the Eastern Himalayas (Cantar et al. 2021). Yaks are well-adapted to the harsh mountain environment, and their grazing activities have long been intertwined with the sustainable use of natural resources. However, as human populations increase and economic pressures grow, the intensity and scale of livestock grazing have intensified, surpassing the carrying capacity of the fragile alpine ecosystems (Alary et al. 2022). The excessive grazing pressure exerted by domesticated yaks alters the natural balance of plant communities in the high-altitude meadows. These meadows are characterized by diverse herbaceous plants, including grasses, sedges, and flowering herbs, which provide food and habitat for a wide range of wildlife species (Newell et al. 2022). However, when overgrazed, these delicate ecosystems undergo a transformation, with unpalatable shrubs becoming dominant. The overgrazing of yaks selectively removes the preferred forage species, leaving behind less palatable vegetation. As a result, the diversity and abundance of desirable grasses and herbs decline while the unpalatable shrubs proliferate. This shift in plant composition negatively impacts the habitat suitability for many herbivorous species, such as musk deer, Himalayan tahr, and various small mammals. These species rely on the availability of nutritious and diverse plant resources to meet their dietary needs and maintain their populations. Furthermore, the dominance of unpalatable shrubs compromises the ecosystem services provided by the high-altitude meadows. These services include soil stabilization, water regulation, and carbon sequestration. Overgrazing disrupts the delicate balance of these ecosystems, leading to increased soil erosion, reduced water infiltration, and diminished carbon storage capacity (Chen et al. 2023). The resulting ecological degradation affects the local biodiversity and compromises the resilience and functioning of the broader landscape. The consequences of overgrazing extend beyond the ecological realm and have profound socio-economic implications for herding communities. Livestock serves as a primary source of sustenance, income, and cultural identity for these communities. However, when overgrazing depletes the forage resources, the health and productivity of the livestock decline (Chen et al. 2023). Herders may face challenges in meeting their basic needs and experience a decline in income from the sale of livestock and livestock products. Addressing the issue of overgrazing requires a holistic approach that combines sustainable land management practices, community-based conservation efforts, and improved livestock management strategies (Darmawan et al. 2023). Implementing rotational grazing systems, where livestock are moved to different grazing areas to allow for recovery and regeneration, can help alleviate the pressures on specific meadows. Supporting herders in adopting sustainable herding practices, such as proper herd size management and diversification of livelihood options, can reduce the dependence on overgrazing as the sole source of income (Ali et al. 2021). Furthermore, engaging local communities in participatory decision-making processes and fostering collaboration between herders, conservation organizations, and government agencies is vital for developing and implementing sustainable grazing management plans. These plans should consider the ecological carrying capacity of the alpine meadows, promote the restoration of degraded areas, and encourage the conservation of critical habitats and biodiversity hotspots (Kumar et al. 2023). By balancing the needs of both human communities and the natural environment, it is possible to mitigate the detrimental effects of overgrazing and ensure the long-term sustainability of the Eastern Himalayas' unique alpine ecosystems. Such efforts can contribute to the conservation of biodiversity, the preservation of cultural traditions, and the resilience of the region's socio-ecological systems (Ahmed and Liquin 2023).

(d) **Commercial trade:**

The commercial trade of plants used in traditional medicine is a significant factor contributing to habitat degradation and biodiversity loss in the Eastern Himalayas (Tan and Omar 2022). This trade involves collecting and extracting plant species valued for their medicinal properties, which threatens specific species and disrupts the natural composition of certain habitats (Xu 2022).

The Eastern Himalayas region is renowned for its rich biodiversity, which includes a vast array of plant species with traditional medicinal uses. Local communities have long relied on traditional medicine for their healthcare needs, and various plant species play a crucial role in these healing practices (Tan and Omar 2022). However, the increasing demand for local and international traditional medicine has led to the commercial exploitation of medicinal plants, often resulting in unsustainable collection practices. The commercial trade of medicinal plants can negatively impact the region's ecosystems. First and foremost, the targeted collection of specific plant species puts pressure on their populations, especially those with slow growth rates or limited distribution. Overharvesting can lead to population declines or even local extinctions, disrupting the delicate balance of plant communities and altering the composition of habitats (Huang and Mabon 2021). Moreover, the collection of medicinal plants often involves destructive harvesting methods. Unsustainable practices, such as uprooting entire plants or excessive stripping of bark, can cause irreparable damage to the targeted species and their surrounding habitats. This not only directly impacts the species being harvested but also has cascading effects on other organisms that depend on these plants for food, shelter, or other ecological interactions The disruption of the natural composition of certain habitats is another consequence of the commercial trade in medicinal plants (Franco et al. 2022). When specific plant species are selectively targeted, it can lead to imbalances within ecosystems, as the removal of key species affects the interactions and dynamics of the surrounding flora and fauna (Chaudhary et al. 2023). This disruption can have far-reaching implications for the overall ecosystem functioning, including nutrient cycling, pollination, and seed dispersal. Furthermore, the commercial trade of medicinal plants can drive habitat degradation indirectly (Mourao and Martinho 2021). The high demand for certain species can incentivize the conversion of natural habitats into cultivation areas to meet market demands. Forests and other natural ecosystems are cleared to make way for plantations, resulting in the loss of biodiversity-rich habitats and the fragmentation of landscapes Addressing the impacts of commercial trade on medicinal plants requires a multifaceted approach (de Alves and de Oliveira 2022). It is crucial to promote sustainable harvesting practices, including the cultivation of medicinal plants in controlled environments or through agroforestry systems. This can help reduce pressure on wild populations and ensure a steady supply of medicinal plant resources. Implementing and enforcing regulations and guidelines for the collection and trade of medicinal plants can also play a vital role in preventing overexploitation and promoting sustainable practice. Furthermore, raising awareness and supporting local communities in sustainable resource management practices is essential (Puangpathumanond et al. 2023). Encouraging the conservation of medicinal plant habitats, promoting the use of alternative plant species, and fostering community-led initiatives for the sustainable management of natural resources can contribute to the long-term conservation of biodiversity and the preservation of traditional healing practices (Mariani et al. 2021). Collaboration between stakeholders, including local communities, governments, conservation organizations, and the herbal medicine industry, is crucial for developing and implementing strategies that balance the socioeconomic benefits of traditional medicine with the conservation of plant species and their habitats (Franco et al. 2022). By promoting responsible trade practices and sustainable resource management, it is possible to mitigate the negative impacts of commercial trade on medicinal plants and ensure their long-term availability while safeguarding the biodiversity and ecological integrity of the Eastern *Himalayas* (Douglas et al. 2022) (Figs. 2 and 3).

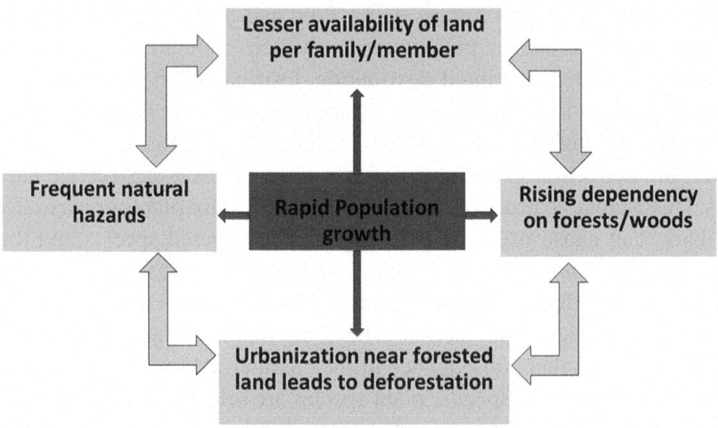

Fig. 2 Flowchart representing inter-relation between the different Himalayan degrading components

Fig. 3 Diagram representing different components of the degradation of the Himalaya Ecosystem

2 Conversion of Forests

The conversion of forests into agricultural lands is a significant driver of ecological change in the Himalayan region (Ahmad et al. 1990). The growing demand for food and the need for additional settlements have led to extensive deforestation, disrupting the delicate balance of the ecosystem, and posing a grave threat to the biodiversity hotspots that the Himalayas are renowned for. Forests in the Himalayas are incredibly diverse, harboring a vast array of plant and animal species (Berreman 1978). They provide critical habitat for numerous endemic and endangered species, playing a crucial role in maintaining the region's ecological integrity (Singh et al. 2020). However, the conversion of forests into agricultural lands results in the loss of habitat availability for these species, pushing them towards the brink of extinction. The removal of forest cover has several detrimental effects on the ecosystem. Firstly, it disrupts the intricate web of ecological relationships that exist within forests. Many species rely on specific forest types and plant communities for food, shelter, and reproduction. When forests are cleared, these vital interactions are disrupted, leading to the decline or disappearance of dependent species and the disruption of ecological processes such as pollination and seed dispersal (Zhang et al. 2011). Moreover, deforestation contributes to soil erosion, a process exacerbated by the steep slopes and monsoon rainfall patterns. The loss of tree cover exposes the soil to the erosive forces of rainfall, leading to increased sedimentation in rivers and streams (Valdiya 1995). This sedimentation affects aquatic ecosystems, disrupting the natural balance and degrading water quality. Additionally, soil erosion reduces the fertility of agricultural lands, requiring increased use of chemical fertilizers and leading to long-term degradation of soil health (Verma and Sharma 2007). Deforestation also diminishes the ability of forests to regulate water flows. Trees play a crucial role in capturing rainfall and facilitating its infiltration into the soil, helping to recharge groundwater reserves and maintain stream flows during dry periods (Lapworth et al. 2015). The loss of forest cover disrupts these natural water regulation mechanisms, leading to reduced water availability in the region and exacerbating water scarcity issues. Furthermore, the conversion of forests increases the vulnerability of the landscape to natural hazards such as landslides. Forests act as natural buffers, stabilizing slopes and reducing the likelihood of soil mass movement. When trees are removed, the integrity of the slopes is compromised, making them more prone to erosion and landslides (Vasudevan and Ramanathan 2016). This poses risks to human settlements, infrastructure, and the safety of communities in the Himalayan region. Addressing the conversion of forests requires a multifaceted approach. Promoting sustainable land management practices that balance agricultural needs with the conservation of forest ecosystems is crucial. This can involve the promotion of agroforestry systems, which integrate tree crops with agricultural cultivation, allowing for the coexistence of food production and forest conservation (Maikhuri et al. 2000). Additionally, there is a need for effective land-use planning and policy implementation that prioritizes the preservation of forest areas and the designation of protected areas to safeguard biodiversity hotspots. Encouraging community participation and engaging local stakeholders in sustainable forest management practices can help foster a sense of ownership and stewardship of forest resources (Koju et al. 2023). Education and awareness programs can play a crucial role in highlighting the importance of forests and the potential consequences of their loss. By promoting the value of intact forest ecosystems for biodiversity, water regulation, soil conservation, and natural hazard mitigation, communities can be encouraged to adopt sustainable practices and support conservation efforts (Negi 1983). Ultimately, safeguarding the remaining forests and restoring degraded forest areas is vital for maintaining the ecological balance and preserving the unique biodiversity of the Himalayan region. By prioritizing sustainable land use and conservation practices, it is possible to mitigate the impacts of forest conversion and ensure a sustainable future for both the local communities and the invaluable natural heritage of the Himalayas (Rosetti et al. 2022).

(f) **Human-wildlife Dispute:**

As human populations continue to grow and expand into wildlife habitats, the frequency of human-wildlife conflict in the Himalayan region escalates. This conflict arises when the needs and activities of humans and wildlife come into direct competition, leading to negative consequences for both sides. The increasing occurrence of human-wildlife conflict is a significant concern as it threatens the well-being of both communities and the region diverse wildlife. Human-wildlife conflict arises from various factors, including habitat loss and fragmentation, competition for resources such as food and water, and the encroachment of human settlements into wildlife territories. As more land is cleared for agriculture, infrastructure development, and human settlements, the natural habitats of wildlife are increasingly fragmented and diminished, forcing animals to venture into human-populated areas in search of resources (Yousuf and Bukhari 2019). This overlap in space and resources can lead to various forms of conflict, including crop depredation, livestock predation, property damage, and even threats to human safety. The disruption of traditional practices further exacerbates human-wildlife conflict. Traditional practices, which have evolved over generations, often incorporated strategies to mitigate conflict and maintain a harmonious relationship between humans and wildlife. However, as these practices are interrupted or abandoned due to changing lifestyles and economic pressures, the tolerance and understanding between communities and wildlife decline. This can lead to heightened levels of conflict and negative perceptions of wildlife, further aggravating the situation. Recognizing the urgency of addressing human-wildlife conflict, conservation organizations such as WWF, along with local communities and governments, are actively working to minimize these conflicts and find sustainable solutions (Chatterjee and Sastry 1997). One example of such efforts is the implementation of community-managed livestock insurance plans. These plans provide compensation to villagers for livestock losses resulting from wildlife predation, such as from snow leopards. By providing financial support, this insurance plan helps reduce the economic burden on communities and promotes a more positive attitude towards wildlife conservation. To mitigate human-wildlife conflict, various strategies are employed. These include the implementation of effective fencing systems to protect crops and livestock enclosures, the development of early warning systems to alert communities about wildlife presence, and the promotion of sustainable land-use practices that create buffer zones and corridors to reduce conflict (Singh et al. 2008) Additionally, raising awareness and conducting education programs that emphasize coexistence and the importance of conservation can foster a better understanding of wildlife and their ecological significance. It is essential to involve local communities in decision-making processes and conservation initiatives to ensure their active participation and ownership (Rahimzadeh 2020). Community-based conservation efforts, where communities are empowered to manage and benefit from wildlife and natural resources, have shown promising results in reducing human-wildlife conflict. Engaging local people as stewards of their environment fosters a sense of responsibility and a shared commitment to finding sustainable solutions that benefit both humans and wildlife (MacRae 2016). Long-term success in mitigating human-wildlife conflict relies on a comprehensive and integrated approach that combines scientific research, community engagement, policy interventions, and conservation strategies. By addressing the underlying causes of conflict and promoting coexistence, it is possible to safeguard both human livelihoods and the precious wildlife, preserving the ecological balance and promoting sustainable development in the Himalayan region.

(g) **Climatic variation:**

The Himalayan region is experiencing significant climatic variations, which have far-reaching impacts on the ecosystem and the well-being of local communities (Dimri et al. 2018). These climatic changes, primarily attributed to global warming, are causing glaciers to melt at an alarming rate, leading to the formation of glacial lakes that are prone to bursting and triggering downstream flooding events (Bajracharya and Mool 2009). Additionally, traditional water springs, which have been relied upon as a vital water source, are drying up, severely limiting the availability of water resources. The consequences of these changing climatic patterns are multifaceted and pose substantial challenges for the agriculture and food security. Farmers who depend on regular and predictable rainfall patterns are facing difficulties as climate change disrupts the timing and distribution of rainfall. Changes in rainfall patterns can result in prolonged dry spells, sudden intense rainfall, or erratic precipitation, all of which adversely affect crop production and yield (Paudel et al. 2016). This uncertainty threatens the livelihoods and food security of the local people who heavily rely on agricultural activities. Rising temperatures and changing humidity levels are also altering the ecological dynamics of the Himalayan region. Warmer temperatures create favorable conditions for the proliferation of insect pests and the spread of diseases in areas where they were previously absent or less prevalent (Yarzábal 2020). These pests and diseases pose significant challenges to agricultural productivity, affecting both crop yields and the health of livestock (Bhattacharyya et al. 2023). In turn, this has implications for food production, economic stability, and the overall well-being of local communities. Furthermore, the melting of glaciers and the formation of glacial lakes are major concerns in the region.

Glacial meltwater serves as a crucial water source for rivers and streams, contributing to the overall water supply and the maintenance of aquatic ecosystems. However, the accelerated melting of glaciers and the formation of glacial lakes increase the risk of catastrophic glacial lake outburst floods (GLOFs) (Bajracharya and Mool 2009). These floods occur when the walls of the glacial lakes breach, releasing enormous volumes of water downstream. GLOFs pose severe threats to human settlements, infrastructure, and the safety of communities in the Himalayan region. Addressing the impacts of climatic variations requires a comprehensive and multi-faceted approach. Mitigation efforts to reduce greenhouse gas emissions globally are crucial for slowing down the pace of climate change. Additionally, adaptation strategies must be implemented at the local level to minimize the adverse effects of changing climatic conditions (Serba et al. 2022). Water management practices must be developed and strengthened to ensure efficient and equitable water distribution. This can include constructing water storage systems, such as reservoirs and tanks, to capture and store rainfall during the wet season for use during dry periods. Implementing water conservation measures, such as rainwater harvesting and efficient irrigation techniques, can also help maximize water availability for agricultural purposes (Brown 2020). Diversification of livelihoods and the promotion of climate-resilient agriculture practices can enhance the resilience of local communities to changing climatic conditions. This can involve the adoption of crop varieties that are more tolerant to erratic rainfall patterns, introducing agroforestry practices to improve soil moisture retention, and implementing sustainable land management practices to mitigate soil erosion and enhance soil fertility (Hessel et al. 2007). Education and awareness programs are vital in enhancing understanding and preparedness for climate change impacts. Empowering local communities with knowledge and skills to adapt to changing conditions can foster resilience and enable them to make informed decisions regarding their livelihoods and natural resource management (Uniyal and Singh 2012). International collaborations, policy interventions, and financial support are essential to address the broader challenges posed by climatic variations. Cooperation between governments, research institutions, non-governmental organizations, and local communities is crucial for effective adaptation and mitigation strategies that protect the Himalayan ecosystem and the well-being of its inhabitants (Tleuken et al. 2022). By taking proactive measures to mitigate greenhouse gas emissions, promoting sustainable water management, fostering climate-resilient agriculture, and strengthening community resilience, it is possible to address the challenges posed by climatic variations in the Himalayan region. Such efforts can contribute to the preservation of vital ecosystems, protect livelihoods, and secure a sustainable future for the people and biodiversity of the Himalayas (Måren and Sharma 2018).

3 Restoring the Terai ARC Landscape

The Terai Arc Landscape (TAL) is a critical conservation area situated in the lowlands of Nepal and India, encompassing a total area of approximately 13,000 sq. km. This region serves as a vital habitat for numerous endangered and charismatic wildlife species, including the Bengal tiger, Indian rhinoceros, and Asian elephant. However, the TAL has faced significant challenges due to habitat fragmentation, encroachment, and unsustainable land-use practices. To address these conservation concerns, the governments of Nepal and India, along with various international organizations and local communities, have initiated efforts to restore and reconnect fragmented habitats within the Terai Arc Landscape (Pohle 1992). One of the key strategies employed is the establishment of protected areas and wildlife corridors to enhance landscape connectivity and facilitate the movement of wildlife across the region. An essential component of the restoration efforts in the TAL is the promotion of participatory conservation and community-based natural resource management. Local communities play a crucial role in protecting and restoring the landscape, as they have traditional knowledge and a deep understanding of the local ecosystems (Negi et al. 2022). Engaging communities in decision-making processes and incentivizing them for sustainable practices contribute to the long-term success of restoration initiatives. Furthermore, restoring and conserving the TAL involves tackling issues such as illegal wildlife trade, habitat degradation, and human-wildlife conflicts. Strengthening law enforcement, enhancing anti-poaching measures, and raising awareness about the importance of wildlife conservation are critical to combat illegal activities that threaten the region, biodiversity (Bhagwat 2012). Restoration efforts also focus on sustainable land-use practices, including agroforestry, to promote biodiversity conservation while meeting local communities' needs. This includes the promotion of sustainable agriculture, establishing community-managed forests, and adopting eco-friendly livelihood options to reduce pressure on natural resources (Farooq et al. 2020). Measuring the effectiveness of restoration initiatives in the TAL is crucial to assess progress and inform future conservation efforts. Monitoring programs employ various techniques such as camera trapping, satellite tracking of wildlife, and ecological surveys to gather data on population dynamics, habitat quality, and connectivity within the landscape (Krogh

2020). The restoration of the Terai Arc Landscape is not only crucial for conserving the region unique biodiversity but also for securing the livelihoods of local communities and promoting sustainable development. The success of these restoration efforts depends on continued collaboration between governments, conservation organizations, local communities, and other stakeholders. By restoring and reconnecting habitats, addressing threats to wildlife, and promoting sustainable practices, it is possible to preserve the ecological integrity and enhance the resilience of the Terai Arc Landscape for future generations. According to the State of Forest Report, 2021, there has been a significant decline in forest in the hill districts of the country, with a loss of 902 sq. km compared to 2019. The Himalayan states have experienced an even more pronounced loss, with an overall decline of 1072 sq. km of forest cover (Schickhoff 1995). Arunachal Pradesh, Manipur, Nagaland, and Mizoram have been particularly affected, losing 257 sq. km, 249 sq. km, 235 sq. km, and 186 sq. km respectively. The consequences of deforestation in the Himalayan region are severe. It accelerates environmental degradation processes such as soil erosion, slope failures, soil fertility depletion, fuelwood and fodder scarcity, increased overland flows, reduced groundwater recharge, and loss of biological diversity. These impacts have long-term implications for the region's ecological health (Reddy et al. 2016). According to the State of Forest Report, 2021, there has been a decline of 902 sq. km in forest cover in the hill districts of the country compared to 2019 (Agrawal 1996). The loss is particularly significant in the Himalayan states, which reported an overall loss of 1072 sq. km of forest cover. Arunachal Pradesh, with 16 hill districts, lost 257 sq. km of forest cover, Manipur lost 249 sq. km, Nagaland lost 235 sq. km, and Mizoram lost 186 sq. km. The loss of forest cover in the Himalayas has various detrimental consequences. It leads to environmental degradation processes such as soil erosion, slope failures, depletion of soil fertility, scarcity of fuelwood and fodder, increased overland flows, reduced groundwater recharge, and loss of biological diversity (Khan et al. 2009). The deforestation of slopes exacerbates these processes, causing accelerated damage to the region's fragile ecosystem. A study by the Scholars Programmed estimated that by 2100, only 10% of the Indian Himalayan land area would be covered by dense forests. The study further projected that the dense forest cover in the Western Himalayas would decrease from 61% in 2000 to 16.8% in 2100, while in the eastern Himalayas, it would decrease from 76.2% in 2000 to 38.7% in 2100 (Bawa et al. 2007). Despite slower deforestation rates, Sikkim is expected to have the least forest cover by 2100 due to 50% of its geographic area lying above the timberline, where forest growth is impossible (Wani et al. 2022). The loss of forest cover in the Himalayas can be attributed to various factors. Unplanned urbanization, commercial timber extraction, capitalistic development projects, mining, and quarrying are identified as troublesome forces of modern development. The activities, driven by economic interests, contribute to the degradation and loss of forested areas in the region (Ali et al. 2005). The vulnerability to soil erosion is higher in the northeastern hills compared to the northwestern hills due to the higher rainfall in the former. The northeastern hills receive annual rainfall ranging from 1500 to 11,500 mm, while the northwestern hills receive rainfall between 350 and 3000 mm annually. This disparity in rainfall makes the northeastern hills more susceptible to soil erosion, affecting 22.3% of the area, compared to 12.6% of the area in the northwestern hills (Hessel et al. 2007). Addressing the challenges of deforestation and forest degradation in the Himalayas requires concerted efforts to promote sustainable land-use practices, conservation measures, and stricter enforcement of environmental regulations. It is crucial to balance development with the preservation of the region valuable ecosystems and biodiversity. By adopting sustainable approaches and considering the long-term ecological implications of development activities, it is possible to mitigate the loss of forest cover and promote the resilience and well-being of the Himalayan region.

4 Himalayan Ecology

The Himalayan region is characterized by its unique mountain ecosystems, which exhibit three—dimensional geo-ecological differentiation, complex topography, and a significant variation in climatic and edaphic conditions (Oulkar et al. 2019). These factors contribute to forming numerous ecological niches and make the Himalayas biodiversity hotspots.

5 Ecosystem Modeling

Correlative modeling approaches have become increasingly popular for predicting species distribution patterns in mountain ecosystems worldwide (Adhikari et al. 2019). These models utilize a combination of historical, current, and projected future data to estimate the potential distribution of various organisms. Among the organisms studied using these models, trees, mammals, herbaceous plants, and birds have received the most attention. In the majority of studies (87%), the widely used WORLDCLIM climate data is employed as the primary predictor variable for modeling species distribution (Schwab et al. 2022). This dataset provides comprehensive global climate data that can be utilized to understand the relationship between species occurrence and factors. Interestingly, nearly half of the studies solely rely on WORLDCLIM data as the main source for their modeling efforts. While the majority of studies predominantly use WORLDCLIM data, a small proportion (2%) of researchers have explored and compared the influence of different climate data sources on model performance (Bobrowski and Udo 2017). These studies recognize the importance of assessing the robustness and reliability of model predictions by considering alternative climate datasets. The interpretation of species-environment relationships in these models varied across studies. Approximately 37% of the studies lacked detailed interpretation, providing limited insight into the ecological needs of the species reflected in the final model. However, 27% of studies included brief information on these relationships, while a more substantial proportion (36%) offered elaborate interpretations that delved into the ecological requirements and associations between the species and the predictor variables (Bobrowski et al. 2021a). Understanding and interpreting these relationships is critical for gaining insights into the current, historical, and future distribution ranges of species, as well as for guiding conservation and management strategies (Vetaas and Chaudhary 1998). By utilizing correlative modeling approaches and integrating various data sources, researchers can enhance our understanding of species-environment relationships and predict species distributions in mountain ecosystems. These models provide valuable insights into how species respond to environmental factors, facilitating effective conservation planning and management efforts. Continued research and exploration of alternative data sources will further refine and improve these modeling approaches, ultimately enhancing our ability to predict and conserve species in mountain ecosystems worldwide (Voigt 1979).

6 Challenges

Studying the Himalayan ecosystem poses several significant challenges, primarily related to large-scale sampling of species occurrences and the acquisition of high-quality, fine-scale environmental predictors (Riaz et al. 2022). The unique characteristics of the Himalayan region, including its rugged terrain, extreme altitudes, and vast geographical extent, present obstacles to conducting comprehensive field surveys and data collection. One of the primary challenges is the difficulty of accessing remote and inaccessible areas within the Himalayas (Singh et al. 2013). The mountainous terrain and often harsh climatic conditions make it challenging for researchers to reach certain locations, particularly in high-altitude regions. These areas, which may harbor unique and specialized species, are often underrepresented in species occurrence data due to their limited accessibility. Additionally, the vast geographical extent of the Himalayan-mountain system poses a challenge in terms of sampling coverage (Ives 2012). The region spans several countries, including India, Nepal, Bhutan, and parts of China, which adds complexity to research coordination and data collection efforts. Achieving comprehensive and representative sampling across such a large area is a logistical and resource-intensive endeavor. Acquiring high-quality, fine-scale environmental predictors is another challenge in studying the Himalayan ecosystem. Fine-scale data on climate, topography, vegetation, and other environmental factors are essential for accurately modeling species distributions and understanding their ecological requirements (Rao et al. 2003). However, such data can be limited or unavailable, particularly in remote areas or regions with limited research infrastructure. Overcoming these challenges is crucial for developing comprehensive ecological model and implementing effective conservation strategies in the Himalayan region (Bobrowski et al. 2021b). Addressing the sampling deficits requires innovative approaches, such as utilizing advanced remote sensing techniques and incorporating citizen science initiatives to gather species occurrence data from diverse locations (Gairola et al. 2013). Collaborative efforts among researchers, institutions, and local communities are essential to enhance access to remote areas and facilitate data collection. Improving the acquisition of high-quality environmental predictors involves integrating data from multiple sources, including satellite imagery, climate models, and field-based measurements. Developing partnerships with local research institutions, government agencies, and international organizations can contribute to the collection and sharing of fine-scale environmental data in the Himalayan region (Batidzirai et al. 2021). By addressing these challenges and filling the gaps in data coverage, researchers can gain a more comprehensive understanding of the Himalayan ecosystem and its unique biodiversity. This knowledge is crucial for informing conservation strategies, identifying priority areas for protection, and promoting sustainable management practices in the face of environmental changes and anthropogenic pressures (Sharma and Sharma 2019).

7 Dispersion of Foreign Species in the Himalayan Ecosystem

The preservation of Himalayan ecology requires comprehensive understanding and management of the unique mountain ecosystems. By incorporating advanced modeling techniques, addressing sampling challenges, and mitigating the dispersion of foreign species, it is possible to safeguard the biodiversity and The Himalayan region, renowned for its rich biodiversity, is experiencing the dispersion of foreign species, posing a significant threat to its unique ecosystem (Campbell 2018). The Indian Himalayan region (IHR) in particular has been a hotspot for introducing and spreading alien plant species (Mehraj et al. 2018). In fact, a comprehensive study identified a total of 771 alien plant species in the IHR, comprising both cultivated and naturalized species. Among the alien plant species present in the IHR, those native to Southern America and characterized by a perennial life span and herbaceous growth form are notably well-represented. These species have successfully established themselves in the region, potentially outcompeting native flora and disrupting local ecosystems (Khuroo et al. 2011). Their presence highlights the significant impact of intentional and unintentional introductions of invasive alien species in the Himalayan ecosystem. The dispersion of foreign species in the IHR exhibits variations in species composition across different parts of the region. Evidence of distance decay of floristic similarity indicates that the species composition becomes more dissimilar as the distance between locations increases. This suggests that the dispersal and establishment of alien plant species are influenced by local environmental factors and geographical constraints within the IHR. Several factors contribute to the richness of alien and naturalized plant species in the IHR (Wangchuk et al. 2014). Average annual rainfall plays a significant role, as regions with higher rainfall tend to have a greater diversity of alien species. Additionally, the length of total traffic, likely associated with human activities and connectivity, is another factor influencing the presence and spread of alien plant species. The accelerated economic activities in the mountain landscapes of the Indian Himalaya have further exacerbated the dispersion of foreign species. Anthropogenic factors such as tourism, urbanization, and development projects have led to intentional and unintentional introductions of invasive alien species (Malik et al. 2016). These introductions, whether through the deliberate planting of non-native species or the inadvertent transport of seeds, have contributed to the alteration of ecosystems and ecological imbalances in the fragile Himalayan region. Preserving the Himalayan ecology necessitates a comprehensive understanding of the unique mountain ecosystems and the implementation of effective management strategies. By incorporating advanced modeling techniques, such as species distribution modeling, researchers can gain insights into the potential distribution patterns and ecological impacts of alien plant species in the region (Khuroo et al. 2011). Addressing the challenges associated with large-scale sampling and acquiring fine-scale environmental data will enhance our understanding of foreign species' dispersal dynamics and impacts. Efforts to mitigate the dispersion of foreign species should include strengthening regulations on importing and cultivating potentially invasive species, promoting awareness about the ecological risks associated with non-native plants, and implementing measures to control and manage invasive species once established (Panda et al. 2018). Additionally, promoting native plant species conservation and restoration initiatives can help restore and maintain the ecological integrity of the Himalayan region (Vasistha et al. 2009). By taking a proactive and holistic approach to addressing the dispersion of foreign species, it is possible to safeguard the biodiversity and ecological balance of the Himalayan ecosystem. This requires collaboration between researchers, conservation organizations, policymakers, and local communities to develop effective strategies and implement sustainable practices for the long-term preservation of the region's unique biodiversity. Ecological integrity of the Himalayan region.

8 Human Effect on the Himalayan Ecology

Overall, the literature on the human effect on the Himalayan ecology highlights the significant impact that various human activities have had on this fragile ecosystem (Sharma and Sharma 2019). The evidence points towards an urgent need for conservation and sustainable management practices to protect the biodiversity and ecological services provided by this region. Ecotourism has emerged as one of the major contributors to the degradation of the Himalayan ecology (Brandt et al. 2019). Numerous studies have documented the negative effects of unregulated tourism activities, such as deforestation, habitat destruction, pollution, and disturbance to wildlife. Additionally, the influx of tourists has also led to increased waste generation and inadequate waste management practices, further degrading the environment. Consequently, the literature emphasizes the importance of implementing sustainable tourism practices, including visitor education and regulations, to mitigate the negative impacts on the Himalayan ecology (Zou et al. 2021). Another key factor affecting the Himalayan ecology is climate change. The literature demonstrates that the region is highly vulnerable to the effects of global warming, resulting in glacial retreat, altered precipitation patterns, and increased frequency of extreme weather events. These changes have profound implications for the biodiversity, freshwater supply, and downstream communities. To address this, further research is needed to understand the specific impacts of climate change on different ecological components of the Himalayas and to develop adaptive strategies that consider the unique vulnerabilities of this mountainous ecosystem (Meenawat and Sovacool 2011). Agricultural activities in the Himalayas also significantly influence the ecology. The literature reveals that traditional farming practices, such as terracing and agroforestry, have had a positive impact on soil conservation and biodiversity. However, with the introduction of modern agricultural practices, such as mechanization, chemical fertilizers, and pesticides, negative impacts such as soil erosion, water pollution, and loss of traditional crop varieties have been observed (Kaushik et al. 2010). To mitigate these effects, it is crucial to promote sustainable agriculture practices that integrate traditional knowledge with modern techniques. Additionally, further research is needed to assess these practices' long-term impacts and develop strategies that enhance food security while minimizing ecological degradation. Lastly, the literature highlights the importance of involving local communities in conservation efforts in the Himalayas. Traditional knowledge and indigenous practices have been found to play a crucial role in maintaining the region's ecological balance. Conservation efforts can be more effective and sustainable by fostering community participation and empowering local stakeholders (Alamoudi et al. 2022). Future research should focus on understanding the socio-economic dynamics and cultural aspects of the indigenous communities living in the Himalayas to develop inclusive conservation strategies that promote both the well-being of local communities and the preservation of the Himalayan ecology. In conclusion, the literature on the human effect on the Himalayan ecology demonstrates the urgent need for conservation efforts and sustainable management practices in the region (Chandel et al. 2019). By addressing issues related to tourism, climate change, agriculture, and local community involvement, society can protect the unique biodiversity and ecological services the Himalayas provides. Further research should explore the specific impacts of these factors on the region, identify innovative solutions, and develop inclusive approaches that ensure the long-term sustainability of this fragile ecosystem (Zhang and Ma 2021).

Conclusion: The Himalayan ecosystem is under pressure from various factors, and urgent action is required to protect its unique biodiversity and ecological services. This article calls for a multidisciplinary and collaborative approach involving researchers, policymakers, conservation organizations, and local communities to develop and implement strategies that ensure the long-term sustainability of this fragile and vital ecosystem. Future research should continue to explore and address the specific challenges and opportunities presented by the Himalayan region.

References

Adhikari D, Tiwary R, Singh PP, Upadhaya K, Singh B, Haridasan KE, Bhatt BB, Chettri A, Barik SK (2019) Ecological niche modeling as a cumulative environmental impact assessment tool for biodiversity assessment and conservation planning: a case study of critically endangered plant Lagerstroemia minuticarpa in the Indian eastern Himalaya. J Environ Manag 243:299–307. https://doi.org/10.1016/j.jenvman.2019.05.036

Agrawal A (1996) The community vs. the market and the state: forest use in Uttarakhand in the Indian Himalayas. J Agric Environ Ethics 9(1):1–15. https://doi.org/10.1007/bf01965667

Ahmad N, Liu Y, Zikovic S, Belyaeva Z (2023) The effects of technological innovation on sustainable development and environmental degradation: evidence from China. Technol Soc 72:102184. https://doi.org/10.1016/j.techsoc.2022.102184

Ahmad A, Rawat JS, Rai SC (1990) An analysis of the himalayan environment and guidelines for its management and ecologically sustainable development. Environmentalist 10(4):281–298. https://doi.org/10.1007/BF02239722

Ahmed S, Liquin P (2023) Socio-ecological challenges of hydroelectric dams among ethnic minorities in northern Laos. Environ Dev 46:100864. https://doi.org/10.1016/j.envdev.2023.100864

Alamoudi AK, Abidoye RB, Lam TYM (2022) The impact of stakeholders' management measures on citizens' participation level in implementing smart sustainable cities. Sustain For 14(24). https://doi.org/10.3390/su142416617

Alary V, Lasseur J, Frija A, Gautier D (2022) Assessing the sustainability of livestock socio-ecosystems in the drylands through a set of indicators. Agric Syst 198:103389. https://doi.org/10.1016/j.agsy.2022.103389

Ali J, Benjaminsen TA, Hammad AA, Dick ØB (2005) The road to deforestation: an assessment of forest loss and its causes in Basho Valley, northern Pakistan. Glob Environ Chang 15(4):370–380. https://doi.org/10.1016/j.gloenvcha.2005.06.004

Ali M, Kirikkaleli D, Sharma R, Altuntas M (2022) The nexus between remittances, natural resources, technological innovation, economic growth, and environmental sustainability in Pakistan. Environ Sci Pollut Res 29(50):75822–75840. https://doi.org/10.1007/s11356-022-21228-4

Ali FH, Liaqat F, Azhar S, Ali M (2021) Exploring the quantity and quality of occupational health and safety disclosure among listed manufacturing companies: evidence from Pakistan, a lower-middle income country. Safety Sci 143. https://doi.org/10.1016/j.ssci.2021.105431

Bajracharya SR, Mool P (2009) Glaciers, glacial lakes and glacial lake outburst floods in the Mount Everest region, Nepal. Ann Glaciol 50(53):81–86. https://doi.org/10.3189/172756410790595895

Barlagne C, Bezard M, Drillet E, Arnaud L, Diman J-L, Alexandre G, Vinglassalon A, Nijnik M (2023) Stakeholders' engagement platform to identify sustainable pathways for the development of multi-functional agroforestry in Guadeloupe, French West Indies. Agrofor Syst 97(3, SI):463–479. https://doi.org/10.1007/s10457-021-00663-1

Batidzirai B, Trotter PA, Brophy A, Stritzke S, Moyo A, Twesigye P, Puranasamriddhi A, Madhlopa A (2021) Towards people-private-public partnerships: an integrated community engagement model for capturing energy access needs. Energy Res Soc Sci 74. https://doi.org/10.1016/j.erss.2021.101975

Bawa KS, Joseph G, Setty S (2007) Poverty, biodiversity and institutions in forest-agriculture ecotones in the Western Ghats and eastern Himalaya ranges of India. Agric Ecosyst Environ 121(3):287–295. https://doi.org/10.1016/j.agee.2006.12.023

Berreman GD (1978) Ecology, demography and domestic strategies in the western Himalayas. J Anthropol Res 34(3):326–368. https://doi.org/10.1086/jar.34.3.3629783

Bhagwat SA (2012) Biodiversity in the Himalayas in a changing world (Himalayan biodiversity in the changing world). Mountain Res Dev 32(3 RU):R141–R142. https://doi.org/10.1659/mrd.mm108.ru

Bhattacharyya PN, Sarmah SR, Somnath R, Sarma B, Nath BC, Bhattacharyya LHH (2023) Perspectives of Beauveria bassiana, an entomopathogenic fungus for the control of insect-pests in tea [Camellia sinensis (L.) O. Kuntze]: opportunities and challenges. Int J Trop Insect Sci 43(1):1–19. https://doi.org/10.1007/s42690-022-00932-1

Bobrowski M, Udo S (2017) Why input matters: selection of climate data sets for modelling the potential distribution of a treeline species in the Himalayan region. Ecol Model 359:92–102. https://doi.org/10.1016/j.ecolmodel.2017.05.021

Bobrowski M, Weidinger J, Schwab N, Schickhoff U (2021a) Searching for ecology in species distribution models in the Himalayas. Ecol Model 458:109693. https://doi.org/10.1016/j.ecolmodel.2021.109693

Bobrowski M, Weidinger J, Schwab N, Schickhoff U (2021b) Searching for ecology in species distribution models in the Himalayas. In: Ecological modelling, vol 458. Elsevier B.V. https://doi.org/10.1016/j.ecolmodel.2021.109693

Brandt JS, Radeloff V, Allendorf T, Butsic V, Roopsind A (2019) Effects of ecotourism on forest loss in the Himalayan biodiversity hotspot based on counterfactual analyses. Conserv Biol 33(6):1318–1328. https://doi.org/10.1111/cobi.13341

Brown T (2020) Pathways to agricultural skill development in the Indian Himalayas. J South Asian Dev 15(2):270–292. https://doi.org/10.1177/0973174120943081

Campbell B (2018) Biodiversity, livelihoods and struggles over sustainability in Nepal. Landsc Res 43(8):1056–1067. https://doi.org/10.1080/01426397.2018.1503241

Cantar NM, Luz Endere M, Laura Zulaica M (2021) The "archaeology" of sustainability in the approach to cultural heritage. Revista De Estudios Sociales 75:71–86. https://doi.org/10.7440/res75.2021.07

Chandel U, Guleria SK, Sudan RS, Kumar D (2019) Genotype by environment interaction and stability analysis for maize hybrids in north western himalayas ecology. Maydica 64(1) https://www.scopus.com/inward/record.uri?eid=2-s2.0-85070062071&partnerID=40&md5=398072f4de04592c547ba911ec97b4c1

Chatterjee S, Sastry ARK (1997) Biodiversity "hotspots" conservation programme of WWF-India. Int J Ecol Environ Sci 23(4):479–483. https://www.scopus.com/inward/record.uri?eid=2-s2.0-0031350189&partnerID=40&md5=d1f2b84b7eb7bba554bd6db73eb9eab4

Chaudhary P, Bawa KS (2011) Local perceptions of climate change validated by scientific evidence in the Himalayas. Biol Lett 7(5):767–770. https://doi.org/10.1098/rsbl.2011.0269

Chaudhary S, Kaur P, Alofysan H, Halberstadt J, Dhir A (2023) Connecting the dots? Entrepreneurial ecosystems and sustainable entrepreneurship as pathways to sustainability. Bus Strategy Environ 32:5935–5951. https://doi.org/10.1002/bse.3466

Chen F, Ali S, Ma J, Arshad S, Ahmad S (2023) Material productivity and environmental degradation: moderating role of environment-related technologies in achieving carbon neutrality. Gondwana Res 117:155–168. https://doi.org/10.1016/j.gr.2023.01.010

Darmawan H, Chang H-L, Wu H-H (2023) A community-based breeding program as a genetic resource management strategy of Indonesian Ongole cattle. Sustain For 15(7). https://doi.org/10.3390/su15076013

de Alves DO, de Oliveira L (2022) Commercial urban agriculture: a review for sustainable development. Sustain Cities Soc 87:104185. https://doi.org/10.1016/j.scs.2022.104185

del Perez-Pena MC, Jimenez-Garcia M, Ruiz-Chico J, Pena-Sanchez AR (2021) Analysis of research on the SDGs: the relationship between climate change, poverty and inequality. Appl Sci-Basel 11(19). https://doi.org/10.3390/app11198947

Dimri AP, Immerzeel WW, Salzmann N, Thayyen RJ (2018) Comparison of climatic trends and variability among glacierized environments in the Western Himalayas. Theor Appl Climatol 134(1–2):155–163. https://doi.org/10.1007/s00704-017-2265-8

Douglas MA, Mollenkopf DA, Castillo VE, Bell JE, Dickey EC (2022) Journeys, not destinations: theorizing a process view of supply chain integrity. J Bus Ethics 181(1):195–220. https://doi.org/10.1007/s10551-021-04906-0

El M'hadi H, Cherkaoui A (2023) Urban planning of business parks (BPs): ecological challenges and commitment to sustainable development, the case study of the technopole ?Casa nearshore? Alex Eng J 67:23–30. https://doi.org/10.1016/j.aej.2022.09.047

Estoque RC (2023) Complexity and diversity of nexuses: a review of the nexus approach in the sustainability context. Sci Total Environ 854:158612. https://doi.org/10.1016/j.scitotenv.2022.158612

Fang X, Wu J, He C (2021) Assessing human-environment system sustainability based on regional safe and just operating space: the case of the Inner Mongolia grassland. Environ Sci Pol 116:276–286. https://doi.org/10.1016/j.envsci.2020.12.007

Farooq Q, Hao J, Liu X, Xiao D, Hao Y (2020) Social and environmental development: fresh concepts and soft measures towards sustainable development. Sustain Dev 28(6):1796–1803. https://doi.org/10.1002/sd.2075

Foggin JM, Lechner AM, Emslie-Smith M, Hughes AC, Sternberg T, Dossani R (2021) Belt and road initiative in Central Asia: anticipating socioecological challenges from large-scale infrastructure in a global biodiversity hotspot. Conserv Lett 14(6). https://doi.org/10.1111/conl.12819

Franco CC, Rebolledo-Leiva R, Gonzalez-Garcia S, Feijoo G, Moreira MT (2022) Addressing the food, nutrition and environmental nexus: the role of socio-economic status in the nutritional and environmental sustainability dimensions of dietary patterns in Chile. J Clean Prod 379(1):134723. https://doi.org/10.1016/j.jclepro.2022.134723

Gairola S, Procheş Ş, Rocchini D (2013) High-resolution satellite remote sensing: a new frontier for biodiversity exploration in Indian Himalayan forests. Int J Remote Sens 34(6):2006–2022. https://doi.org/10.1080/01431161.2012.730161

Hessel R, Gupta MK, Datta PS, Van Den Elsen E, Sharma SD (2007) Rainfall, soil moisture content and runoff in a small catchment in the Indian Himalayas. Int J Ecol Environ Sci 33(2–3):115–128. https://www.scopus.com/inward/record.uri?eid=2-s2.0-60149091380&partnerID=40&md5=a92516a2afa7ba3d784b50ceb3467a87

Huang Y-C, Mabon L (2021) Coastal landscapes, sustainable consumption and peripheral communities: evaluating the Miramar resort controversy in Shanyuan Bay, Taiwan. Marine Policy 123:104283. https://doi.org/10.1016/j.marpol.2020.104283

Ives JD (2012) Environmental change and challenge in the Himalaya. A historical perspective. Pirineos 167:29–68. https://doi.org/10.3989/Pirineos.2012.167003

Jalil A, Rauf A, Sikander W, Yonghong Z, Tiebang W (2021) Energy consumption, economic growth, and environmental sustainability challenges for belt and road countries: a fresh insight from Chinese going global strategy. Environ Sci Pollut Res 28(46):65987–65999. https://doi.org/10.1007/s11356-021-15549-z

Jiang Z, Mahmud AR, Maneengam A, Nassani AA, Haffar M, Cong PT (2022) Non-linear effect of biomass, fossil fuels and renewable energy usage on the economic growth: managing sustainable development through energy sector. Fuel 326:124943. https://doi.org/10.1016/j.fuel.2022.124943

Kamble T, Bahadure SP (2022) A sustainability assessment framework for population density in central Indian cities. Archnet-Ijar Int J Archit Res 16(1):134–153. https://doi.org/10.1108/ARCH-05-2021-0142

Karaeva A, Ionescu G, Cioca LI, Tolkou A, Katsoyiannis I, Kyzas G (2023) Environmental sustainability for traditional energy small and medium enterprises. Environ Sci Pollut Res 30:47822–47831. https://doi.org/10.1007/s11356-023-25718-x

Kattel GRR, Wu C (2023) Reconfiguration of ecohydrology as a sustainability tool for Himalayan waterways. Ecohydrology 17. https://doi.org/10.1002/eco.2522

Kaushik A, Sharma HR, Jain S, Dawra J, Kaushik CP (2010) Pesticide pollution of river Ghaggar in Haryana, India. Environ Monit Assess 160(1–4):61–69. https://doi.org/10.1007/s10661-008-0657-z

Khan K, Almas AS, Chaudhry MN (2009) Environmental assessment of land degradation: a case study of NW Himalayas, Pakistan. In: Proceedings of the IASTED International Conference on Environmental Management and Engineering, EME 2009, pp 183–188. https://www.scopus.com/inward/record.uri?eid=2-s2.0-74549138309&partnerID=40&md5=89635f2dd66cf540f559b440dac01a6f

Khanal P, Dhakal R, Khanal T, Deepak P, Devkota NR, Nielsen MO (2022) Sustainable livestock production in Nepal: a focus on animal nutrition strategies. Agriculture-Basel 12(5). https://doi.org/10.3390/agriculture12050679

Khuroo AA, Weber E, Malik AH, Reshi ZA, Dar GH (2011) Altitudinal distribution patterns of the native and alien woody flora in Kashmir Himalaya, India. Environ Res 111(7):967–977. https://doi.org/10.1016/j.envres.2011.05.006

Koju U, Karki S, Shrestha A, Maraseni T, Gautam AP, Cadman T, Sherpa AP, Lama ST (2023) Local stakeholders' priorities and perceptions towards forest ecosystem services in the Red panda habitat region of Nepal. Land Use Policy 129. https://doi.org/10.1016/j.landusepol.2023.106657

Krogh DS (2020) Landslides: monitoring, susceptibility and management. https://www.scopus.com/inward/record.uri?eid=2-s2.0-85144658896&partnerID=40&md5=c8c8ec985c067d644316969a68b1b057

Kumar S, Gopinath KA, Sheoran S, Meena RS, Srinivasarao C, Bedwal S, Jangir CK, Mrunalini K, Jat R, Praharaj CS (2023) Pulse-based cropping systems for soil health restoration, resources conservation, and nutritional and environmental security in rainfed agroecosystems. Front Microbiol 13. https://doi.org/10.3389/fmicb.2022.1041124

Lapworth DJ, MacDonald AM, Krishan G, Rao MS, Gooddy DC, Darling WG (2015) Groundwater recharge and age-depth profiles of intensively exploited groundwater resources in Northwest India. Geophys Res Lett 42(18):7554–7562. https://doi.org/10.1002/2015GL065798

MacRae G (2016) Himalayan agricultures, ecologies and local food sovereignties. Anthropol Forum 26(3):262–275. https://doi.org/10.1080/00664677.2016.1192989

Maikhuri RK, Semwal RL, Rao KS, Singh K, Saxena KG (2000) Growth and ecological impacts of traditional agroforestry tree species in central Himalaya, India. Agrofor Syst 48(3):257–271. https://doi.org/10.1023/A:1006344812127

Malik ZA, Pandey R, Bhatt AB (2016) Anthropogenic disturbances and their impact on vegetation in Western Himalaya, India. J Mt Sci 13(1):69–82. https://doi.org/10.1007/s11629-015-3533-7

Måren IE, Sharma LN (2018) Managing biodiversity: impacts of legal protection in mountain forests of the Himalayas. Forests 9(8). https://doi.org/10.3390/f9080476

Mariani M, Casabianca F, Cerdan C, Peri I (2021) Protecting food cultural biodiversity: from theory to practice. Challenging the geographical indications and the slow food models. Sustainability 13(9). https://doi.org/10.3390/su13095265

Meenawat H, Sovacool BK (2011) Improving adaptive capacity and resilience in Bhutan. Mitig Adapt Strateg Glob Chang 16(5):515–533. https://doi.org/10.1007/s11027-010-9277-3

Mehraj G, Khuroo AA, Qureshi S, Muzafar I, Friedman CR, Rashid I (2018) Patterns of alien plant diversity in the urban landscapes of global biodiversity hotspots: a case study from the Himalayas. Biodivers Conserv 27(5):1055–1072. https://doi.org/10.1007/s10531-017-1478-6

Mourao PR, Martinho VD (2021) Discussing a challenging document focused on land use: the first bibliometric analysis of Laudato Si. Land Degrad Dev 32(16):4680–4692. https://doi.org/10.1002/ldr.4070

Murtagh N, Scott L, Fan J (2020) Sustainable and resilient construction: current status and future challenges. J Clean Prod 268:122264. https://doi.org/10.1016/j.jclepro.2020.122264

Nandy Rekha Pant Rao SK (n.d.) Indian Himalaya a demographic database environmental information system on Himalayan ecology G.B. Pant Institute of Himalayan Environment And Development

Negi SS (1983) Need for a conservation-based rural development strategy in the Himalayas. J Rural Dev 2(1):138–143. https://www.scopus.com/inward/record.uri?eid=2-s2.0-0021044275&partnerID=40&md5=b0dbebc6af8ea5440f9713d359b79ba3

Negi VS, Pathak R, Dhyani V, Durgapal M, Joshi RK, Bhatt ID (2022) Land restoration in the Himalayan region: steps towards biosphere integrity. Land Use Policy 121:106317. https://doi.org/10.1016/j.landusepol.2022.106317

Newell R, Dale A, Lister N-M (2022) An integrated climate-biodiversity framework to improve planning and policy: an application to wildlife crossings and landscape connectivity. Ecol Soc 27(1). https://doi.org/10.5751/ES-12999-270123

Oulkar S, Siingh D, Saha U, Kamra AK (2019) Distribution of lightning in relation to topography and vegetation cover over the dry and moist regions in the Himalayas. J Earth Sys Sci 128(7). https://doi.org/10.1007/s12040-019-1203-9

Panda RM, Behera MD, Roy PS (2018) Assessing distributions of two invasive species of contrasting habits in future climate. J Environ Manag 213:478–488. https://doi.org/10.1016/j.jenvman.2017.12.053

Parmar P, Kumar R, Neha Y, Srivatsan V (2023) Microalgae as next generation plant growth additives: functions, applications, challenges and circular bio-economy based solutions. Front Plant Sci 14. https://doi.org/10.3389/fpls.2023.1073546

Paudel B, Gao J, Zhang Y, Wu X, Li S, Yan J (2016) Changes in cropland status and their driving factors in the Koshi River basin of the Central Himalayas, Nepal. Sustainability (Switzerland) 8(9). https://doi.org/10.3390/su8090933

Pohle P (1992) Adaptation to the environment and economic change in the Himalayas of Nepal. The example of Manangki; [Umweltanpassung und okonomische Wandel in Nepal-Himalaya. Das Beispiel der Manangki]. Geogr Rundsch 44(7–8):416–425. https://www.scopus.com/inward/record.uri?eid=2-s2.0-0027041535&partnerID=40&md5=31f468cb305790b1cbfae27d804d6373

Puangpathumanond S, Qiu Q, Lew TTS (2023) Engineering plants as sustainable living devices. MRS Bull 48:1086–1095. https://doi.org/10.1557/s43577-023-00505-8

Qin Y, He J, Wei M, Du X (2022) Challenges threatening agricultural sustainability in Central Asia: status and Prospect. Int J Environ Res Public Health 19(10). https://doi.org/10.3390/ijerph19106200

Rahimzadeh A (2020) Fraternal polyandry and land ownership in Kinnaur, Western Himalaya. Hum Ecol 48(5):573–584. https://doi.org/10.1007/s10745-020-00181-1

Rana SK, Rawal RS, Dangwal B, Bhatt ID, Price TD (2021) 200 years of research on Himalayan biodiversity: trends, gaps, and policy implications. In: Frontiers in ecology and evolution, vol 8. Frontiers Media S.A. https://doi.org/10.3389/fevo.2020.603422

Rao KS, Semwal RL, Maikhuri RK, Nautiyal S, Sen KK, Singh K, Chandrasekhar K, Saxena KG (2003) Indigenous ecological knowledge, biodiversity and sustainable development in the Central Himalayas. Trop Ecol 44(1):93–111. https://www.scopus.com/inward/record.uri?eid=2-s2.0-0141685909&partnerID=40&md5=951ced7c80d2172c82d9a912b94623cb

Reddy CS, Satish KV, Jha CS, Diwakar PG, Murthy YVNK, Dadhwal VK (2016) Development of deforestation and land cover database for Bhutan (1930–2014). Environ Monit Assess 188(12):658. https://doi.org/10.1007/s10661-016-5676-6

Riaz MT, Basharat M, Pham QB, Sarfraz Y, Shahzad A, Ahmed KS, Ikram N, Waseem MH (2022) Improvement of the predictive performance of landslide mapping models in mountainous terrains using cluster sampling. Geocarto Int 37(26):12294–12337. https://doi.org/10.1080/10106049.2022.2066202

Rietveld AM, Groot JCJ, van der Burg M (2022) Predictable patterns of unsustainable intensification. Int J Agric Sustain 20(4):461–477. https://doi.org/10.1080/14735903.2021.1940731

Rosetti I, Bertrand Cabral C, Pereira Roders A, Jacobs M, Albuquerque R (2022) Heritage and sustainability: regulating participation. Sustain For 14(3). https://doi.org/10.3390/su14031674

Schickhoff U (1995) Himalayan forest-cover changes in historical perspective: a case study in the Kaghan Valley, northern Pakistan. Mt Res Dev 15(1):3–18. https://doi.org/10.2307/3673697

Schwab N, Bürzle B, Bobrowski M, Böhner J, Chaudhary RP, Scholten T, Weidinger J, Schickhoff U (2022) Predictors of the success of natural regeneration in a Himalayan Treeline ecotone. Forests 13(3). https://doi.org/10.3390/f13030454

Serba DD, Hejl RW, Burayu W, Kai U, Bushman BS, Williams CF (2022) Pertinent water-saving management strategies for sustainable Turfgrass in the desert US Southwest. Sustainability 14(19). https://doi.org/10.3390/su141912722

Sharma A, Sharma A (2019) Impact of anthropogenic activities on Himalayan ecosystem services and their management/sustainability: a review. ~ *361* ~. J Pharm Innov 8(1):361–365. www.thepharmajournal.com

Singh G, Chandra N, Lingwal S, Bisht MPS, Tiwari LM (2020) Distribution and threat assessment of an endemic and endangered species Angelica glauca in high ranges of Western Himalaya. J Herbs Spices Med Plants 26(4):394–404. https://doi.org/10.1080/10496475.2020.1748783

Singh RP, Dubey CS, Singh SK, Shukla DP, Mishra BK, Tajbakhsh M, Ningthoujam PS, Sharma M, Singh N (2013) A new slope mass rating in mountainous terrain, Jammu and Kashmir Himalayas: application of geophysical technique in slope stability studies. Landslides 10(3):255–265. https://doi.org/10.1007/s10346-012-0323-y

Singh K, Maikhuri RK, Rao KS, Saxena KG (2008) Characterizing land-use diversity in village landscapes for sustainable mountain development: a case study from Indian Himalaya. Environmentalist 28(4):429–445. https://doi.org/10.1007/s10669-008-9164-6

Tan YY, Omar R (2022) Green practices and innovations of traditional Chinese medicine (TCM) industry in Singapore: idea worth sharing. Sustain For 14(18). https://doi.org/10.3390/su141811588

Theodora Y, Spanogianni E (2022) Assessing coastal urban sprawl in the Athens' southern waterfront for reaching sustainability and resilience objectives. Ocean Coastal Manage 222:106090. https://doi.org/10.1016/j.ocecoaman.2022.106090

Tleuken A, Tokazhanov G, Jemal KM, Shaimakhanov R, Sovetbek M, Karaca F (2022) Legislative, institutional, industrial and governmental involvement in circular economy in Central Asia: a systematic review. Sustain For 14(13). https://doi.org/10.3390/su14138064

Uniyal SK, Singh RD (2012) Natural resources assessment and their utilization: analyses from a Himalayan state. Environ Monit Assess 184(8):4903–4919. https://doi.org/10.1007/s10661-011-2311-4

Valdiya KS (1995) Proterozoic sedimentation and pan-African geodynamic development in the Himalaya. Precambrian Res 74(1–2):35–55. https://doi.org/10.1016/0301-9268(95)00004-O

Vasistha HB, Negi M, Rawat M (2009) Restoration of degraded forests: a case study of Kujapuri Hills of Garhwal Himalaya, Uttarakhand. Int J Ecol Environ Sci 35(2–3):267–272. https://www.scopus.com/inward/record.uri?eid=2-s2.0-73949126924&partnerID=40&md5=1f7e6f44e28349600dca18861f6a20b5

Vasudevan N, Ramanathan K (2016) Geological factors contributing to landslides: case studies of a few landslides in different regions of India. IOP Conference Series: Environ Earth Sci 30(1):012011. https://doi.org/10.1088/1755-1315/30/1/012011

Verma S, Sharma PK (2007) Effect of long-term manuring and fertilizers on carbon pools, soil structure, and sustainability under different cropping systems in wet-temperate zone of Northwest Himalayas. Biol Fertil Soils 44(1):235–240. https://doi.org/10.1007/s00374-007-0207-9

Vetaas OR, Chaudhary RP (1998) Scale and species-environment relationships in a central Himalayan oak forest, Nepal. Plant Ecol 134(1):67–76. https://doi.org/10.1023/A:1009768122692

Voigt W (1979) The management of mountain ecosystems (Himalayas). Nepal Research Centre, Publication, pp 2–3. https://www.scopus.com/inward/record.uri?eid=2-s2.0-17144463077&partnerID=40&md5=0ea5436133bb973a708573c317fd8853

Vujanovic M, Wang Q, Mohsen M, Duic N, Yan J (2021) Recent progress in sustainable energy-efficient technologies and environmental impacts on energy systems. Appl Energy:283. https://doi.org/10.1016/j.apenergy.2020.116280

Wangchuk K, Darabant A, Rai PB, Wurzinger M, Zollitsch W, Gratzer G (2014) Species richness, diversity and density of understory vegetation along disturbance gradients in the Himalayan conifer forest. J Mt Sci 11(5):1182–1191. https://doi.org/10.1007/s11629-013-2942-8

Wani ZA, Samant SS, Pant S (2022) Diversity, utilization pattern and representativeness of dye yielding plants in North Western and Western Himalaya, India: an untapped source for bioprospection. Environ Dev Sustain 24(4):4493–4510. https://doi.org/10.1007/s10668-021-01664-x

Xu X (2022) The impact of natural resources on green growth: the role of green trade. Resour Policy 78:102720. https://doi.org/10.1016/j.resourpol.2022.102720

Yadav A, Verma P, Raghubanshi AS (2022) An overview of the role of plant functional traits in tropical dry forests. In: Research anthology on ecosystem conservation and preserving biodiversity. IGI Global, pp 92–117. https://doi.org/10.4018/978-1-6684-5678-1.ch006

Yarzábal LA (2020) Perspectives for using glacial and periglacial microorganisms for plant growth promotion at low temperatures. Appl Microbiol Biotechnol 104(8):3267–3278. https://doi.org/10.1007/s00253-020-10468-4

Yousuf M, Bukhari SK (2019) Impact of geological and environmental complexities over infrastructure development in Kashmir Himalaya. Disaster Advances 12(4):29–40. https://www.scopus.com/inward/record.uri?eid=2-s2.0-85066606173&partnerID=40&md5=62ee9cc361b82317c28bae0bb8e8b092

Zahraee SM, Golroudbary SR, Shiwakoti N, Stasinopoulos P, Kraslawski A (2020) Water-energy nexus and greenhouse gas-sulfur oxides embodied emissions of biomass supply and production system: a large scale analysis using combined life cycle and dynamic simulation approach. Energy Convers Manage 220. https://doi.org/10.1016/j.enconman.2020.113113

Zeigermann U (2021) Scientific knowledge integration and the implementation of the SDGs: comparing strategies of sustainability networks. Politics Governance 9(1):164–175. https://doi.org/10.17645/pag.v9i1.3630

Zhang Z-Q, Kress WJ, Xie W-J, Ren P-Y, Gao J-Y, Li Q-J (2011) Reproductive biology of two Himalayan alpine gingers (Roscoea spp., Zingiberaceae) in China: pollination syndrome and compensatory floral mechanisms. Plant Biol 13(4):582–589. https://doi.org/10.1111/j.1438-8677.2010.00423.x

Zhang J, Ma L (2021) Urban ecological security dynamic analysis based on an innovative emergy ecological footprint method. Environ Develop Sustain 23(11):16163–16191. https://doi.org/10.1007/s10668-021-01341-z

Zheng Z-W, Chou R-J (2023) The impact and future of edible landscapes on sustainable urban development: a systematic review of the literature. Urban For Urban Green 84. https://doi.org/10.1016/j.ufug.2023.127930

Zhong L, Wang J, Zhang X, Ying L (2020) Effects of agricultural land consolidation on ecosystem services: trade-offs and synergies. J Clean Prod 264:121412. https://doi.org/10.1016/j.jclepro.2020.121412

Zou Y, Meng F, Bi J, Zhang Q (2021) Evaluating sustainability of cultural festival tourism: from the perspective of ecological niche. J Hosp Tour Manag 48:191–199. https://doi.org/10.1016/j.jhtm.2021.06.009

Part IV

The Possible Way Forward

The Himalayan Ecosystem and Development Paradigm: A Sustainability Perspective

Achyutananda Mishra

1 Introduction

The Himalaya is the highest of mountain system of earth (Rana et al. 2021). Himalaya is a treasure of resources and a source of livelihood for millions. Human induced developmental activities has a taken a toll on the health and ecosystem of Himalaya. Hindu Kush Himalaya (HKH) spans about 3500 km in length from Afghanistan to Myanmar covering parts or all of Pakistan, India, China, Nepal, Bhutan, and Bangladesh, housing unique cultures, hugely diverse and distinct landscapes, and all of the world's peaks above 7000 m. Due to climate change Cryosphere (glaciers, snow, permafrost) in HKH is experiencing large scale irreversible changes unknown to human history. The impacts like increased temperature in higher elevations, accelerated melting of glaciers, increased permafrost thaw, declining snow cover, and more erratic snowfall patterns are visible (ICIMOD 2023).

Traditionally, Himalaya was protected and preserved by the communities inhabiting it. It was a source of survival for the communities. They were able to balance economy and ecology. The colonial rulers ignored such age-old traditional way of protecting Himalaya and focused on the pure economic aspect of Himalayan resources which were followed by the successive governments after independence leading to degradation and destruction of Himalaya ecosystem.

Himalayan ecology is very delicate and fragile, and has very limited capacity to adapt to changes which has been and is being damaged by anthropogenic (developmental) activities. Due to human activities ecological concern of Himalaya has been grossly ignored which is quite unsustainable for the present as well as future. To save the Himalayan ecology from destruction, local communities have initiated

A. Mishra (✉)
Christ University, Bangalore, India
e-mail: achyuta.mishra@christuniversity.in

A. Borthakur, P. Singh (eds.), *The Resilience of Traditional Knowledge Systems for a Sustainable Future*, Advances in Asian Human-Environmental Research,
https://doi.org/10.1007/978-3-031-56858-9_11

activist actions, mass movements and litigated before courts. Regardless of the actions Himalayan ecology continues to suffer due to the inadequacy of the prevalent development model. To save it from further deterioration it is imperative that a model which takes into account the concern of Himalayan ecology need to be adopted. Further, it should cater to the need of the present as well as future generation into account and promote sustainable development for all.

2 Mountain Ecosystem and Sustainability

2.1 Defining Ecosystem

Ecosystem involves the symbiotic relationship of organisms amongst themselves and their environment. According to United States Council of Environmental Quality, ecology represents an intricate web of relationship between living organism and their living and non-living surroundings which constitute an ecosystem. While interacting, which is countless, plants and animals respond to each other and to their surrounding making the changes in the ecosystem continuous and uninterrupted. In the process they adapt to the inevitable changes. But nature has limited such adaptable capacity by a genetic composition beyond which species cannot adapt and perish. Interactions are so continuous and countless that they form a chain of reactions which, eventually, can be perceived throughout the system. It means a small change in one part of the ecosystem can be felt across the system (Divan and Rosencranz 2001).

Dramatic changes can be seen in the ecosystem if the course of the nature is altered artificially. For example, if a dam was built on a river only to generate electricity, it will produce electric power but at the same time it will affect the fish population and increase disease-bearing aquatic snails. The stability, resilience and endurance of an ecosystem bank on its diversity (Divan and Rosencranz 2001).

2.2 Mountain Ecosystem

Mountains have a very significant role for the human race; they play a very crucial role in the lives of millions of people all over the world by providing a source of water and food. They are a perennial source of water. Due to orographic reason mountains precipitate water/rain from the atmosphere/clouds. Their natural forest coverage transforms seasonal rain water into perennial water sources (Shiva 1991). They are a hub of rich bio-diversity and cultural diversity. They are a home to socio-economic practices, agricultural practices, cultural identities, arts, languages, governance, and music of the mountain population which majorly consist of diverse groups. Mountains are "water towers" (Mastrojeni 2022; Wester et al. 2019) of the world providing for planets 60–80% of fresh water resources. Around half of the population of the world rely on the services of mountain eco system, especially water. They provide habitation to more than 85% of world's bird, amphibian, and mammal species. They are house to nearly 1 billion people of the world. They have local, national and global significance. They are the 'water tower' for consumption, land use, and growing energy needs; the weather makers for major parts of the world; a major source of valuable natural resources which come from the rich biological diversity they nurture; and a rejuvenating place for the body, mind and spirit (Stone 1992).

Mountains are the key sources of energy, water and biological diversity. They are a rich source of minerals, forest products, and agricultural products. Further, they serve as a major source of recreation for human society. Mountains are a major ecosystem containing complex and interrelated ecology of the planet. They are very essential for the survival of the global ecosystem (Wester et al. 2019). Mountains are one of the fragile ecosystems of the world. They are highly vulnerable to human action and natural ecological imbalances. They are very sensitive to climate change. They are susceptible to soil erosion, landslides, loss of habitats and genetic diversity. Further, they inhabit poor mountain inhabitants and experience loss of indigenous knowledge. Globally, mountains are experiencing environmental degradation (United Nations 1992a, 1992b).

Mountain ecosystems are sensitive and susceptible to small changes, and consequences of such changes are irreversible in nature. Damage to natural ecosystem and ecological diversity of mountains are a serious human and international concern. (Ahmad et al. 1990).

2.2.1 Sustainable Development Goals (SDG)

Mountains play a key role in the global environmental landscape. Realising this, the United Nations brought to sharp focus the mountain ecology in its global agenda for sustainable development. SDGs are a global road map for the future of the world. It places high priority on the terrestrial and mountain ecosystem as they contribute majorly to the development and sustenance of the human society.

SDG 15 aims at Protecting, restoring and promoting sustainable use of terrestrial ecosystems, managing forests, combating desertification, and halting and reversing land degradation and biodiversity loss.

Further, Goal 15.1 seeks to ensure the conservation, restoration and sustainable use of terrestrial and inland freshwater ecosystems and their services, in particular forests, wetlands, mountains and drylands, in line with obligations under international agreements by 2020. Goal 15.2 targets to promote the implementation of sustainable management of all types of forests, halt deforestation, restore degraded forests and substantially increase afforestation and reforestation globally by 2020.

Goal 15.4 intends to ensure the conservation of mountain ecosystems, including their biodiversity, in order to enhance their capacity to provide benefits that are essential for sustainable development by 2030 (UN 2015).

3 Himalayan Ecosystem and its Significance

Himalaya(s), a great mountain system of Asia consisting of parallel mountain ranges, touches Afghanistan, Bangladesh, Bhutan, China, Myanmar, Nepal, Pakistan, and India; it has a huge significance for the people of these countries specifically. It is the world's most populous and fragile mountain system. It inhabits people of diverse language, culture and race. It is ground for many medicinal plants used in famous medicinal systems like Ayurveda, Yunani and Tibetan medicine. It is a source to all three major river systems, i. e. the Indus, the Yangtze and the Ganga-Brahmaputra, of the Asian continent and more than 19 major rivers. It is one of the youngest mountain ranges of the planet. After Antarctica and the Arctic Himalayas have the third largest deposit of ice and snow in the world, Glaciers play an indispensible role in feeding the Himalayan rivers. It is a host to one-fourth of all or parts of global biodiversity hotspots supporting diverse flora and fauna (Britannica 2023; Chatterjee and Bishop 2023).

The Himalaya being a great climatic divider—regulating air and water circulation—determine the meteorological conditions in the Indian subcontinent in the south and Central Asian highlands in the north. Due to its height and location it obstruct the passage of continental cold air from north to south in winter and also forces the south-westerly rain bearing monsoon winds to leave most of their moisture before crossing to north resulting in heavy precipitation on the Indian side and arid conditions in Tibet (Chatterjee and Bishop 2023).

80% irrigated land in Indus Basin receive water from Hindu Kush Himalayas where 180 million people's food is grown (Mastrojeni 2022). 1.3 billion People across Asia rely on Himalayan ecosystem for their livelihood.

Himalayan ecosystem has a special significance to the ecology of India. It provides forest cover, feed perennial rivers which are a source for irrigation, drinking water and hydropower, conserve biodiversity, provide a valuable ground for agriculture, and a magnificent and picturesque landscape for tourism. "Natural geological wealth, forestry, wild life, flora, fauna and biodiversity, snow, ice and water bodies, traditional knowledge and mountain agriculture which characterize the region are special" (GoI 2010).

Some of the major geophysical changes following the formation of the Himalaya include the inception of the monsoon system, the development of glaciers, and the birth of a wide network of river. The ecosystem and biodiversity services for the local communities in the Himalaya include timber, medicinal plants, spices and condiments, fodder and pastures for livestock rearing, and irrigation and manure for agriculture (Pandit et al. 2014). Himalayan forests plays crucial role as a sink of CO_2 and help mitigate climate change impact (Negi 2022).

Himalaya is a house to many crops of the world, natural wealth as well as indigenous societies and knowledge system. It is a rich repository of plant and animal wealth providing a source of livelihood to millions of people in the Asian region. 10,000 angiosperm species are found in Himalaya out of which 1800 are medicinal and 700 are of edible in nature. Its significance lies in the fact that it has been recognised as one among the 34 global biodiversity hotspots.

Himalayan ecology is very delicate and carries limited capacity to adapt to change (Stone 1992). The Indian Himalayan Region is one of the rich biodiversity regions of the world. It houses over 10,000 plants, 977 birds, 300 mammals, 269 fishes, 281 herpetofauna, and several species of invertebrates and microorganisms. Many of such species and organisms have high global conservation significance (Singh et al. 2021).

3.1 Himalayan Ecosystem: The Traditional Way of Preservation

Indigenous practices of development in India were always sensitive to natural ecosystem. Pre-colonial indigenous economic processes were not destructive of environment rather they were protective of it. Himalayan eco-sensitive codes were guiding the visits to key pilgrim centres. Social norms were prevalent to utilise resources in an environmentally safe manner (Shiva 1991). Since time immemorial socio-cultural-environmental practices have been guiding norm of development in Himalaya (NITI 2018). Indian Mountain system, including Himalayan, is a complex socio-ecological system embodying cultural and biological diversity. Maintenance of mountain by traditional societies was guided by ecological prudence, it was a holistic one. It was a process of maintenance of natural resources and biodiversity. Sustainable livelihood was the norm. Changing human population dynamics has altered the pattern bringing disruption in the relation between man and nature in mountains (Ramakrishnan 2007).

4 Himalaya: The Locus of Conflict in Retrospect

Himalaya, a reservoir of rich resources, has been a life sustaining source to the communities inhabiting there since generations. It had a very robust system to manage forest and mountain resources. Traditionally forests resources were under common ownership and management, the entire community used to be the owner and preserver of those resources. There was practice in those communities that anyone cuts a tree must plant another tree in its place. Colonial forest policy making way for commercialisation of forestry and prioritising forest revenue collection deeply eroded the traditional forest management and conservation practices and systems.

In 1850 one Englishman named Mr. Wilson was granted a lease for exploitation of forest resources of Bhagirathi Valley in the Garhwal Himalayas at an annual rent of Rs. 400/–. He axed and destroyed many valuable Deodar and Chir forest in pursuance to his timber business. Inspired by his flourishing business the British rulers took a lease of 20 years for exploitation of forest resources. European settlements in the area led to the large scale cutting of oak forests for cultivation of food crops (Shiva 1991).

4.1 Chipko Movement

The conservation of the forest was never on the agenda for colonial rulers. Conflict arose due to colonial rulers disregard and indifference towards the conservation of nature and the harmonious relationship between nature and the communities. The indifference towards the need of the locals led to accelerated deforestation. The colonial forest policy of denying the ancient rights of the people/communities to forest resources inspired popular opposition leading to unrest in the state of Tehri Garhwal in 1906. In 1930 the people of Garhwal resorted to non-cooperation movement to highlight the right of people to forest resources. Further, to protect their forests people in Himalaya resorted to one of the popular movement called as Chipko Movement/Andolan (hugging of tree against it cutting and felling). It was a very prominent movement of locals for forest conservation against the commercialisation and overexploitation of forest resources. It was a monumental environmental movement sparking protest across the Himalaya and beyond. "It was a response to the alarming signals of rapid ecological destabilisation in the hills (Shiva 1991). The growth imperative inspired commercial exploitation of forest resources leading to conflict between development and ecology.

The conflict between people's right and the revenue requirements of the state remained unsolved and continued even after the departure of Britishers from India.

4.2 Doon Valley Case

The resource rich Himalaya faced denudation, deforestation, accelerated soil erosion and blockage of underground water channels due to mining and quarrying activities granted by the government. Such activities were destroying Himalayan ecosystem and altering the nature of the valley. Concerned citizens approached the Government of Uttar Pradesh to stop such unscrupulous destruction but it was in vain. Then the Rural Litigation and Entitlement Kendra, Dehradun and a group of concerned citizens filed petition in Supreme Court for remedy and stoppage of such mining activities. The Supreme Court in this case (popularly known as Doon Valley case or the Mussoorie Hills case) said, it is a first of its kind flagging issue concerning environment and ecological balance. The questions raised are of grave significance involving not only people of Mussoorie Hill range in Himalaya but also the people in general across the country and their welfare (Supreme Court 1985).

Further, the court remarked:

> But there can be no gainsaying that lime stone quarrying and excavation of the lime stone deposits do seem to effect the perennial water springs. This environmental disturbance had however to be weighed in the balance against the need for extraction of limestone for industrial purposes… (Supreme Court 1985, p. 434)

5 Anthropogenic Activities in Himalayas

To pursue economic development anthropogenic activities are on the rise across the globe and the process of globalisation and industrialisation has exacerbated it exponentially. Such activities have deeply impacted the ecological balance and created environmental problems worldwide. In the age of Anthroposcene, which is characterised by the dominance of human activity over the climate and environment, mountain ecosystems, including Himalaya, are facing a grave threat. Mountain ecosystems are very sensitive to small changes and disturbances. The consequences of any such disturbance are often irreversible (Ahmad et al. 1990).

Himalaya witnessed intensive human activities in the 1950s when countries sharing common boundaries in Himalaya improved accessibility to remote mountain areas for military reasons (Stone 1992). Such accessibility facilitated by roads and few air routes led to other impacts. Such increased access, of course, integrated the Himalayan terrain to the richer and resourceful mainland territory, at the same time exposed its rich resources to mainland territory and people. Accessibility facilitated trade and commerce routes. Resources like forest, hydro-power, minerals, biodiversity and advantageous climate conditions, hitherto used by locals, began to be exploited in an unprecedented scale in the name of development and economic transformation (Stone 1992). It is doubtless that integration of the Himalaya to mainland territory has immensely benefitted the people inhabiting in it like the availability of food supplies in times of scarcity, possibility of health care services and reduction of death rates due to better transport facility, general increase in socio-economic condition etc. But all this happened without looking into the fragility and vulnerability of Himalayan ecology.

> Forests became denuded, roads and mines created enormous land instabilities, the intensification of agriculture led to soil degradation and erosion, pastures were damaged by overexploitation while natural course of rivers were dammed and all these factors together created a massive sedimentation problem. Tourist concentrations created pollution even in the "abode of gods". Much of the blame for environmental degradation in the Himalaya was directed at the growing local population, whose indigenous modes of natural resource management were, however, made increasingly redundant by the pace of socio-economic and cultural transformation and state intervention (Stone 1992).

Flooding in Bangladesh is linked to the environmental degradation in Himalaya's catchment area (Stone 1992). Undoubtedly climate change, which is an outcome of human activities, is impacting the Himalayan ecology, at the same time development-induced impact on the Himalayan ecology is widely noticeable.

Unplanned urbanization, haphazard establishment of infrastructure in risky locations, construction of large dams and modification of natural waterways are responsible for the happening of disaster in Himalaya (Satyala et al. 2017).

Criticising the sorry state of affairs of Himalaya, noted historian, environmentalist, author and public intellectual Ramachandra Guha says, successive governments have assaulted the Himalaya and its population in four ways, i. e. promoting commercial forestry, open-cast mining, large hydel projects, and unregulated tourism.

These activities have resulted in accumulation of toxic wastes, rise in air pollution, degeneration of forests and loss of biodiversity, degradation of water resources, landslides and floods. After this he added the fifth one i. e. massive road-building project, which is devastating the already damaged landscape (Guha 2020).

Mindless human activities have been a cause of the degradation of mountain ecosystems which is a global problem. Himalayas faces serious environmental problems due to human-induced development. Himalaya has been place of human-induced development and a sight of unplanned activities like, major engineering constructions, construction of highways, unplanned land use, over-exploitation of forest resources, tourism and recreation (Wester et al. 2019). It is a sight of development taking a toll on the ecology of the Himalaya; development taking center stage and environment taking back seat.

5.1 Tourism

Due to rich in biodiversity, glaciers, water resources, cultural diversity, and suitable conditions for recreational activities, adventure sports and religious pilgramages the Himalayan region attracts tourists from across the world. Tourism provides a source of livelihood to the people of the region. At the same time it creates the problem of pollution having detrimental effect on the Himalayan mountain ecosystem (Kuniyal 2023; Singh et al. 2021).

Tourism in Himalaya is comparatively advantageous and growing faster. Commenting on the spread of tourism in Himalaya, NITI Aayog in a report observes:

> Specific negative impacts linked to the current form of tourism in IHR (Indian Himalayan Region) include the replacement of traditional eco-friendly and aesthetic architecture with inappropriate, unsightly and dangerous construction, poorly designed roads and associated infrastructure, inadequate solid waste management, air pollution, degradation of watersheds and water sources, and the loss of natural resources, biodiversity and ecosystem services (NITI, 2018, p. 30).

5.2 Hydroelectric Projects and Dams

Himalaya has lot of potential for hydroelectric generation. To harness such potential the process began in 1950s and a large scale multi-purpose project was initiated in Bhakra-Nangal on the Sutlej River in the Outer Himalayas. Similar projects were initiated on other rivers by India as well as China and Nepal (Chatterjee and Bishop 2023).

Reckless dam building is another threat to Himalaya. Himalayan nations are on a dam building spree on the Himalaya which is causing huge loss to the biodiversity in it. "More than 600 large dams have been built or are in some stage of construction or planning in the geologically active Himalayan Mountains… (Mukerjee 2015). It is projected that in the next few decades Himalaya will have highest average dam densities in the world, one dam in every 32 kilometres of the river channel. It will cause species loss, loss of forests by submergence, loss of tree species richness, reduction of tree density and in general it will cause habitat fragmentation.

According to the Central Electricity Authority the Himalayan region in India, by the end of November, 2022, had 81 large hydropower projects, 26 projects under construction, and another 320 large projects in the pipeline. This has aggravated the already existing risk of landslide due to climate and seismic activities in the river valleys of Himalaya. Hydropower projects have become a constsnt threat to Himalayan region leading to frequent disasters in recent years. A committee of experts examined 24 hydropower projects concluded that 23 projects had "irreversible impact" on the ecology of the region (Shagun 2023).

5.3 Char Dham Project

In December 2016 The Ministry of Road Transport and Highways, Government of India announced a programme called 'Chardham Mahamarg Vikas Pariyojna' constructing highways connecting four shrines of Hindu religion i. e. Yamunotri (NH94/134 up to Janki Chatti), Gangotri (NH-108), Kedarnath (NH-109, up to Sonprayag), Badrinath (NH-58) and the Tanakpur-Pithoragarh stretch of the Kailash Mansarovar Yatra route (NH-125) all situated in the lap of Himalaya. These shrines use used to be located in glaciers in entirety and now in paraglacial zones and considered to be ecologically sensitive. The project was conceptualised to improve access to these shrines by widening the existing roads, and to make travel smooth, safe, and fast. It will widen the existing roads to double lane with 16 bypasses, realignments and tunnels, 101 small bridges, 3516 culverts, and 15 flyovers (SCI 2021).

The Ministry's decision was challenged in Supreme Court of India. The court after discussing the vulnerability and fragility of Himalaya, and also, at length the imperative of sustainable development prioritised the infrastructural need of the state. The Court accepted the infrastructural requirement of the Armed Forces and refused to interfere with policy choice of the government and allowed the project to continue (SCI 2021).

The judgment did not consider the massive crisis of climate change, and rewarded the violations of environmental norm instead of establishing a range of accountability; Loss of Himalaya is a loss to India too (Patel 2021). With this Himalayan ecology lost a chance of preservation, survival and revival. The concern of the Himalayan ecology, which has far reaching consequences for humanity, was overlooked. The Himalayan ecology gave way to the economic and security interest of the nation.

6 Himalayan Ecology: A Victim of Anthropogenic Activities

Loss of or damage to Himalayan ecology is quite unsustainable for the planet earth. It is consistently facing the pressure due to human activities which has deeply impacted the ecology of the Himalaya. It has become a victim of human action and paying a heavy price for it. Man-made activities, which include unplanned land use, cultivation on steep slopes, overgrazing, major engineering activities (road construction, mining; dams, reservoirs, irrigation canals, quarrying, hydroelectric projects, etc.), over-exploitation of community forests, etc. have been accelerating the degradation of the Himalayan mountain ecosystem and are threatening its equilibrium. (Ahmad et al. 1990).

The plight of Himalaya is its degeneration of forest cover openingup gates for diverse problem like depletion of soil fertility, soil erosion, paucity of fuel wood and fodder, slope failures, accelerated siltation of river beds in lowlands areas, amplified overland flows and abridged ground water recharge hammering biological diversity (Singh et al. 2021).

"Despite significant ecological, hydrological and biological values, the fragile mountain ecosystems in the IHR are seriously threatened due to increasing anthropogenic pressures, mainly development" (Singh et al. 2021, p. 1645).

Impacts of Human Activities on Himalaya:

- Soil Impoverishment and loss
- Landslide
- Landscape transformation
- Melting/recession of glaciers
- Loss of biodiversity
- Migration
- Loss of water sources
- Loss of livelihood
- Destruction of vegetation cover
- Desertification

7 Development Paradigm: Need for Reorientation

Post World War II marked a rapid industrialization to end poverty and unemployment, and witnessed development across. This era termed as "age of ecological innocence" did not consider ecological concern as a factor while persuing the agenda of development (Guha 2020). The process of development planning between 1950s and 1970s was primarily represented by large infrastructure, modernisation of industrial and agricultural projects, and mobilisation of new technologies. There was very little concern about impacts of such policies on the environment (Elliot 2013).

Growing evidence suggests that mankind has caused incalculable harm to the natural environment. Due to man-made activities unprecedented level of pollution has been caused to the air, water, earth and living being. To prevent such catastrophic consequences for the present and the posterity, and to achieve a better life, a goal needs to be set to protect and preserve the natural resources of the earth. In pursuance to this a responsibility was put on mankind/states to balance between socio-economic development and preservation of natural resources including wildlife (UN 1972).

Since the days of industrialisation development thinkers have often disagreed as to the meaning of the word 'development' and keeps debating. Oftentimes it is associated with the gross domestic product (GDP) and GDP became a measure of development. Thinking on development keeps hovers around economic growth, application of scientific and technical knowledge. Although use of GDP to measure development of economy is useful, it has limitations in measuring human welfare and individual well-being. It was primarily in currency till 1970s. Prof. Amartya Sen designated the identification of development only with growth with gross national product (GNP) or with rise in the individual incomes or rise in industrialisation or advances in technology or modernisation of society as narrow. Development, he concludes, in addition to the above, should be inclusive of socio-economic interest and rights (Sen 1999).

"Deep Ecology", a strand of American Environmentalism, argues for *biocentrism* which believes in intrinsic worth of all living being regardless of their utility to human needs; humans reserve no right to diminish the richness and diversity except to the extent of satisfying their vital needs; human interference in the nonhuman world is disproportionate and worsen the situation. Deep Ecologist calls for a fundamental change in the dominant human thinking and adopt an ecological friendly approach in politics, society, technology, economy and ideology which will lead to a global state of ecologically sustainable development Guha 2020).

There is necessity of new approaches to re-establish the harmony between development and conservation. Economic interest of exploitation of resources needs to be in sync with the imperative of ecological conservation (Rana et al. 2021).

Over a period of time and faced with multiple problems across the world a general consensus has emerged the development to be measured by people's welfare and quality of life and all 17 sustainable goals are a proof to it highlighting a shift in the paradigm od development discourse. Development paradigm took a major shift in the post Stockholm Declaration on Human Development, 1972 followed by Brundtland Report, 1987 entitled 'Our Common Future', Earth Summit, 1992, Millenium Development Goals and Sustainable Development Goals.

8 Sustainable Development: A New Approach

Environmental stress can be noticed due to increased demand on scarce resources and the pollution caused by the rising standard of living by the relatively affluent masses and widespread poverty. Led by the growth of the technologies quality of life and living standards of man have substantially improved since mid-1950s which has entailed substantial pollution across the globe. The scale of the impact of such development on environment is such that human history had never ever witnessed it before (Brundtland 1987).

The economic approach towards the development is inherently antithetical to the protection and preservation of environment. Environment belongs to the whole of humanity and sustainable development demands that it must be preserved so that not only the present generation but also all future generations should enjoy it without any limitation. It underlines the necessity of bringing an end to conflict between development and environment. "Sustainable development is development that meets the needs of the present without compromising the ability of future generations to meet their own needs (Bruntland, 1987). It emphasise on the socio-economic needs of people and the ability of the environment to remain stable to meet the needs of all without compromising the interest of the future generation.

In essence, sustainable development is a process of change in which the exploitation of resources, the direction of investments, the orientation of technological development; and institutional change are all in harmony and enhance both current and future potential to meet human needs and aspirations (Brundtland 1987, Chap. 2). It moots the idea of the maintenance of development over time. It represents three pillars, i. e. economy, environment and society (Elliot 2013). The idea of sustainable development put human beings at the center. It envisages a "healthy and productive life in harmony with nature" for all human beings. It calls for a global partnership for the conservation, protection and restoration of the health and integrity of the planet's ecosystem (UN 1992a, 1992b).

Sustainable development is a multi-dimensional concept which encompasses economy, social interest, individual interests, environmental interest, and a balanced development focusing on inter and intra-generational equity. The UN Declaration on the Right to Development, 1986 recognises that development is a comprehensive process involving politics, society, economy, and culture exclusively aiming at improving the well-being of the entire population and all individuals on the basis of their free, active and meaningful participation in the process of development and fair distribution of benefits resulting therefrom.

9 Himalayan Ecology and Sustainable Development

Growth in Himalaya led by unfettered demand-driven economy has spurred unplanned urbanisation, environmental degradation, and risks and vulnerabilities. Any road map towards sustainable development of Himalayas must approach through the environmental, socio-cultural and economic arena to further the well-being of the people inhabiting the region and beyond (NITI 2018).

Sustainable development requires utilisation of ecological system in a way which satisfies the need of the present generation without compromising the interest of the future generations. Sustainable development strategies must be premised upon reliable, authentic and broad-based data on socio-economic-cultural resources, as well as on the prevailing environmental set-up (Singh 2006).

A successful sustainable development strategy must identify as a priority "sustainable land use practices, promotion of on-farm activities, value addition to all resources and adoption of environment-friendly technologies, restoration of degraded ecosystems, biodiversity conservation, water resource and hydro-power development, promoting community-based management, upgrading infrastructure, improving quality education and capacity building to ensure benefits, to improve livelihoods, income and environment in the Himalaya" (Singh 2006).

10 National Mission for Sustaining the Himalaya Eco-System

Recognising the necessity to protect the Himalayan ecology from vulnerability posed by poorly conceived anthropogenic activities; to conserve of biological diversity of Himalayan ecology; to promote sustainability of Himalayan Ecosystem and as a part of National Action Plan on Climate Change (NAPCC) the government of India established the National Mission for Sustaining Himalayan Ecosystems (NMSHE). The Mission aims at scientifically understanding the complex processes affecting Himalayan eco-system, frame appropriate policy framework for its, including glaciers, sustainable management and safeguarding, and implement scientifically backed policy with help of the states in the region. The Mission is tasked with involving all stakeholders/institutions involved in exploration, conservation and management of natural resources in Indian part of Himalaya.

The basic principles which guide the course of the Mission in achieving the goals are:

- Protection of the vulnerable sections of the society.
- Development of human resources.
- Enhance ecological sustainability.
- Deployment of technologies.

To achieve the goals the Mission undertakes the responsibility of collection of all relevant knowledge and to document them; build all human and institutional capacities for observance and assessment of all changes in the ecosystem and correlating them to likely causes for change to assist in the formulation of policies and strategies for effective management; establishing a State of the Art National Center for Himalayan Glaciology for promotion of research; assisting the Ministry of Environment and Forests in formulating policies and strategies for management; and assisting the states in implementing sustainable actions in the Indian Himalayan Region.

11 Conclusion

Nature nurtures human race and it is the duty of the race to nurture and conserve it as a trust. Economic development and improvement of quality of life people needs to go hand in hand with environmental protection and regeneration. Due to anthropocentric approach of development mountains are facing grave threats. Developmental activities may be having immediate benefit to the states surrounding Himalaya but it will cause irreparable harm and deprive the future generation of the benefit of Himalaya which is against the very idea of sustainable development. Himalaya is extremely critical to the lives and livelihoods of millions of people and also it has significance for the global climate condition and sustainable development.

Given the ecological fragility of Himalaya the relationship between the environment and development has been grossly neglected in the policy planning arena. Undertaking development activities in ecologically fragile Himalayan region is highly risky and unsustainable; it goes against the basic tenets of sustainable development. Undeniably, people in the region need development; they need education, electricity, health care, road, employment and options of livelihood. But, it is equally to be noted that the development they need shall be short-lived in a depleting Himalaya. Development at the cost of Himalayan ecology is deadly and quite unsustainable.

Anthropogenic activities have brought and continue to bring manifold changes to the Himalaya. Legislative and Policy initiatives appears to be deficient; and judicial approach is not so encouraging. To save the Himalaya from irresponsible and careless human action a concerted international efforts and action (policy as well as legislative) is required. As it has a wider significance to a large segment of Asian population regional efforts on the same line is equally called for mandating the concerned states to bring in legislation with mandatory standards for the protection and preservation of Himalaya. Multilateral and bilateral cooperation among the Himalayan nations is highly necessary to conserve the biodiversity of Himalaya. Further, interest and involvement of the local communities in the process should be prioritised.

NITI Aayog has recommended for the establishment of a 'Himalayan Authority' to for coordination among the Himalayan states for the sustained development of the Himalayan region. In line with that an 'Association for Regional Cooperation of the Development of Himalaya' with an exclusive focus on the interest of Himalaya is required among the Himalayan nation.

References

Ahmad A, Rawat JS, Rai SC (1990) An analysis of the Himalayan environment and guidelines for its management and ecologically sustainable development. Environmentalist 10(4):281–298

Brundtland GH (1987) Our common future: report of the world commission on environment and development. United Nations

Chatterjee SP, Bishop BC (2023) Himalayas. www.britannica.com/place/Himalayas

Divan S, Rosencranz A (2001) Environmental law and policy in India: cases, materials and statutes. Oxford University Press, New Delhi

Elliot JA (2013) An introduction to sustainable development. Routledge, London and New York

Encyclopaedia Britannica, Inc (Britannica) (2023) Himalayas. https://www.britannica.comhttps//www.britannica.com/place/Himalayas

Government of India (GoI) (2010) National mission for sustaining the Himalayan eco-system. https://dst.gov.in/sites/default/files/NMSHE_June_2010

Guha R (2020) The Wilful destruction of Himalaya must stop. https://www.ndtv.com/opinion/killing-the-himalaya-one-cut-at-a-time-by-ramachandra-guha-2287551

International Centre for Integrated Mountain Development (ICIMOD) (2023) Water, ice, society, and ecosystems in the Hindu Kush Himalaya: an outlook. International Centre for Integrated Mountain Development

Kuniyal JC (2023) Safety from numbers: tourism in the Himalayas is not uniform; its impacts though are detrimental. Down to Earth

Mastrojeni G (2022) Why mountains matter. https://www.oneearth.org/why-mountains-matter

Mukerjee M (2015) The impending dam disaster in the Himalayas. https://www.scientificamerican.com/article/the-impending-dam-disaster-in-the-himalayas/

Negi GCS (2022) Trees, forests and people: the central Himalayan case of forest ecosystem services. Trees For People 8(2022):100222. https://doi.org/10.1016/j.tfp.2022.100222. https://www.sciencedirect.com/science/article/pii/S2666719322000322?via%3Dihub

NITI Aayog (2018) Contributing to sustainable development in the Indian Himalayan region

Pandit MK et al (2014) Dancing on the roof of the world: ecological transformation of the Himalayan landscape. BioScience 64(11):980–999. https://doi.org/10.2307/90006977

Patel P (2021) When the Himalayas lose: assessing char Dham Pariyojana judgment. Down to Earth

Ramakrishnan PS (2007) Sustainable mountain development: the Himalayan tragedy. Curr Sci 92(3)

Rana SK, Rawal RS, Dangwal B, Bhatt ID, Price TD (2021) 200 Years of research on himalayan biodiversity: trends, gaps, and policy implications. Front Ecol Evolution 8

Satyala P, Shrestha K, Ojha H, Vira B, Adhikari J (2017) A new Himalayan crisis? Exploring transformative resilience pathways. Environ Develop 23:56

Sen A (1999) Development as freedom. Alfed A Knopf, New York

Shagun (2023) Himalayan plunder: how hydropower projects in the hills have increased disasters there. Down To Earth, 1–15

Shiva V (1991) Ecology and politics of survival: conflicts over natural resources in India. Sage Publications India Pvt Ltd. And United Nations University Press

Singh C, Chauhan N, Upadhyay SK, Singh R, Rani A (2021) The Himalayan natural resources: challenges and conservation for sustainable development. J Pharmacogn Phytochem 10(1):1643–1648

Singh JS (2006) Sustainable development of the Indian Himalayan region: linking ecological and economic concerns. Curr Sci 90:6

Stone PB (1992) The Himalaya: prospects for and constraints on development. In: The state of the World's mountains: a global report. Zed Books Ltd., London and New Jersy. https://archive.org/details/stateofworldsmou0000unse

Supreme Court of India (SCI) (1985) Rural Litigation and Entitlement Kendra, Dehradun v. State of U.P. (1985) 2 Supreme Court Cases 431

Supreme Court of India (SCI) (2021) Citizens for Green Doon & Ors v. Union of India & Ors

United Nations (1972) Declaration of the United Nations conference on the human environment, Stockholm

United Nations (1992a) Agenda 21. Rio de Janerio, Brazil. https://sdgs.un.org/sites/default/files/publications/Agenda21.pdf

United Nations (1992b) Rio declaration on environment and development, Rio de Janeiro

United Nations (2015) Transforming our world: the 2030 agenda for sustainable development. https://sdgs.un.org/sites/default/files2023/04SDG/Progress/Report/Special/Edition/2023ADVANCE/UNEDITED/VERSIO

Wester P et al (2019) The Hindu Kush Himalaya assessment: mountains, climate change, sustainability and people. Hindu Kush Himalayan Monitoring and Assessment Programme (HIMAP)

Revitalizing Roots: Harnessing Traditional Socio-Cultural Institutions for Sustainable Development of the Himalayas

Sachin Kumar and Amit Shoshta

1 Introduction

Mountains embody the multifaceted challenges of our era, grappling with ecological fragility, economic disparities, and social crises. The robustness of these landscapes underscores our collective duty to preserve the delicate equilibrium between nature, economies, and the mountain communities. Consequently, SDG 15, Target 1 explicitly recognizes the imperative to conserve, restore, and sustainably use mountain ecosystems in alignment with international agreements (UN 2015). The Himalayas, akin to other mountain systems, demand comprehensive policies and interventions that not only rejuvenate their ecology but also enhance economic vitality and nurture social cohesion, propelling them toward sustainable development. Achieving sustainability necessitates robust institutions at all levels (e.g., Mitra and Moldavanova 2018; Pfahl 2005; Spangenberg 2007), elucidated by SDG 16's Target 6 and Target 7, focusing on effective, transparent institutions and inclusive, participatory decision-making, respectively. Traditional socio-cultural institutions (TSCIs) operating locally, especially in mountain regions, have a rich history of engaging in sustainable practices. These institutions, known for their accountability and effectiveness, facilitate responsive and participatory decision-making. This chapter contends that these inherently sustainable TSCIs merit recognition, respect, and seamless integration to guide the Himalayas toward a sustainable future. The argument emphasizes that development interventions should align with these time-tested institutions to ensure greater acceptance, improved

S. Kumar
Department of Geography, Government Degree College, Shahpur, Himachal Pradesh, India

A. Shoshta (✉)
Department of Geography, Government Degree College, Shahpur, Himachal Pradesh, India

Department of Geography, Government Degree College Rampur Bushahar, Shimla, India

A. Borthakur, P. Singh (eds.), *The Resilience of Traditional Knowledge Systems for a Sustainable Future*, Advances in Asian Human-Environmental Research,
https://doi.org/10.1007/978-3-031-56858-9_12

implementation, and ultimately, effective outcomes. The subsequent section provides a conceptual overview of TSCIs and articulates the rationale for promoting and integrating them into sustainable development, accompanied by a concise review of relevant literature. The chapter then presents two illustrations of TSCIs contributing to SDG achievement in the Himachal Himalayas, India. The first narrates the conservation efforts for the Himalayan Birch by *Praja Mandal*, a customary institution in Pangi Valley. The second recounts the promotion of organic farming inspired by a Buddhist Gompa in a village in Hangrang Valley. Subsequently, the chapter addresses potential challenges in the integration process. The final section proposes a pragmatic way forward for the effective incorporation of TSCIs into broader development initiatives. This chapter draws upon a foundation of secondary sources and supplemented with a series of comprehensive interviews conducted with key informants, including development practitioners actively engaged at the grassroots level.

2 Conceptualising TSCIs

Defining institutions becomes intricate, particularly when extending the investigation beyond formal organizational structures to scrutinize socio-cultural institutions (Tirrell 2017). As Tirrell (2017) succinctly summarizes, socio-cultural institutions have been viewed through various lenses: as ingrained thought habits (Veblen 1969), as relational norms (Parsons 1990), as societal game rules (North 1990), and as decision-making mechanisms (Kiser and Ostrom 1982). Scholars generally concur on their enduring and resistant nature. These institutions play a pivotal role in establishing accepted standards and enforcing contracts, thereby facilitating transactions that might otherwise not occur. Scholars categorize institutions into three approaches based on formality, hierarchical levels, and the scope of analysis (Elobeid 2012). As North (1990) defines, institutions encompass constraints shaping human interaction, encompassing formal rules such as constitutions and laws alongside informal norms. Alternatively, Williamson (2000) proposes an institutional classification based on hierarchical levels, spanning social embeddedness to governance structures and resource allocation mechanisms. A third classification approach by Jutting (2003) distinguishes institutions based on the area of analysis, encompassing economic, political, legal, and social institutions.

In light of this diverse conceptualization, TSCIs emerge as socially embedded, informal entities whose constraints are adhered to in culturally appropriate manners. These time-honoured societal structures and practices shape behaviour, norms, and values within a community, passed down through generations. Examples of such institutions include family units, religious organizations, community councils, and indigenous governance systems.

3 TSCIs and Sustainable Development in Himalayas: The Rationale

The sustainable development of the Himalayas is intricately tied to the preservation and revitalization of TSCIs that have thrived in this region for centuries. These institutions, deeply rooted in the cultural fabric of Himalayan communities, can play a pivotal role in fostering sustainability on various fronts.

Firstly, TSCIs in the Himalayas often possess a profound understanding of local ecosystems and natural resources. Indigenous knowledge passed down through generations equips these communities with insights into sustainable agricultural practices, water management, and biodiversity conservation. By integrating such traditional wisdom with modern sustainable practices, the Himalayan region can benefit from a harmonious balance between human activities and the fragile mountain environment. Moreover, these institutions serve as guardians of cultural heritage and identity, fostering a sense of community and collective responsibility. In the face of contemporary challenges, such as climate change and globalization, the preservation of cultural values becomes paramount. Traditional socio-cultural institutions act as custodians of rituals, festivals, and traditional lifestyles, instilling a deep connection to the land. This connection, in turn, can inspire a commitment to sustainable living practices, as individuals recognize their role as stewards of a rich cultural and environmental legacy.

In terms of economic sustainability, TSCIs often support local economies through practices like community-based tourism and handicrafts. By promoting these traditional economic activities, the Himalayan communities can achieve sustainable development without compromising their cultural integrity. This approach not only generates income but also helps in the equitable distribution of resources within the community, reducing dependence on environmentally detrimental practices.

Furthermore, TSCIs socio-cultural institutions are known for their communal decision-making structures. These community-driven governance models, often based on consensus-building, can be harnessed for sustainable development initiatives. When communities actively participate in decision-making processes, there is a higher likelihood of adopting environmentally responsible practices that align with local needs and aspirations.

TSCIs are typically local institutions and as Uphoff (1992) points out, local institutions can play a vital role in sustainable development for several reasons. They are essential for mobilizing and regulating resources, ensuring a long-term foundation for productive activities. Local institutions possess location-specific knowledge, enabling efficient and sustainable resource utilization. Involving local communities in monitoring resource changes accelerates adaptive decision-making. While local institutions may not always resolve conflicts, their absence results in slower and less suitable resolutions at higher levels. Preserving environmentally sound practices requires more than individual incentives; it necessitates adherence to community norms. Institutions foster a long-term perspective, creating common expectations and cooperative bases that extend beyond individual interests, enhancing compliance without excessive inducements or sanctions.

However, in the face of modernization and external pressures, these institutions face the risk of withering away. However, their revitalization and active integration into contemporary development strategies offer a promising pathway for sustainable development in the Himalayas. Recognizing the importance of these institutions is not merely a nod to tradition but a strategic approach to address the unique challenges faced by the region.

4 Contribution of TSCIs for Sustainable Development: A Brief Review

Social institutions are integral components of societies worldwide, and the scholarly exploration of these institutions spans diverse dimensions. Studies conducted across the globe have delved into various facets of social institutions, encompassing conflict resolution, maintenance of social solidarity, administration, natural resource management, societal welfare, and nature conservation. Historically, scholars have scrutinized the roles of these institutions in different contexts, shedding light on their significance in addressing societal challenges and contributing to development.

Examining conflict resolution, early studies by Majumdar (1962) and later by Tafese (2016) have offered insights into how social institutions mediate disputes, promoting harmony within communities. Additionally, research by Rajpurohit and Prakash (2015) and Singh et al. (2021) has explored the crucial role of these institutions in fostering social solidarity, emphasizing their capacity to bind communities together.

Administrative functions of social institutions have been investigated by Pur and Moore (2010) and Aye and Sangtam (2018). Their studies contribute to our understanding of how these institutions play a vital role in governance structures, managing internal affairs, and ensuring the effective functioning of societies. Furthermore, the management of natural and common resources, such as water, grazing grounds, and forest produce, has been a focal point of research. Agrawal and Gibson (1999) and Aggarwal (2008) have examined the intricate relationships between social institutions and sustainable resource management, highlighting the potential for these institutions to contribute to environmental conservation.

Recent studies have expanded the scope of inquiry, investigating the involvement of social institutions in developmental activities. Lashley's work in New Zealand (2000) and Muyinda et al.'s research on sex education in Uganda (2004) exemplify how these institutions are actively engaged in implementing treaties, policies, and programs, contributing to societal progress.

In the context of disaster risk reduction and resilience, Trosper (2002) in NW America and Rumbach and Foley (2014) in Samoa have explored the role of social institutions in enhancing community resilience and mitigating the impact of disasters. Meanwhile, Chuluu's study in China (2020) delves into how these institutions contribute to social and environmental justice.

While various studies globally have enriched our understanding of social institutions, the focus on specific regions and communities remains uneven. In India, scholars have explored facets of *Khumri* and *Sayana* (Majumdar 1962), rural indigenous institutions (Pur and Moore 2010), *Khap panchayats* (Rajpurohit and Prakash 2015), and the institution of Sumi tribal chief (Aye and Sangtam 2018). Despite the rich tapestry of social institutions in Himachal Pradesh, scholarly exploration in this domain remains notably sparse. Existing studies, such as those on the *Praja* system of Pangi (Kapoor et al. 2020) and the involvement of monks in community organization during the Covid-19 pandemic (Singh et al. 2021), touch upon the roles of these institutions in social organization and conflict resolution. However, these works fall short in delving into the broader implications of these social institutions, particularly in relation to the attainment of SDGs. There exists an unexplored terrain regarding the potential contributions of these institutions towards broader societal advancements, a gap that warrants further scholarly attention for a more comprehensive understanding of their significance in the pursuit of SDGs.

Furthermore, the study of social institutions within tribal communities is relatively limited, indicating a gap in our understanding of these communities' organizational structures and dynamics. Additionally, although some works have explored monasteries in Himachal Pradesh (Harcourt 1972; Bajpai 1987; Handa 1987; Jahoda 2008; Lal 2017), these primarily focus on religious functions, emergence, and architectural elements. There is a noticeable dearth of research exploring the socio-economic and non-religious roles that monasteries play in Spiti, highlighting a potential avenue for future investigation (Table 1).

In conclusion, the extensive body of research on social institutions globally has provided valuable insights into their diverse functions and contributions to societal well-being. However, the uneven distribution of studies across regions and communities, particularly in Himachal Pradesh and within tribal contexts, underscores the need for further exploration. Addressing these gaps will contribute to a more comprehensive understanding of the multifaceted roles that social institutions play in shaping societies.

5 Illustrations of TSCI's Contribution for a Sustainable Himalayas

5.1 Case 1: Conservation of Himalayan Birch by *Praja Mandal*, a TSCI in the Pangi Valley of Himachal Himalaya

5.1.1 The Locale

Nestled between the Pir Panjal Range and the Great Himalayan Range, the Pangi Valley occupies a crucial position in the Chandra-Bhaga basin. Its geographical coordinates lie between 32° 12′ 41″ to 32° 47′ 59" N latitude and 76° 13′ 56″ to 76 ° 47′ 48″ E longitude (Shahnawaj 1999). The valley shares boundaries with Jammu & Kashmir to the north, Churah tehsil of Chamba district to the west and southwest, and Lahaul-Spiti district to the south and east. Positioned in the semi-arid zone of

Table 1 TSCI and sustainable development: a summary of literature review

Region	Area/Country	Institution/ community studied	Aspect covered Findings/contribution	Reference
Himalayan region	Uttrakhand (Jaunsar)	Khumri, Sayana	As a local decision making and dispute resolving body	Majumdar (1962)
	Indian Himalaya	Indigenous institutions	Forest management & forest produce distribution	Gupta (2006)
	Mt. Kailash region	Institution of migratory pastoralism	Factors affecting vulnerability of migratory pastoralism	Pandey et al. (2017)
	India	Sumi tribal chief (Nagaland).	Customary laws and traditional system of administration	Aye and Sangtam (2018)
	Himachal Pradesh (Pangi)	*Praja*	Village council's quasi-judicial role and maintaining social harmony	Kapoor et al. (2020)
	Himachal Pradesh	Spitian	Monks as community organisers during Covid-19	Singh et al. (2021)
Rest of India	India	Indian indigenous institutions,	Natural resource management	Aggarwal (2008)
	India	Rural indigenous institutions of South India	Analysed traditional governance and local democracy	Pur and Moore (2010)
	India	Khap panchayat	Legitimacy, reality and reforms	Rajpurohit and Prakash (2015)
Rest of Asia	Kazakhstan	Informal health institutions	Health services after retrenchment of state led health services	Sharipova (2016)
	China	Mangolian indigenous institutions	Environmental justice	Chuluu (2020)

(continued)

Table 1 (continued)

Region	Area/Country	Institution/ community studied	Aspect covered Findings/contribution	Reference
Africa	Ghana	Indigenous institution	Environmental assessment	Appiah-opoku and Mulamoottil (1997)
	Ethiopia	Indigenous institutions of Borana	Role of indigenous institutions in development	Watson (2003)
	Ethiopia	Indigenous institutions of the Awi people	Role of indigenous institutions in socioeconomic affairs management	Genet (2021)
	Uganda	Community sengas (father's sister)	Utilizing sengas for community sex education	Muyinda et al. (2004)
	E-Africa (Ethiopia & Sudan)	Anyuaa society and Nyieye (nobleship) and Kwaaro (headman) institutions	Solving social issues and conflict management	Tafese (2016)
Oceania	New Zealand	Indigenous institution	Implementing treaty settlement	Lashley (2000)
	Samoa	Indigenous institutions of Fa'a Samoa or the culture of Samoa.	Role of indigenous institutions in disaster risk reduction and resilience	Rumbach and Foley (2014)
Americas (N & S)	NW America	Indigenous institutions of native Indians	Resource management, resilience and sustainability	Trosper (2002)
	Columbia	Rural informal institutions	Impact on entrepreneurial activity in rural and urban area	Escandón-barbosa et al. (2019)
Europe	Hungary	Indigenous institutions of chamber system	Economic development	Ingleby (1996)
	Norway	Fishing communities	Fisheries management	Tirrell (2017)
	–	–	Role of community in natural resource management	Agrawal and Gibson (1999)

the inner Himalayas, Pangi faces the climatic challenges typical of a rain shadow area, where imposing mountain peaks obstruct the monsoon winds, resulting in harsh conditions. The predominant form of precipitation is snowfall, often accompanied by avalanches. As part of the cold arid region, Pangi Valley is characterized by rugged terrain, meagre rainfall, severe weather conditions, and substantial snowfall (Shahnawaj 1999).

Elevated in both altitude and remoteness, Pangi Valley is recognized as a tribal area renowned for its rich floral and faunal biodiversity. The high-altitude passes remain impassable for a significant portion of the year (October to June) due to heavy snowfall, isolating the Valley from the rest of the country. Falling under the cold and dry zone in terms of agro-climatic conditions, Pangi Valley spans an area of 1601 sq. km, with a population of 18,868 individuals according to the 2011 census. Notably, among all the Fifth Schedule areas in the state, designated by the Government of India as tribal territories, Pangi exhibits the lowest literacy rate at 71.02%, with women's literacy standing at 52.27%, and the percentage of main workers at a modest 28.9%.

5.1.2 The Institution

The challenging climate, rough terrain, and isolated geographical location of Pangi and its sub-valleys prompted the local population to create a self-help group, known as *Praja* or *Praja Mandal*, to ensure the survival of individuals and their families. Moreover, the primary objective was to foster cohesion within the community. *Praja*, a Sanskrit term signifying 'people' or 'subject,' holds significant cultural importance for the *Pangwala* people. It has evolved into a highly respected customary village council, functioning as a social organization dedicated to the region's development in the Pangi tehsil.

Given that villages in the area have small populations with shared needs and interests, and their livelihoods and social interactions intersect, the necessity for a local body with firsthand knowledge arises to maintain order and facilitate daily activities. Originally tasked with addressing socio-cultural, developmental, and legal matters, *Praja* has transformed into a quasi-judicial entity with the advent of modern legal and political systems. Notably, it remains an apolitical organization with obscure origins, though folktales suggest its inception in Mahiliat Village, where 25 families initiated an institution to resolve issues and act as a disciplinary body.

Praja serves as an assembly by the people, for the people, and comprised of the people (Kapoor et al. 2020). Its membership is exclusively patriarchal, with only men eligible to join. The eldest male member from each family in the village represents their household in *Praja*. The members collectively elect a head known as the *Prajapati*, who serves on a rotational basis for a year. In this democratic body, every member has equal rights and a voice in general meetings, where disputes are amicably resolved, often in the presence of *Praja* to ensure transparency and unbiased judgments.

The core functions of *Praja* encompass social welfare, fostering social solidarity, organizing events, resolving disputes, imposing fines, and complementing the Gram Panchayat, the local governance unit (Kapoor et al. 2020). *Praja* actively contributes to the welfare of the people by engaging in various development-related activities and taking responsibility for organizing local fairs, festivals, and contributing to the construction and maintenance of temples. The decisions made by *Praja* are binding on all members, and failure to attend meetings or non-compliance with decisions may result in fines. Notably, the association between members of Gram Panchayat and *Praja* enhances their understanding of villagers' needs and problems.

5.1.3 The Initiative

In Pangi valley, *Betula utilis*, commonly known as the Himalayan Birch or *Bhojpatra* is one of the important floral species. Apart from its ornamental appeal, it serves various purposes, religious, economic, ethnobotanical and environmental—true to its Latin epithet *utilis* meaning 'useful. Its lightweight wood is used for crafting tools and utensils, and extracts from its bark utilized in traditional medicine (Sarswat et al. 2022) Playing a crucial role in environmental conservation, it stabilizes soil and prevents erosion. Thriving at elevations up to 4500 m, this resilient, multi-branched species reaches 20 m in height, with high freezing tolerance, forming treelines in the Himalayas (Sarswat et al. 2022). Enduring up to 400 years, it dominates sub-alpine altitudes, preserving the ecosystem despite critical endangerment from deforestation and anthropogenic activities.

Nestled along the left bank of Sural Nalla, a tributary of the Chandrabhaga River, lies a significant expanse of Himalayan Birch, spanning over 10 square kilometres. However, the once-thriving woodland had faced the relentless onslaught of human activity, with the birch trees being extensively lopped for fuel, fodder, and medicinal purposes. This widespread exploitation had led to a gradual thinning of the forest, raising concerns among the local villagers. Recognizing the urgent need for conservation, the concerned villagers convened in the *Praja Mandal* of Sural Valley, a venerable customary institution. In a collective effort to safeguard the birch forest, a pivotal decision was made—the woodland would be closed to the public, except for a limited period during the summer, ensuring its undisturbed regeneration. Moreover, strict guidelines were established: only dry wood and Non-Timber Forest Products (NTFPs) could be collected, and any form of green felling was strictly prohibited. To enforce these regulations, a democratic process determined the quota of forest usage, with vigilant monitoring to ensure compliance.

The success of this initiative hinged on the commitment of the local community to adhere to the mutually agreed-upon strictures. In cases of violation, a comprehensive punishment framework was instituted. Initial transgressions incurred monetary fines, while repeated offenses could result in banishment or boycott from the community, entailing exclusion from social functions. This robust system effectively curbed reckless and uncontrolled usage, allowing the forest to rejuvenate naturally.

Presently, the local villagers not only utilize the forest sustainably but have also emerged as stewards of its conservation. Their profound respect for the rules established by traditional institutions of local governance has been pivotal in ensuring the success of this conservation model. The unintended consequence of this meticulous conservation effort has been the transformation of the regenerated forest into a magnet for tourists. It stands as a unique spectacle within the valley, boasting a rich and dense expanse of Himalayan Birch, a sight that has become a draw for nature enthusiasts and wanderers alike.

This harmonious blend of ecological preservation and traditional governance exemplifies the potential for community-led initiatives to restore and safeguard fragile ecosystems. The narrative of the Sural Nalla birch forest serves as a beacon, illuminating the transformative power of collective responsibility and community-driven conservation, where the roots of tradition intertwine seamlessly with the branches of sustainable development.

5.2 Case 2: Promotion of Organic Farming by a Buddhist Gompa (Monastery) in the Shalkar Village, Hangrang Valley of Himachal Himalaya

5.2.1 The Locale

Shalkar village is situated in the Hangrang valley, a trans-Himalayan region in Himachal Pradesh, India. Despite being administratively part of Kinnaur district, it shares geographical and cultural ties with the Spiti Valley. Surrounded by Spiti tehsil on three sides and separated from Kinnaur district by tehsil Pooh on the others, Hangrang also shares an international boundary with China to the east and north-east, while the Spiti River and its tributaries drain the valley. Hangrang experiences dry temperate to arctic climatic conditions with a short, dry summer from late May to September and a prolonged winter from November to April, characterized by sub-zero temperatures and heavy snowfall. The region is in a rain-shadow zone during the southwest monsoon, resulting in scanty summer rainfall but substantial snowfall in winter due to western disturbances. The area's vegetation is sparse and of dry alpine nature, with mountainsides devoid of vegetation except along river-banks and near springs. Lower reaches of the valley boast large woody trees. Spread over 727 sq. km, this area is a designated tribal land with a population of 4659 people as per the 2011 census, Classified as cold desert, the region faces drought-like conditions due to low rainfall and low humidity.

5.2.2 The Institution

The predominant religious affiliation in this region is Tibetan Buddhism, and its profound impact is evident in shaping the cultural, social, and spiritual landscape of the Valley. The enduring presence of Buddhism is manifested through numerous ancient monasteries, locally referred to as Gompas, which have served as the focal points for various facets of socio-cultural and economic life (Handa 1987; Lal 2017). These Buddhist monasteries stand as pivotal centers with significant cultural, spiritual, and historical importance, intricately woven into the local fabric and drawing global visitors. Operating as spiritual hubs, these monastic institutions facilitate religious rituals, meditation, and the study of Buddhist philosophy, contributing to the spiritual enrichment of the community.

As custodians of cultural heritage, exemplified by revered monasteries like Tabo, they safeguard ancient art, scriptures, and artifacts, providing a glimpse into the historical evolution of Buddhism in the Himalayas. Furthermore, these monasteries function as educational hubs, imparting Buddhist knowledge and rituals to monks, fostering intellectual growth, and ensuring the transmission of traditions to future generations. Vibrant Buddhist festivals, such as *Losar* and *Hemis Tsechu*, celebrated in these monastic settings, play a pivotal role in fostering community bonds. Beyond their cultural significance, these monasteries attract tourists, bolstering local economies and offering serene locations for meditative retreats, making them multifaceted contributors to the identity, well-being, and broader understanding of Buddhism in the region.

Monks hold considerable respect among the local population, treating them with high esteem. Considered cultural leaders, monks exert significant influence over the people. This region has produced highly venerated monks, and Reverend T. K. Lochen Tulku Rinpoche stands as one of them. Regarded as the reincarnation of Rinchen Zangpo, a great monk of the tenth century credited with constructing 108 monasteries in the Himalayan region and playing a significant role in the spread of Buddhism in Ladakh, Spiti, and Tibet, Lochen Tulku Rinpoche currently holds the position of Spiritual Director at Kee Monastery, the largest monastery in Himachal Pradesh. Born in Shalkar village and having received instructions from His Holiness the Dalai Lama, Lochen Tulku Rinpoche, as a respected teacher and leader, has amassed thousands of disciples and followers across the Indian Himalaya and in overseas monasteries.

5.2.3 The Initiative

In this resource-deficient region, the primary occupations revolve around agriculture, horticulture, and animal husbandry. The cultivation of cash crops like apples, peas, and potatoes is notable, primarily aimed at the market rather than self-consumption. However, the labor and input-intensive nature of these crops led to the widespread use of chemical insecticides and pesticides, negatively impacting the fragile local ecology. This practice contradicted the fundamental tenets of Buddhism, emphasizing non-violence and compassion for all living beings, and posed significant ecological threats.

Concerned about this environmental degradation, Reverend Lochen T. Rinpoche intervened approximately 8–9 years ago. His influential guidance prompted locals to abandon chemical practices in favour of traditional farming methods. Rinpoche advocated the use of buttermilk and cow dung to protect crops and enhance soil fertility. While some argue that Rinpoche did not explicitly address insecticides/pesticides, villagers voluntarily refrained from their use to avoid harm to insects.

Initially, farmers faced challenges during the first and second years of the transition. However, with the blessings of His Holiness Lochen T. Rinpoche, these challenges gradually subsided. An annual religious ceremony, *Hawan*, conducted by Rinpoche in October, involves the distribution of sanctified ash to each household. Mixed with fireplace cinders, this ash is then sprinkled in fields, believed to enhance productivity and protect crops. Villagers established a mechanism to enforce this practice, imposing a 21,000 Rupee penalty for violators. Adherence has been voluntary, with no penalties imposed, indicating successful adoption of traditional methods. Now, this practice is being replicated by farmers in nearby villages.

Coincidentally, during the 2017–2022 state government regime in Himachal Pradesh, a natural farming program was initiated. Trained personnel were deployed to educate farmers on utilizing local resources for increased yield and crop protection. Practices involving cow urine, such as *jeevamrut* and *ghanjeevamrut*, were taught as organic manure and foliar spray, significantly reducing input costs and boosting net profits. Simultaneously, the NGO HIMOARD played a crucial role in aiding farmers to obtain organic farming certificates, allowing them to command premium prices by selling their products as organic. This initiative not only reduced production costs but also positioned the village as an organic farming model for the entire Spiti Valley. Farmers, benefiting from better market prices, have witnessed an improvement in income levels.

This shift towards organic farming has not only contributed to ecological balance and produced healthier food but has also bolstered the community's self-esteem. Importantly, the success of initiatives by both the state government and non-state actors can be attributed to the community's cultural preparedness to practice non-violence, instilled by socio-cultural institutions and leaders. Shalkar village's journey towards sustainable and organic farming exemplifies how such institutions can catalyse positive change, aligning with the broader goal of sustainable development.

6 Integrating TSCIs for Development: Key Challenges

As the Himalayan region embarks on the path of sustainable development, the involvement of Traditional Socio-Cultural Institutions (TSCIs) becomes imperative. However, this journey is not without its challenges, necessitating a nuanced understanding and proactive measures for effective integration into development initiatives. One significant hurdle lies in the resistance exhibited by traditional institutions towards changes or interventions perceived as disruptive to established cultural practices. The apprehension often stems from concerns about the erosion of traditional values and practices deeply rooted in the community's identity.

Compounding this challenge is the limited formal recognition of traditional institutions in legal and governmental structures. This lack of acknowledgment hampers their influence in decision-making processes, constraining their ability to access essential resources and support. Language and communication barriers pose another obstacle, creating potential misunderstandings between government officials and traditional institution members. Overcoming this challenge is crucial for conveying development goals and strategies effectively.

Moreover, traditional institutions may face limitations in administrative and technical skills required for participation in complex development projects. Capacity-building efforts become imperative to enhance their ability to engage meaningfully in sustainable development initiatives. Within traditional institutions, existing power dynamics may hinder inclusive and equitable decision-making. Dominance by certain groups or individuals in discussions can marginalize other community members, posing a challenge to fostering truly representative development.

Differing priorities and interests within traditional institutions and among communities may lead to conflicts, requiring a delicate balance to align with sustainable development goals. The inadequacy of resources, both financial and infrastructural, further impedes their active participation in and contribution to development projects. The diversity of cultural practices across different villages adds complexity to creating standardized development approaches. Striking a balance between tailoring initiatives to local customs while adhering to a broader sustainable development framework demands careful consideration.

External economic, social, or political pressures may influence traditional institutions in ways misaligned with sustainable development goals. Thus, balancing external influences with local needs and aspirations becomes paramount. Given the ecologically sensitive nature of the Himalayan region, traditional practices may inadvertently contribute to environmental degradation. Addressing this challenge requires finding a delicate equilibrium between preserving cultural practices and promoting environmentally sustainable solutions.

Additionally, traditional institutions may lack access to information about modern sustainable practices, necessitating targeted efforts to bridge the information gap and provide relevant knowledge in an accessible manner. In navigating these challenges, a thoughtful and inclusive approach is crucial to harness the potential of Traditional Socio-Cultural Institutions in driving sustainable development in the Himalayas.

7 The Way Forward for Development Practitioners

- Foster a collaborative approach that respects and integrates traditional values. Work with communities to identify innovative solutions that align with cultural practices, ensuring a sense of ownership and cooperation.
- Advocate for formal recognition of traditional institutions in legal and governmental frameworks, empowering them to actively participate in decision-making processes. Establish mechanisms for resource allocation and support to enhance their role in sustainable development.
- Implement training programs and workshops aimed at bridging language and communication gaps. Develop a framework for effective dialogue between government officials and traditional institutions, fostering mutual understanding of development goals and strategies.
- Prioritize capacity-building initiatives to enhance administrative and technical skills within traditional institutions. Collaborate with educational institutions, NGOs, and government agencies to provide tailored training programs, ensuring active and informed participation.
- Establish inclusive governance structures within traditional institutions, ensuring diverse representation. Facilitate community dialogues to address power imbalances, fostering an environment where all voices are heard in decision-making processes.
- Implement conflict resolution mechanisms within and between traditional institutions. Foster a participatory approach to development planning that considers diverse interests, promoting consensus-building and minimizing conflicts.

- Develop mechanisms to secure funding and resources for traditional institutions. Facilitate partnerships with government agencies, NGOs, and private sector entities to ensure sustainable funding for development projects.
- Invest in basic infrastructure, such as communication networks and transportation, to enhance the engagement of traditional institutions in development planning and implementation. Strengthening connectivity will enable more effective collaboration.
- Embrace the diversity of cultural practices by adopting a flexible and context-specific development approach. Tailor initiatives to suit local customs while maintaining overarching sustainable development goals, allowing for adaptability to varied needs.
- Develop guidelines to navigate external economic, social, and political pressures. Encourage traditional institutions to prioritize local needs while maintaining a balance with external influences, fostering sustainability aligned with community aspirations.
- Collaborate with traditional institutions to identify and promote environmentally sustainable practices. Balance the preservation of cultural practices with ecological sensitivity, ensuring that development initiatives contribute positively to environmental conservation.
- Bridge the information gap by providing targeted and accessible knowledge about modern sustainable practices. Establish channels for continuous learning and information exchange, empowering traditional institutions with the tools to make informed decisions.

8 Concluding Thoughts

In contemporary discourse, there is a growing acknowledgment of institutions' pivotal role in guiding communities and nations toward sustainable development. Scholars assert that these institutions, along with their configurations, constitute a crucial link revealing variations in growth rates and developmental trajectories among developing countries (Elobeid 2012). Extending this perspective to the Himalayas, Traditional Socio-Cultural Institutions (TSCIs) transcend their role as mere guardians of heritage; they emerge as catalysts for sustainable development, intricately weaving tradition into the fabric of a resilient and harmonious future (e.g., Gupta 2006). The future development of the Himalayan region hinges on recognizing and empowering these institutions. Their wealth of knowledge, cultural significance, and community-centric approaches form a robust foundation for crafting a sustainable future that respects both the environment and the people calling the Himalayas home. This importance amplifies in light of scholarly reports indicating a gradual weakening of these institutions (e.g., Pandey et al. 2017). The aim of this chapter is to build a case for identifying, involving, and leveraging these institutions, outlining the challenges that accompany this agenda. Overcoming these challenges demands a collaborative and adaptive approach, involving continuous dialogue, capacity-building initiatives, and a genuine commitment to understanding and respecting local cultures and traditions. Effectively engaging with TSCIs necessitates addressing these challenges to ensure inclusivity, cultural sensitivity, and shared ownership of sustainable development goals. Simultaneously, one must recognize the Himalayan region's broader challenges, including climate change, population pressure, and modernization, which can impact the continued viability of TSCIs. Therefore, the key challenge in the pursuit of sustainable development in the Himalayas lies in balancing tradition with adaptation to changing circumstances. This delicate equilibrium is essential to navigate the complex landscape of the region and foster a future that is both rooted in tradition and responsive to the evolving needs of the community and the environment.

References

Aggarwal A (2008) Indigenous institutions for natural resource management : potential and threats. Econ Polit Wkly 43(23):21–24

Agrawal A, Gibson CC (1999) Enchantment and disenchantment : the role of Community in Natural Resource Conservation. World Dev 27(4):629–649

Appiah-opoku S, Mulamoottil G (1997) Indigenous institutions and environmental assessment : the case of Ghana. Environ Manag 21(2):159–171

Aye KV, Sangtam KV (2018) Customary laws and traditional system of administration with special reference t o Sumi tribal chief. Int J Human and Soc Sci Invent 7(10):30–36

Bajpai SC (1987) Lahaul–Spiti: a forbidden land in the Himalayas. Indus Publishing Company, New Delhi

Chuluu KE (2020) The Tongpo case : indigenous institutions and environmental justice in China the Tongpo case : indigenous institutions and environmental. Crit Asian Stud 0(0):1–17. https://doi.org/10.1080/14672715.2020.1854616

Elobeid D (2012) The role of institutions in sustainable development: the experience of Sudan economy. OIDA Int J Sustain Dev 4(5):53–68

Escandón-barbosa DM, Urbano D, Hurtado-ayala A, Paramo JS, Dominguez AZ (2019) Formal institutions, informal institutions and entrepreneurial activity: a comparative relationship between rural and urban areas in Colombia. J Urban Manage, June, pp 1–14. https://doi.org/10.1016/j.jum.2019.06.002

Genet A (2021) Indigenous institutions and their role in socioeconomic affairs management: evidence from indigenous institutions of the Awi people in Ethiopia. Int Soc Sci J 1–17:225–241. https://doi.org/10.1111/issj.12297

Gupta HK (2006) Cultural significance of indigenous institutions and Forest management practices in the Indian Himalayas: implications for policy and sustainable livelihoods. In Survival of the commons: mounting challenges and new realities, the Eleventh Conference of the International Association for the Study of Common Property, June 19-June 23, 2006 Bali, Indonesia

Handa OC (1987) Buddhist Monasteries in Himachal Pradesh. Indus Publishing Company, New Delhi

Harcourt AFP (1972) Himalayan District of Kooloo, Lahoul and Spiti. Vivek Publishing House, New Delhi

Ingleby SJ (1996) The role of indigenous institutions in the economic transformation of Eastern Europe: the Hungarian chamber system–one step forward or two steps back? J Eur Publ Policy 31(March):102–121. https://doi.org/10.1080/13501769608407020

Jahoda C (2008) Political space and socio-economic Organization in the Lower Spiti Valley (early nineteenth to late twentieth century). J Int Assoc Tibetan Stud 4

Jutting, J. (2003). Institutions and development: a critical review. OECD development Centre, Working paper no. 210; Paris: OECD

Kapoor K, Mehta J, Sinha AK, Sahani R (2020) PRAJA : a Customary Village Council of the Pangwala of Himachal Pradesh. J Anthropol Survey India 69(1):82–92. https://doi.org/10.1177/2277436X20920979

Kiser L, Ostrom E (1982) The three worlds of action: a meta-theoretical synthesis of institutional approaches. In: Ostrom E (ed) Strategies of political inquiry. Sage Publications, Beverly, Hills, pp 179–222

Lashley ME (2000) Implementing treaty settlements via indigenous institutions: social justice and detribalization in New Zealand. Contemp Pac 12(1):1–55

Lal R (2017) Climate change, livelihood options and adaptability in Spiti valley Himachal Pradesh. An Unpublished Ph.D. Thesis,. Department of Geography, Himachal Pradesh University Shimla

Majumdar DN (1962) The Himalayan polyandry: structure, functioning and culture change, A field study of Jaunsar-Bawar. Asia Publishing House, Bombay

Mitra R, Moldavanova A (2018) Social institutions and sustainability: a transdisciplinary research agenda. Crit Sociol 44(2):275–283

Muyinda H, Nakuya J, Whitworth JAG, Pool R (2004) Community sex education among adolescents in rural Uganda: utilizing indigenous institutions. AIDS Care 16(1):69–79. https://doi.org/10.1080/09540120310001633985

North D (1990) Institutions, institutional change and economic performance. Cambridge University Press, Cambridge

Pandey A, Pradhan N, Chaudhari S, Ghate R (2017) Withering of traditional institutions? An institutional analysis of the decline of migratory pastoralism in the rangelands of the Kailash sacred landscape, western Himalayas. Environ Sociol 3(1):87–100. https://doi.org/10.1080/23251042.2016.1272179

Parsons T (1990) Prolegomena to a theory of social institutions. Am Sociol Rev 55(3):319–333

Pfahl S (2005) Institutional sustainability. Int J Sustain Dev 8(1–2):80–96

Pur KA, Moore M (2010) Ambiguous institutions: traditional governance and local democracy in rural South India. J Dev Stud 46(4):603–623

Rajpurohit GS, Prakash A (2015) Khap panchayat in India: legitimacy, reality and reforms. IJAPRR Int Peer Rev Ref J, II III:81–90

Rumbach A, Foley D (2014) Indigenous institutions and their role in disaster risk reduction and resilience : evidence from the 2009 tsunami in American Samoa. Ecol Soc 19(1)

Sarswat N, Dasila K, Lata R, Samant SS, Pandey A, Singh M (2022) Policy brief on Himalayan birch: Betula utilis D. Don. Mohal-Kullu. G B Pant National Institute of Himalayan Environment, Himachal Regional Centre, Himachal Pradesh, India

Shahnawaz (1999) Environmental development interface in mountain eco-system: a case study of Pangi valley in western Himalayas. An Unpublished Ph.D. Thesis, Jawaharlal Nehru University New Delhi

Sharipova D (2016) State retrenchment and informal institutions in Kazakhstan: people's perceptions of informal reciprocity in the healthcare sector. Central Asian Survey 34(3):310–329. https://doi.org/10.1080/02634937.2015.1044199

Spangenberg JH (2007) The institutional dimension of sustainable development. In: Hák T, Moldan B, Dahl AL (eds) Sustainability indicators: a scientific assessment. Island Press, Washington, DC, USA, pp 107–124

Singh TP, de Schweinitz PA, Khanna I, Norbu C (2021) Buddhist monks as community organizers: an indigenous response to COVID-19 in the Spiti Valley of northern India. Crit Public Health 00(Online):1–7. https://doi.org/10.1080/09581596.2021.1931662

Tafese T (2016) Conflict management through African indigenous institutions: a study of the Anyuaa community. World J Soc Sci 3(1):22–32. https://doi.org/10.5430/wjss.v3n1p22

Tirrell A (2017) Sociocultural institutions in Norwegian fi sheries management. Mar Policy 77(July 2016):37–43. https://doi.org/10.1016/j.marpol.2016.12.002

Trosper RL (2002) Northwest coast indigenous institutions that supported resilience and sustainability. Ecol Econ 41:329–344

United Nations (2015) Transforming our world: the 2030 agenda for sustainable development. (ARes/70/1). Retrieved from: http://www.un.org/ga/search/view_doc.asp?symbol=A/RES/70/1&Lang=E United Nations Department of Economic and Social Affairs [online]
Uphoff N (1992) `local institutions and participation for sustainable development', IIED gatekeeper series, no. 3. IIED, London
Veblen T (1969) Veblen on Marx, race, science, and economics (the place of science in modern civilization and other essays). Capricorn Books, New York
Watson EE (2003) Examining the potential of indigenous institutions for development: a perspective from Borana, Ethiopia. Develop Change 34(2):287–309
Williamson OE (2000) The new institutional economics: taking stock, looking ahead. J Econ Lit 38(3):595–613

The manufacturer's authorised representative in the EU is Springer Nature Customer Service Centre GmbH, Europaplatz 3, 69115 Heidelberg, Germany. If you have any concerns regarding our products, please contact ProductSafety@springernature.com

Printed and bound by CPI Group (UK) Ltd, Croydon, CR0 4YY
01/10/2026
02225910-0009